Wireless Networks

Series Editor

Xuemin Sherman Shen, University of Waterloo, Waterloo, ON, Canada

The purpose of Springer's Wireless Networks book series is to establish the state of the art and set the course for future research and development in wireless communication networks. The scope of this series includes not only all aspects of wireless networks (including cellular networks, WiFi, sensor networks, and vehicular networks), but related areas such as cloud computing and big data. The series serves as a central source of references for wireless networks research and development. It aims to publish thorough and cohesive overviews on specific topics in wireless networks, as well as works that are larger in scope than survey articles and that contain more detailed background information. The series also provides coverage of advanced and timely topics worthy of monographs, contributed volumes, textbooks and handbooks.

** Indexing: Wireless Networks is indexed in EBSCO databases and DPLB **

More information about this series at http://www.springer.com/series/14180

Michael Mao Wang • Jingjing Zhang

Machine-Type Communication for Maritime Internet-of-Things

From Concept to Practice

Springer

Michael Mao Wang
School of Information Science
and Engineering
Southeast University
Nanjing, China

Jingjing Zhang
School of Internet of Things
Nanjing University of Posts and Telecomm
Nanjing, China

ISSN 2366-1186 ISSN 2366-1445 (electronic)
Wireless Networks
ISBN 978-3-030-77907-8 ISBN 978-3-030-77908-5 (eBook)
https://doi.org/10.1007/978-3-030-77908-5

This Springer imprint is published by the registered company Springer Nature Switzerland AG
The registered company address is: Gewerbestrasse 11, 6330 Cham, Switzerland

This book is dedicated to my wonderful children, to the memory of my beloved parents, and to all my diligent students.

Michael Mao Wang

Preface

Internet of Things (IoT) can be viewed as a global infrastructure for the information society, enabling applications and services in the fields such as energy, health, and transportation by seamlessly interconnecting and integrating "things" into the information network. The IoT technology is expected to burgeon in the coming decade. The same situation abounds in the maritime domain, as manifested by the recent effort by the United Nations chartered International Maritime Organization (IMO) to modernize the maritime industry under the name "*e*-Navigation," involving the harmonized collection, integration, exchange, presentation, and analysis of maritime information to enhance berth-to-berth navigation and related services for safety and security at sea as well as the protection of the marine environment.

The backbone of IoT is the machine-type communication technology that provides wireless connectivity for massive IoT devices or "things." It can be deemed the framework that the IoT network is hinged on. Machine-type communication (MTC), also known as machine-to-machine communication (M2M), is automated data communication among machines or things and, in principle, does not need human intervention. The communication may occur between an IoT device and an application server or directly between two or more IoT devices. A distinguishing characteristic of MTC from traditional human-type communication (HTC) is that the communication destination is typically not the "person of interest" but the "service of interest." Maritime MTC is a special type of MTC with a specific application in maritime IoT. Like in the general IoT scenarios, most maritime applications and services require little or no human involvement and work even in the absence of human operators, particularly in adverse conditions. Compared to most terrestrial MTC systems, maritime MTC is more complex in coverage, deployment environment, and radio spectrum allocation.

However, it is not an overstatement to say that maritime communication technology has been literally "crushed" by its land counterparts in the past few decades, in large part because of their mass markets and profits that have lured investments and research efforts in information and communication technology (ICT), virtually unchallenged for decades, depleting genuinely experienced maritime ICT professionals. Today's reality is that maritime communication technology has lost traction and fallen far behind, struggling to meet the emerging maritime service demand. This book is intended to fill this gap by pointing out the key challenges and requirements in maritime IoT, presenting a maritime MTC concept to address the challenges and requirements, and providing a comprehensive design example of a practical maritime MTC system, focusing on network architecture, air interface, and radio spectrum. The primary readers of this book include maritime communication engineers, maritime IoT professionals, maritime academia, and the general MTC and IoT communities.

Overall, this book can be partitioned into two parts. The first half, consisting of Chaps. 1, 2, 3, 4 and 5, identifies the challenges and requirements of maritime MTC, delineates an MTC concept—a portrait of a maritime MTC system from an architectural perspective, and describes an MTC framework along with definitions of the fundamental components and functionalities. In the second half, Chaps. 6, 7, 8, 9 and 10, we tie together all discussed in the first part and demonstrate how to apply it on a realistic radio spectrum. Particularly, we deal with more specific aspects through a practical design example based on an international maritime mobile spectrum, which stresses the unified approach introduced in the first part to address the various requirements. Throughout the book, we focus on describing the ideas behind the design using explanatory arguments and simple mathematical treatments to guide readers to an easier understanding of the provided material. A sequential reading of these chapters is recommended, although each chapter is reasonably self-contained.

Specifically, Chap. 1 introduces the Internet of Things and machine-type communication with a focus on maritime applications. It briefly reviews the maritime communication history and describes the challenges and requirements of maritime MTC. Chapter 2 overviews the maritime MTC system architecture and describes the associated network topology and components. It then examines the characteristics of various wireless communication techniques and networks pertinent to maritime MTC, such as proximity, terrestrial, and satellite communication networks. Chapter 3 defines the protocol layers, protocol channels, and layer interfaces for a maritime MTC system. Chapter 4 describes the network structure and the control and application planes to enable service-centric networking and interworking with the Internet protocol or the IP-based network. Chapter 5 reviews the elements of radio communication techniques most relevant to MTC and serves as a basis for the following chapters. It explains the basic concepts of the radio spectrum, modulation, baseband and passband signals, bandwidth, propagation, transmission, and reception to provide the reader with a technical foundation to understand the radio communication techniques used in the maritime MTC air interface design.

For materialization of the concept portrayed in the first half, Chap. 6 introduces a VHF maritime mobile band, recently assigned to *e*-Navigation by ITU. It then examines the individual frequency bands devoted to different maritime MTC scenarios regarding regulations, sharing, and interference mitigation. Chapter 7 deals with the proximity air interface design for proximity-based maritime IoT services operating on this VHF maritime mobile band. Chapter 8 provides an in-depth treatment of a terrestrial air interface for nearshore maritime services. Chapter 9 deals with the issues unique to a satellite air interface for global coverage of maritime services. Finally, Chap. 10 considers the open issues and highlights the potential pitfalls to watch out for in ongoing and future development and standardization of maritime MTC technologies.

This book is an outgrowth of our teaching activities on wireless networking and research work on the maritime Internet of Things. In writing this book, many people have contributed in many ways, and we acknowledge those whose efforts should not go unnoticed. Credit is due to Professor Xiaohu You of Southeast University for originally suggesting and funding the research that uncovered the material in this book. The authors are grateful to Professor Cong Shen of the University of Virginia and Professor Min Dong of Ontario Tech University for thoughtfully and constructively sharing their perspectives, insights, and expertise. We want to thank Mr. Ross Norsworthy, a delegate from the United States to ITU and IEC, for sharing his considerable knowledge of writing and maintaining technical standards for maritime communications and navigation, which has enriched Chap. 6. We also want to pay special thanks to Mr. Hans Christian Haugli of Space Norway, Mr. Yoshi Miyadera of Japan Radio Company (JRC), Dr. Ronald Raulefs of German Aerospace Center, and Dr. Krysztof Bronk of the Polish National Institute of Telecommunications for many valuable discussions and suggestions. Several gifted young minds—the graduate students of the authors have also helped with simulations and critical reading or proofing. In this regard, the efforts of Tingting Xia and Lei Wang and their continual assistance are invaluable and greatly appreciated.

San Diego, California, USA

Michael Mao Wang
Jingjing Zhang

Contents

Chapter 1
Introduction

Abstract This chapter introduces the concept of maritime Internet of Things (IoT) and machine-type communication (MTC) and explains how MTC is related to maritime IoT. We begin with examples of maritime IoT applications and services and motivate the study of maritime MTC by highlighting the key areas in maritime IoT where maritime MTC is sorely needed. But before diving into the maritime MTC topic, we examine the history of maritime communications, debatably the most intriguing episode in the modernization of human society. We then provide the definition of MTC and its strong tie with IoT. Finally, we present the foremost challenges facing maritime MTC, rooted in complex maritime environments, diverse service requirements, and scarce radio frequency spectrum and accentuated by the ever-growing maritime IoT applications and services in the past decade.

Keywords Maritime Internet of Things (IoT) · Maritime communication · Machine-type communication (MTC) · Maritime IoT applications and services · Maritime MTC challenges and requirements · Maritime information network · Maritime cloud · *e*-Navigation

1.1 Maritime Internet of Things

The earth is a water planet, where the oceans cover more than 70% of the surface. These vast, expansive, and interconnected bodies of water represent a prime international domain for maritime transportation. Most remarkably, international freight shipping is responsible for the carriage of about 90% of world trade. This reality creates a new paradigm and perfect application for the Internet of Things (IoT), that is, maritime IoT. In the broadest sense, the IoT encompasses everything connected to the Internet so that information can be shared, analyzed, and acted upon—all happening under the hood of what is commonly called the *application* (or simply an *app*), software for performing specific functions or services. The main objective of introducing IoT into the maritime domain is to achieve better efficiency in operations, enable automation, minimize human errors and potential risks, and enhance overall service quality and security. The maritime domain consists of all

M. M. Wang, J. Zhang, *Machine-Type Communication for Maritime Internet-of-Things*, Wireless Networks, https://doi.org/10.1007/978-3-030-77908-5_1

areas and "things" of, on, under, and adjacent to seas, oceans, and waterways, including related infrastructure, marine equipment/devices or sensors, cargo, vessels, and other conveyances.

A concept similar to the maritime IoT was developed initially to modernize the maritime industries by the United Nations chartered International Maritime Organization (IMO) under the name "*e*-Navigation" to support various types of navigational services [1, 2]. This concept is extended to a more technically formalized, full-fledged maritime IoT solution in [3], under which all things associated with the maritime domain are interconnected and ultimately linked to the *maritime information network* or maritime services network through a unified *machine-type communication* (MTC) system for undisrupted maritime services worldwide. Examples of such maritime IoT applications and services include search and rescue, salvage and intervention, ship operations, environment monitoring and protection, vessel traffic services, aids to navigation, route exchange, as well as exchanging time-sensitive maritime safety information, including distress alerts in an emergency to improve situational awareness and navigational safety.

1.1.1 Search and Rescue

In search and rescue (SAR), an MTC terminal or device embedded in SAR equipment (such as life vests/rafts, maritime survivor locating devices, or man-overboard devices) enables the communication between the equipment and maritime SAR coordination center, or the ships in the vicinity, providing the precise location, weather condition, and other information that helps the SAR operation. MTC also allows the SAR coordination center to poll ships in the vicinity to ask for their SAR capabilities in an automated manner [4]. Apparently, this type of device should require minimum human intervention to operate due to either lack of knowledge or physical incapacitation.

The Maritime Safety Committee of IMO divides the oceans into 13 SAR regions, in each of which the countries have delimited SAR areas for which they are responsible. Wherever a distress signal is received, the rescue is coordinated by a SAR organization and by cooperation between neighboring SAR organizations when necessary.

1.1.2 Aid to Navigation

In general, an *aid to navigation* (AtoN), also known as a *navigational aid*, is any marker or indicator that helps the mariner navigate. It is a device external to a vessel specifically intended to assist navigators in determining their positions or safe course or warn them of dangers or obstructions to navigation in areas such as dangerous coastlines and channels and hazardous shoals and reefs.

There are primarily two types of AtoN devices: buoys and beacons. A buoy is a float typically anchored to the seabed, anything from a floating device to a lightvessel, whereas a beacon is a structure permanently fixed to the seabed or land, ranging from a single-pile pole to a lighthouse. Traditional buoys and beacons provide a variety of navigation information via shape, color, light, and audible signals.

Although still widely in use, this type of traditional AtoN is limited by human visual range and the visibility of the AtoN, which is determined by its elevation and affected by weather, hence susceptible to human errors. Conversely, AtoN devices equipped with MTC utilizes radio signals (radio beacons) to provide precision piloting to passing ships via messages known to "machines," for example, MTC receivers aboard the ships. It extends the human visual perception and is less prone to weather extremes and human errors.

An AtoN system based on physical buoys and beacons is costly to maintain since it is easily damaged by weather and passing traffic, not to mention the installation and replacement in a harsh marine environment. An extension of the AtoN concept under MTC is to go virtual. Unlike a traditional AtoN, a virtual AtoN does not physically exist but comprises a radio signaling system, simply put, a virtual buoy or beacon, presented in a form only visible to a machine or an app. It can be easily deployed or broadcast to any location and maintained by an AtoN service located elsewhere in a maritime cloud or cloudlet without being physically present in that specific location.

Furthermore, the AtoN concept can be extended beyond its original meaning, that is, a marker. With the benefit of a maritime IoT network, the maritime safety information services provide a vessel, throughout its voyage, with more current navigational warnings, meteorological forecasts, and hydrographic services, among other safety-related information [5].

1.1.3 Ship Reporting

In ship reporting, a ship periodically broadcasts (1) its static and voyage-related information, such as the ship's identification, draught, vessel type (e.g., a passenger ship, a pilot vessel, a SAR vessel, a tugboat, a port tender, a law enforcement vessel), intended destination, and estimated time of arrival or (2) its dynamic information, such as position, speed over ground, course over ground, and navigational status (e.g., underway using the engine, at anchor, not under command). These types of information not only help collision avoidance but also allow long-range identification and tracking of vessels at sea, as well as monitoring maritime devices worldwide, such as relaying warnings from tsunami buoys in the open ocean and even monitoring the safety status of individuals working in the most remote areas, for example, the polar regions of the earth.

Other reports include the deviation report, the dangerous goods report, the marine pollutants report, and much more.

1.1.4 Vessel Traffic Service

The purpose of a vessel traffic service is to provide active monitoring and navigational advice for vessels in a limited geographical area, particularly ports and waterways—similar to the air traffic control near an airport. The service is implemented by port authorities, which requires interacting with the traffic and responding to traffic situations developing in the service area. In the United States, it is established and operated by the United States Coast Guard Navigation Center, encompassing a wide range of techniques and capabilities aimed at preventing vessel collisions and groundings in the port and waterways, as well as synchronization and optimization of ports of call to increase transportation system efficiency, expedite vessel movements, and improve all-weather operating capability.

Under the maritime IoT concept, the MTC network forms the basis of these primary services, through which transiting vessels report their identity, course, speed, and other data to the vessel traffic service provider and are, in turn, informed with accurate, complete, and timely navigational safety information. It empowers the vessel traffic service to play a more effective role in marine traffic management, thereby decreasing vessel congestion, critical encounter situations, and the probability of a marine casualty that may result in environmental damage.

1.1.5 Container Tracking

Shipping container tracking allows for geo-locating, tracking, and monitoring a specific container onboard a cargo vessel, providing full visibility of marine assets, logistics, trade flows, and a host of related matters. All parties involved in the shipping process (e.g., commodity traders, insurers, liners, and logistic providers) can reap the benefits of knowing the whereabouts and conditions of the assets when making an oceanic voyage where there is no shortage of uncertainty. Typically, a battery-operated tracking device is attached to the exterior of a shipping container, whose location data is sent either directly through the MTC network to the application provider or relayed by a local area network (e.g., Bluetooth [6], RFID [7]) and then backhauled through the MTC network to the provider.

Indeed, with the current state of global affairs, supply chains are now more critical than ever to ensure the continual running of essential services and secure the global movement of essential supplies for all aspects of daily life, including food, healthcare supplies, and medical equipment. It creates a landscape where commercial companies, financial institutions, and governments must carefully monitor their supply chains and maritime domains for illegal shipping practices and criminal activities. Hence, real-time cargo tracking and tracing have become prevalent and vital for today's maritime service operators.

1.1.6 Route Exchange and Collision Avoidance

Autonomous shipping may be the ultimate way forward; route exchange, however, is the coveted and viable solution to vessel collisions. Route exchange allows ships to be informed of other ships' intentions, foresee possible dangerous situations, and reduce route detours due to traffic conditions. Ships in close proximity (e.g., within the *closest point to approach* or CPA of 0.25 nautical miles) coordinate and optimize their routes autonomously so that close-quarter situations can be predicted and avoided at an early stage, especially when navigating across a shipping lane and in poor weather.

1.1.7 Smart Navigation and Environment Protection

Smart navigation triggers the growing need for better visibility of operations at sea. With maritime IoT, shipowners and fleet operators monitor fuel consumption and machinery performance to improve maintenance and vessel operations, prevent equipment failures and deficient performance, and reduce downtime. Vast amounts of data coming from heterogeneous sources are made available for technologies like artificial intelligence, big data, and blockchain to harness and leverage the full potential of data, turning them into trusted and actionable intelligence for making data-driven decisions and optimizing operational efficiency at all levels. Such examples include fleet management, voyage planning, weather and traffic-related routing, ship speed and engine configuration, and delay reduction at ports to improve navigational efficiency and reduce fuel consumption and emissions.

How to lessen the impact on the marine environment has been another key focus of the maritime industry. Stricter regulations are being enforced recently by IMO to reduce the maritime industry's environmental footprint. Maritime IoT provides a valuable tool to address global climate concerns, allowing operators to monitor their compliance with environmental regulations.

1.1.8 Marine Environment Monitoring and Sensor Network

Indeed, ocean activities and processes have direct environmental impacts and human implications. Yet, the understanding of the ocean system is still far from complete. There is a pressing need for ocean research infrastructure to support both fundamental research and societal priorities. Such infrastructure commonly consists of (1) land-based facilities (e.g., laboratories and control centers) where research and management are conducted, (2) remote platforms where observations and monitoring are performed, and (3) information and communication systems for data transfer and sharing where maritime IoT is expected to play an essential role.

A typical use case is the marine meteorological and oceanographic information collection via maritime sensor networks for monitoring, studying, and protecting the marine environment, where one of the most common data acquisition and collection methods (still in use on many occasions) has been using oceanographic and hydrographic research vessels and aircraft. Moored buoys equipped with wireless communication modules are deployed in coastal and offshore waters, forming a quasi-permanent array of ocean observing stations to collect data on the state of the ocean itself and the surrounding lower atmosphere. Here, a sensor module is used to sense and monitor environmental parameters, such as barometric pressure, wind direction, speed and gust, seawater temperature, salinity, turbidity, oxygen density, and chlorophyll levels.

A power supply module usually provides power to the sensing and communication modules through primary batteries, necessitating replacement when depleted. The replacement is costly or may even be unfeasible in these applications, especially for remote offshore deployments. Low energy consumption is thus critical. In this regard, energy-harvesting technologies have shown to be promising in extending the battery lifetime. However, the energy-harvesting device captures low levels of ambient energy, such as solar, wind, tide, or seawater, and typically generates a limited amount of power. Therefore, unlike onboard vessels, the transmit power for these stations' communication modules is rather limited, and hence the support of these types of energy and power-limited systems becomes another key to maritime IoT.

1.2 A Brief History of Maritime Communications

Disparate maritime elements are brought together under the maritime IoT concept. At the heart of this concept is the machine-type communication technology since to enable maritime IoT, there is a need to establish wireless communication between "things" and the maritime IoT service providers.

There are various ways to incorporate "wireless" functionality into a communication system. One of these is the use of electromagnetic radiation, which is the basis for radio communication. However, electromagnetic radiation is not unique in its ability to extend communication into the wireless domain. In human history, man had relied on fire, smoke, and drum to encode information over a limited geographic area. The uncanny ability of doves and pigeons to find their way back to their nests was also utilized to extend the communication distance by attaching messages to them and releasing them to fly back to their nest, the destination.

Indeed, human attempts at distance communication were extremely limited before the advent of radio communications at the end of the nineteenth century. The ability to communicate between ships was not at all trivial. Communicating with ships (from shore) and between ships at sea relied on signals that humans could hear or see. The former, like horns, is via sound wave—a pressure wave, whereas the latter, like semaphores and flags, is via light wave—an electromagnetic wave; both are still very much relevant today. For example, one short blast of a horn indicates, "I intend to pass you on my port (left) side," and two short blasts mean, "I intend to pass you on my starboard (right) side." Shutter semaphore uses optical shutters (e.g., two vertical rows of three on a white [bright] background) to spell out words and sentences for communication between ships by changing the pattern of the black (dark) and white stripes of the shutters. The flag semaphore is an alphabet signaling system involving the positions of two hand-held flags; each position and motion represent a letter or number. Waving a pair of hand-held flags in particular patterns signals letters of the alphabet and numbers. This type of communication made it easy for fleets to communicate and was all very well for ships and shore as long as they were within sight of each other, but what happened when they were not? This limitation was indeed a problem at that time since communication between ships and from ship to shore is vital for survival on the high seas.

Like communication through visible light, radio communication (or simply radio) is also the transmission of signals via electromagnetic waves but at frequencies well below visible light, that is, invisible by human eyes. Hence, special transmitters and receivers are needed at communicating ends (details in Chap. 5). Radio communication brought about a radical transformation in communication at sea because of the ability beyond sight. The very first commercial use of radio occurred at the end of the nineteenth century. In 1879, at the Royal Society in London, David Edward Hughes demonstrated his invention, where he transmitted and received radio waves using a detector made of carbon, which was later recognized as the now-popular term "wireless signaling" or "wireless communication," albeit that communications through sound and light are both "wireless." Practical experiments began to be made in the 1890s by many inventors like Guglielmo Marconi [8]. In 1896, Marconi sent his first long-distance wireless transmission over a 2-km distance, which earned him a place in radio communication history as the man who gave the world the first radio. Radio, known as "wireless telegraphy," was born. Its primary application was maritime communication, where it was deemed a real breakthrough in safe shipping and control of naval ships. Through radio waves, ships can send messages across vast ocean stretches to communicate with each other and shore. The first-ever radio distress call was transmitted from the East Goodwin Lightship on March 17 in 1899 and received by the radio operator on duty at the South Foreland Lighthouse, who was able to summon a lifeboat [9]. The range of radio signaling was then increased by Marconi, who made a one-way transatlantic transmission in 1901.

From the early years of the twentieth century, ships started fitting radio for communicating distress alerts among ships in the neighborhood and to the shore. Shore stations were soon established in most maritime nations. Radiotelegraphy using Morse code became popular for maritime communication. However, unregulated transmissions and intense commercial competition resulted in interference. Thus, the regulations governing the use of maritime radio became necessary. The first International Radiotelegraph Conference was held in Berlin in 1906, administered by the International Telegraph Union (ITU). It took its present name, the International Telecommunication Union, in 1932, and in 1947 became a specialized agency of the United Nations. The 1906 conference produced the International Radiotelegraph Convention with an annex containing the first regulations in this field, which were later expanded and revised by numerous conferences to become what is now known as the Radio Regulations. Since then, radio telecommunication at sea had undergone a sea-change in the last century. Radio communication technology had advanced at an unprecedented pace, improving how the ships communicate with each other and the shore, owing to its superiority in communication range, mobility, and cost. Channel 16 (156.8 MHz) in the VHF maritime mobile band is probably the best-known radio frequency, designated as an international distress frequency by ITU, for distress and safety priority calls [10]. It was and still is monitored 24/7 by coast guards around the world.

Since satellite use had been made permissible in the 1960s, satellite communication extended ship-to-shore communication and was increasingly used onboard ships for commercial and safety-related communications. As opposed to terrestrial communication systems, satellite services typically need the help of geostationary satellites for relaying signals through a ground station (or teleport), where shore stations cannot reach. The 1970s was a big leap forward for maritime communications. IMO (then known as the Intergovernmental Maritime Consultative Organization) began to consider using mobile satellite communication for improving distress and safety of life at sea (SOLAS) communications and efficiency of ship management. The International Maritime Satellite Organization was established in 1976 to provide marine vessels with global communication for mariners to stay connected with shore. Mobile satellite telephones had since become more common. Maritime communication between ships or with the shore was carried with onboard systems through both shore and satellite stations. Some satellite systems provide two-way communications, and some are limited to the reception of signals from the emergency position (e.g., position-indicating radio beacons).

While ship-to-ship voice communication was brought about by radio, digital selective calling (DSC), introduced by ITU, came up with remote control commands to transmit or receive distress alerts, urgent or safety calls, or regular messages. These messages are pre-defined and transmitted in a more efficient "digital" format via the medium-frequency (MF), high-frequency (HF), and very-high-frequency (VHF) maritime radio systems, allowing mariners to instantly send an automatically formatted distress alert to the rescue authority anywhere in the world. This service also allows mariners to initiate or receive distress, urgency, safety, and routine radiotelephone calls to or from any similarly equipped vessel or shore station.

In the 1990s, primarily for collision avoidance and in support of vessel traffic information service, ITU introduced a VHF data transmission system, coined automatic identification system (AIS), which uses two VHF channels to provide vessel identification, position reporting and tracking [11]. AIS supports up to 64 pre-defined messages and integrates a VHF transceiver fitted with a global positioning system, such as a GPS receiver. Along with other electronic navigation sensors gyrocompass or rate of turn indicator, a ship automatically (with minimal human intervention) communicates a ship's static and dynamic information to other ships or shore-based vessel traffic services, allowing an AIS receiver to portray a ship's location and other information on a presentation device, such as an electronic chart display and information system (ECDIS), and hence reduces the risk of human errors. Before then, navigators and shore stations relied mostly on visual navigation, analog radar, and voice communications to mitigate collisions. In that sense, AIS could be deemed a primitive version of the so-called machine-type communication, as will be formally introduced in the following section.

During the last two decades, the IMO has popularized AIS for exchanging navigational data between ships and shore stations to improve situational awareness over VHF voice, radar, and sight. Nonetheless, one of the many limitations of AIS is the lack of a well-defined protocol structure that most modern communication systems rely on to interwork with a broad range of IP-based applications. It imposes a strong tie to a specific application developed only for AIS and limits its applicability to new maritime IoT applications. This is an inherent structural limitation, as will be elaborated in the following chapters. Another significant inherent weakness of AIS is the wireless interface that essentially adopted a *wired* interface design, thereby rendering AIS inherently unreliable and power and spectrally inefficient in a maritime IoT environment. Not surprisingly, it has been showing incompetence in dealing with the emerging maritime applications and has been hollering for more spectrum to offload the maritime service traffic. However, more spectrum may not be the ultimate solution to all the problems and ever-increasing challenges.

1.3 Maritime Machine-Type Communication

The history of radio communication amply demonstrates the never-ending push for progress in communication technology. Undoubtedly, the key to IoT is connectivity between machines or things, and thus the maritime IoT concept brings about a paradigm shift in maritime communications and unfolds a brand new chapter for maritime communications.

Although the infancy and youth of radio communication were mainly linked to maritime services, maritime communications have been relegated to a lower priority tier for many decades because of the limited market and slim profits, resulting in depletion of genuinely experienced professionals in the field of modern wireless communications and networking. Consequently, advancements in maritime communication technologies are severely lagging behind their land counterparts at every level since the advent of the mobile internet and IoT. While terrestrial wireless communication technologies have undergone revolutionary developments and breakthroughs in the past decades [12], maritime communication has been mostly relying on traditional voice communications and the legacy AIS for exchanging information for the safety of navigation and has stalled since AIS. Progresses in the maritime domain for the modernization and mobilization of maritime-related businesses, as well as maritime safety and traffic management continue to challenge the legacy maritime communication systems, which have gradually shown their incompetence in meeting the ever-increasing demand from maritime IoT services, hindering the broad use of these systems for IoT. This situation has accentuated the need and accelerated the timetable for complete maritime communication technology and infrastructure transformation, and at the heart of this transformation is the need for machine-type communication technology.

Indeed, the terrestrial communication network landscape has undergone revolutionary progress since the emergence of the IoT concept in the late 2000s, witnessing a surge of new developments driven by the exploding IoT applications and services. The most notable paradigm shift is from the traditional network-centric network architecture toward the service-centric architecture, followed by the emerging machine-type communication concept, replacing the conventional network-centric architecture that human-type communication has been relying on.

Machine-type communication or MTC (also known as machine-to-machine communication) is a form of automated data communication among machines or things, that is, the IoT devices under the IoT concept, which is autonomous from humans, and hence does not need human interaction or intervention. The communication may occur between an IoT device (i.e., the application host) and an application server or directly between two or more MTC devices. A distinguishing characteristic of MTC from traditional human-type communication is that the communication destination is typically not the "person of interest" but, more often than not, the "service of interest," which has a profound impact on the system design.

Maritime MTC is a special type of MTC with a specific application in maritime IoT. Like in the general MTC use cases, most maritime applications and services require little or no human intervention and work even in the absence of human operators, particularly in adverse conditions. We have enumerated such use cases in the previous section.

A central aspect of maritime MTC is the heterogeneity of IoT devices in cost and power constraints, which co-exist in the same network. This coexistence of diverse IoT devices affects the communication behavior and ergo the design of the transceiver and communication protocols, contrasting the human-type communication that has a sole goal in mind, that is, pursuit of the highest data rate or the best human experience, by taking advantage of the high-cost, high-performance transceiver, and high-capacity battery—the luxury that is not available for all IoT devices. Above all, the unique characteristics of maritime MTC from any other communication technology, which present unprecedented challenges to the maritime MTC design, are (1) vast and distinctive deployment environments, from nearshore across high seas and to the polar regions and (2) global coverage, containing different nations and regions.

Maritime MTC will certainly trigger another wave of the maritime communication revolution. It is expected to incorporate advanced wireless networking and mobile communication technologies to facilitate ubiquitous communication between a broad range of maritime "things" at sea and their service providers to support a wide range of maritime IoT applications and services. The "things" include not only the traditional "common devices" aboard a vessel (like GPS, compass, steering, radar, sonar, and fuel monitoring and management systems) but also maritime devices, like SAR and AtoN devices, as well as marine sensor networks, with interconnection to shore-side maritime information networks or maritime clouds. It is not an overstatement to say that the maritime MTC system constitutes the "backbone" of maritime IoT. Without it, maritime IoT will remain a mirage.

1.4 MTC Requirements and Challenges

The maritime IoT concept involves a comprehensive communication network for information collection and distribution, processing and real-time analytics, and control and actions. Rather than an incremental change, it calls for a complete rethinking of maritime communication and exploring new communication and networking technologies, which creates the most serious challenges that the maritime communication community has ever faced. For that, we need a clearly defined maritime MTC concept and its associated requirements per the maritime IoT use cases iterated in the previous sections.

1.4.1 Ubiquitous Connectivity and Service Continuity

First and foremost, a maritime MTC system is required to provide ubiquitous connectivity between vessels (including marine equipment) and shore on a global scale, especially over open oceans, including the most remote areas of the world (like the polar regions), to ensure unbroken and consistent existence of maritime services. Currently, the presence of services in offshore settings is limited by the absence of adequate information and communication infrastructures. This crucial lack of infrastructure poses a unique challenge. Unlike in terrestrial wireless communication, where wide-area wireless coverage is provided via the mass deployment of base stations interconnected with the core network, it is obviously impractical (if not impossible) to cover the open oceans with such base stations.

Moreover, today's maritime service networks are deployed in a "campus-style." Depending on the vessel's national, *cross-region* continuity of maritime service remains inconsistent and even absent. Service continuity between even neighboring countries is still in question. Indeed, ownership or governance remains the main stumbling block to a unified global maritime service network. Even though the undeniable appeals of full data authority and infrastructure control are driving the adoption of privately managed connectivity across maritime service networks, ultimately, a unified cooperative service network is essential for global maritime IoT to provide the seamless roaming ability with undisrupted services across organizational, regional, and national boundaries, especially in times of crisis.

1.4.2 Traffic Nonuniformity

Despite its global nature, marine traffic is highly unevenly distributed. Heavy traffic concentration is typical in ports, nearshore, and waterways. For instance, coastal shipping accounts for more than 50% of the total maritime transport of goods to and from the main ports of the coastal nations, where the cargo ships primarily follow the routes that are set close to the shore wherever possible while sailing between trading ports. Besides, leisurely activity, like recreational boating on coastal waters, is also a factor in the traffic densification in nearshore areas. Quite the contrary, maritime traffic on the high seas is mainly from intercontinental transportation or deep-sea shipping and is relatively sparse in density. Maritime MTC thus calls for an effective solution to cope with this type of traffic characteristics effectively, a scenario that the legacy systems like AIS do not have a readymade solution other than asking for more spectrum.

1.4.3 Service-Centricity and Adaptability

Unlike traditional mobile networks where services are built around the network architecture, the maritime MTC network must be service-oriented, encompassing network functions like service routing, service composition, and dynamic resource management to support efficient provisioning, discovery, address resolution, and execution of service requirements of various types of maritime application and service components distributed over the maritime information network or maritime cloud network in order to reap the full benefits of moving to maritime IoT.

As aforementioned, maritime IoT applications and services vary from simple periodic reporting to route exchange and remote control (for autonomous shipping, for instance) and remain in flux due to the ongoing maritime IoT evolution. In this regard, a maritime MTC system is not confined to a one-time design and deployment. Instead, it is expected to offer fluidity in distributing network functions and provide amorphous services that adapt to a wide variety of maritime IoT-specific needs and match-changing demands. Hence, both network configuration and communication resources must be made flexible and adaptive to the offered specific service.

In traditional maritime communication systems, components from networks, hardware, middleware to applications, and services are customized and intrinsically coupled for the specific requirements of a specific system. Consequently, service clients are burdened with the tremendous cost of maintaining the services and receiving new services. Every new service comes with a new set of communication equipment, and the same goes even for switching service providers for the same service, at which time it is often necessary to literally tear down the "old" equipment and replace it with the one provided by the new service provider. This situation often presents a conundrum for an ocean-faring vessel to install all kinds of communication equipment (and antennas) to receive various services, much less to maintain and feed the power to all the equipment. The reasons for this situation can be traced to many sources; however, it is undoubtedly true that this situation is largely rooted in the absence of unified system architecture and protocols for dealing with diverse maritime applications and heterogeneous end systems, rendering the traditional maritime communication technologies and systems no longer up to the task or adequate for this requirement.

1.4.4 Device Heterogeneity

To serve a myriad of maritime IoT applications, maritime MTC must support a diverse set of MTC terminals or IoT devices from the low-end or low-cost category with reduced functionality to the high-end category with full functionality.

A low-cost IoT device is typically used for mass deployment like disposable sensors and buoys that generate a small amount of data periodically and are mostly power or energy-limited, while the high-end device is for large vessels, like a ship that boasts high power and high-performance transceivers and even encompasses a premises network, that is, a local area network with multiple maritime application hosts. Consequently, these devices may have transmit powers varying from a few mW (e.g., 5 mW, harvested from random ambient energy sources) to more than 1 W (from continuous mains power onboard large ships), antenna gains from less than 0 dB to tens of dB, and synchronization sources from tens of ppm to less than 1 ppm. Therefore, a serious challenge that faces maritime MTC is the high degree of heterogeneity in terms of the device's communication capability, including hardware, power supplies, interoperability, and protocols, which calls for an efficient MTC system with the capability of absorbing the heterogeneity under a *unified* framework.

1.4.5 Simplicity and Robustness

Simplicity has traditionally been the overall system design criterion for maritime communication systems. It must be taken into consideration as we strive to meet these goals and requirements since a simpler system is faster to develop, cheaper to manufacture, easier to maintain, greater in longevity, and most importantly, more reliable and robust under complex marine environments, which is of paramount importance to marine equipment and systems. Indeed, maritime communication devices must endure harsh weather and environmental conditions, such as wind and waves, precipitation, seawater corrosion, extreme temperature, under which simplicity can bring performance benefits to any maritime system.

Cost is also an essential factor that cannot be overlooked. In order to modernize the maritime industry and improve the safety of navigation, the MTC system will eventually be mandated on all vessels by the international maritime authority like IMO. Therefore, cost-efficiency and free loyalty become essential to the selection/development of MTC technology. It is a misconception to suppose that low cost means low reliability. On the other hand, in some IoT applications, devices are deployed on a massive scale, like mass disposable sensors, and hence must be cheap to manufacture.

These properties associated with general maritime IoT devices strongly contrast with the human-type communication devices (like cell phones) that are built to "shine," whose complexity and cost are much less of an issue.

1.4.6 Capacity and Scalability

It is foreseeable that the demand to handle the ever-growing maritime traffic will place high stress on the MTC system and drive the need for higher capacity. Efficiency is vital to maximizing the system capacity constrained by the extremely scarce communication resources (i.e., the radio spectrum). Physical and higher layer protocols of the MTC system are hence to be optimized to enable spectrally efficient communications.

Furthermore, the system capacity must be scalable with future growth in response to the growing demand for capacity and the increasing bandwidth needs. This concept connotes an MTC system's ability to handle growing volumes of workload gracefully or expand when resources are added to accommodate that growth without the need for major changes to the architecture or the need for reengineering the existing infrastructure. The scalability of an MTC system subject to the growing demand is crucial to its long-term success.

1.4.7 Interoperability

It is essential for a vessel or equipment in a maritime IoT system to receive services from different systems or networks. The inherent dilemma of receiving such services is their interoperability due to the usage of distinct protocols, interfaces, software instances on top of different IoT infrastructures and resources in diverse physical environments.

Interoperability is the ability of different maritime applications to exchange and integrate information flexibly, effectively, consistently, and cooperatively. These applications may very well be provided by various organizations and businesses across the maritime industry. Interoperability is then at the very center of maritime IoT's promises and is fundamental to its ecosystem. For that reason, maritime MTC is expected to offer the ability for different maritime IoT applications and services to access the network seamlessly both within and across network boundaries and to provide portability of information efficiently and securely across the complete spectrum of maritime IoT services without effort from the end-user, regardless of its manufacture or origin.

In fact, maritime service providers are already beginning to adopt internet-based application interfaces as a foundational component of their information technology (IT) platform. Therefore, maritime MTC must migrate from the traditional maritime communication framework to a newer network and protocol architecture to take on this new challenge for the maritime IoT to succeed. Nonetheless, it takes a big, effortful leap of information and communication technology (ICT) moving from the legacy AIS platform to the modern ICT arena. It is a leap that the current maritime community is not ready to make for the reason mentioned in Sect. 1.3.

1.4.8 Radio Spectrum Internationality

Last, but by no means least, another fundamental element for a maritime MTC system is the radio communication spectrum. The radio spectrum is, without doubt, the most critical component for any wireless communication system, and even more so for maritime MTC because of its global coverage nature. However, the radio spectrum is a natural resource, and natural resources within the nation's geographic boundaries are generally owned by that nation, much like freshwater, land, forests, and mineral deposits. This fact increases the difficulties in overcoming the inherent barriers that prevent the globalization of the maritime MTC, ergo the transformation and revitalization of the world's maritime information technology, ultimately breaking the promise of more effective and efficient use of the world's precious maritime resources.

Indeed, this ultimate goal of globalizing the maritime MTC will not be possible without adequate global spectrum access to support it. For maritime MTC to be successfully deployed worldwide, it is imperative to have appropriate international standards and regulations to enable the proper operation and sharing of the system.

However, new wireless communication technologies and systems, fueled by exploding mobile applications (government and commercial), have created an unprecedented demand for the radio spectrum, and emerging countries deploy wireless systems to modernize communication infrastructure on land and at sea. This reality, plus the pervasive presence of legacy radio systems, makes an extremely competitive environment for worldwide spectrum access. Therefore, international spectrum allocations and regulatory challenges must be addressed by the international standards and regulatory bodies, as well as the world's maritime community and institutions.

1.5 Summary

The requirements and challenges that we have identified so far will serve as a reflection and long-term measure of maritime MTC technology needs and will play a crucial role in backing the renaissance of maritime IoT and determining a maritime MTC system's success. It has become evident that the traditional maritime communication methodology is no longer up to the task, and an up-to-date wireless networking concept is required to take up the new challenges. The fundamental question is what to do about these requirements and challenges. Notably, we need (1) a service-centric maritime MTC architecture, (2) radio/air interfaces that accommodate diverse maritime service requirements, and (3) an internationally authorized radio spectrum. These are the focus of the remainder of this book.

References

1. *Strategy for the Development and Implementation of e-Navigation*, MSC 85/26/Add.1 Annex 20 (IMO, London, 2008)
2. J.E. Hagen, *Implementing e-Navigation* (Artech House, Norwood, 2017)
3. T. Xia, M.M. Wang, J. Zhang, L. Wang, Maritime Internet of Things: challenges and solutions. IEEE Wireless Commun. **27**(2), 188–196 (2020, April)
4. *International Convention for the Safety of Life at Sea (SOLAS)* (IMO, London, 2013)
5. M. Cho, H.R. Choi, C. Kwak, A Study on the method of IoT-based navigation aids management. Adv. Sci. Technol. Lett. **98**, 55–58 (2015)
6. Sig Bluetooth, "Specification of the Bluetooth system v4. 0" (2014)
7. IEEE-USA Communications & Information Policy Committee (CCIP), *RFID: The State of Radio Frequency Identification (RFID) Implementation and Policy Implications* (IEEE, 2005)
8. *The history of telecommunication*. https://www.mitel.com/articles
9. *Overview of ITU's History*. http://search.itu.int/history/HistoryDigitalCollectionDocLibrary
10. T. Konrad, *First Radio Distress Signal?* (2009, October 9). https://gcaptain.com/first-radio-distress-signal
11. *Technical Characteristics for an Automatic Identification System Using Time Division Multiple Access in the VHF Maritime Mobile Frequency Band*, document ITU-R M.1371-5 (ITU, Geneva, 2014, February)
12. M. Agiwal, A. Roy, N. Saxena, Next-generation 5G wireless networks: a comprehensive survey. IEEE Commun. Surveys Tuts. **18**(3), 1617–1655 (2016, 3rd Quarterly)

Chapter 2
System Architecture

Abstract Motivated by the maritime IoT applications and requirements presented in the previous chapter, this chapter identifies and addresses the commonalities and differences of these applications and requirements and introduces a maritime MTC system architecture comprising three types of communication networks tailored to the various service requirements and the unique maritime IoT deployment environments. We shall examine the distinct characteristics of this envisioned hybrid maritime MTC system and explain how it differentiates itself from its land counterparts. This chapter lays out an architectural foundation that leads to an ultimate solution to the maritime IoT requirements and challenges.

Keywords System architecture · Ad hoc communication · Infrastructure-based communication · Satellite communication · Terrestrial communication · Proximity communication · Maritime information network · Maritime cloud · Distributed medium access · Centralized medium access · Full-duplex · Half-duplex · Frequency-division-duplexing · Time-division-duplexing

2.1 Architecture

A maritime MTC system pertinent to maritime IoT is envisioned in Chap. 1. It possesses unique architectural characteristics, such as colossal network scope, dynamic topological structures, trans-regional mobility, and highly non-uniform traffic distribution. These distinct characteristics signify a fundamental transformation of the traditional maritime communication concept and call for a distinctive architectural design from the legacy maritime systems and their land counterparts.

In particular, as we recall from Sect. 1.4 of Chap. 1, the first and foremost requirement of maritime MTC is the provision of ubiquitous connectivity between vessels and the maritime information network over open oceans to ensure maritime service continuity. This requirement poses an unprecedented challenge since, unlike in terrestrial communication where wide area wireless coverage is provided by establishing an infrastructure, comprising of mass deployment of base stations as

M. M. Wang, J. Zhang, *Machine-Type Communication for Maritime Internet-of-Things*, Wireless Networks, https://doi.org/10.1007/978-3-030-77908-5_2

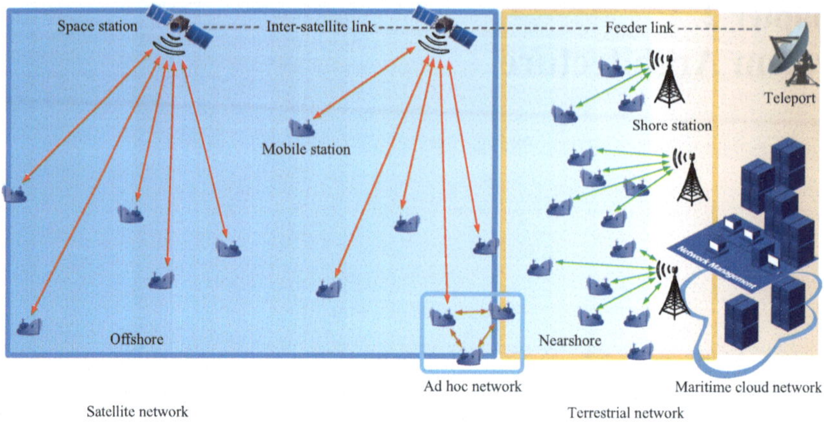

Fig. 2.1 Maritime MTC system architecture, where a mobile station (or simply a mobile) is an MTC terminal onboard a vessel or embedded in any maritime equipment. The feeder link is a backhaul radio link from a teleport (i.e., earth station) at a fixed earth location to a space station or vice versa. Inter-satellite link is a backhaul radio link interconnecting satellite space stations operating in a constellation, which can be either an RF beam or optical beam. The backhaul network (e.g., a fiber-optic network) interconnecting shore stations and maritime cloud networks is not shown

network access points; it is clearly unrealistic to cover the open oceans with such base stations considering the struggles to construct, operate, and maintain such colossal infrastructure in harsh maritime environments, not to mention the environmental implications. A strategic architecture is needed to provide global coverage and, at the same time, to ensure the highest robustness, best efficiency, and greatest flexibility to facilitate new maritime IoT services and incorporate emerging technological advances and yet minimum environmental footprints.

The primary way we seek to achieve these ends is to implement an operationalized strategy that should comprise three integrated elements. Figure 2.1 paints a picture of such a conceptual architecture. First, both satellite and terrestrial communication networks are needed; second, both infrastructure and ad hoc communication topologies are present, where the infrastructure consists of deployments of *control stations*, *space stations*, and *shore stations*, interconnected through backhaul networks; and third, a centralized network controller is in charge of integrating and managing both space and terrestrial segments.

Here, a *mobile station* is an MTC terminal aboard a vessel or embedded in marine equipment (e.g., a SAR transponder, a buoy, or a lighthouse). More generally, it provides wireless interworking between the maritime MTC network and an on-premises network. Such an on-premises network can be a sensor network or a local area network aboard a ship with one or more end-hosts running diverse applications, such as the global navigation service, meteorological service, marine geoscience data service on board a ship, and a container tracking system onboard a cargo vessel. A mobile station is also capable of engaging in direct communication with other mobile stations to form a self-organized ad hoc local network for maritime proximity-based services (e.g., route exchange, aids-to-navigation).

Within the maritime MTC network, a station is uniquely identified or addressed by the Maritime Mobile Service Identity (MMSI) or any uniform resource name that can uniquely and persistently identify any maritime resource on a global scale. MMSI was initially issued by ITU in 1982 to uniquely identify ship stations and shore stations or a group of stations [1]. It is adopted by AIS and also this book as the physical address to identify a maritime MTC station.

An MMSI is a series of nine digits, including the 3-digit Maritime Identification Digits (MID) ranging from `201` to `775`, denoting the country or geographical area of the country responsible for the ship so identified. For instance, the MMSIs under MID `338` are for various registering organizations in the United States, and those under `247` are for Italy. The MMSI for an individual mobile station begins with an MID, i.e., `MIDXXXXXX`, where `X` is any figure from `0` to `9`. An MMSI with the first figure equal to `0` is designated to identify a group of mobile stations using the format `0MIDXXXXX`.

Both satellite and terrestrial infrastructures are present in the architecture to provide mobile stations with ubiquitous, adaptive, and seamless connectivity to the maritime information network or the *maritime cloud*—a trusted platform for providing various maritime services and applications with the highest computational and storage capacity in the maritime IoT framework. The federated network management combines the salient features of service provision, management, and orchestration with dynamic network and communication resource management, consolidation, service resolution, and forwarding mechanisms, through which the infrastructure assets and physical resources are maximally shared among service providers and are fine-tuned to meet the individual service requirements. Furthermore, it enables interconnectivity between cloud-based or cloud-enabled applications, services, solutions, and ultimately service-centric networking.

Under this envisioned architecture, the nearshore communication is through dense geographically distributed shore stations, whereas offshore connectivity (e.g., on the high seas) is provided by large-footprint space stations. These control stations act as access points to the maritime cloud for mobile stations. The space station is mainly for offshore communication via satellite ground station (teleport), thereby facilitating communication beyond the reach of the shore stations, especially over open oceans, including the most remote areas of the world. The nearshore traffic is hyper-dense, whereas offshore is much sparser. Thus, the MTC system makes use of both "micro-coverage" to handle the hyper-dense nearshore traffic by exploiting spatial spectrum reuse and "macro-coverage" to accomplish the global coverage. Besides, an ad hoc or self-organized network can be quickly and dynamically established to facilitate direct communication among mobile stations, particularly for proximity-based maritime IoT services.

2.2 Communication Topologies

The communication topology defines how the stations are interconnected to create networks for information exchange. The topology has a direct impact on how the communicating stations share the communication resources. We focus on the two most commonly used topologies in wireless communications that best serve the maritime MTC architecture presented previously.

2.2.1 Ad hoc Communication

Self-organizing or *ad hoc* communication follows a "point-to-point network" topology, as graphically illustrated in Fig. 2.1. It is a natural topology for a decentralized wireless communication network composed of individual mobile stations communicating directly, bypassing or negating a central control station (i.e., a space station or shore station). The network is "ad hoc" because it does not rely on a pre-existing infrastructure that consists of control stations connected to a core network that provides network management.

In a practical communication system, users or stations share the communication medium or radio resources based on *multiplexing*. A medium access method or a multiple access method allows more than two mobile stations connected to the same transmission medium to transmit over it and share its capacity. Without the benefit of centralized management, the communication resource or medium is shared among mobile stations through a *distributed* or *autonomous medium access control* mechanism. Such accessing schemes can be classified into two main categories: *random medium access* and *sensing-based medium access*.

2.2.1.1 Random Medium Access

Random medium access or *random access* is the simplest medium access strategy by which each transmitter sharing the medium transmits independently at random times. In the context of MTC in this book, the medium is the radio frequency (RF) channel or *radio channel* of a given bandwidth at a given frequency. There is no need for a transmitter to abide by a specific timeline or check whether the radio channel is occupied, and hence no need for transmitters to set up a common timing mechanism with their receivers. When a transmitter is up and ready to transmit signals, it just goes ahead and transmits without coordination with other transmitters sharing the same channel. Therefore, the synchronization burden is all on the receiver side, which needs to continuously detect the intended transmission. This transmission strategy is commonly known as the *Aloha* random access [2].

To get an idea of the effectiveness of the random-access strategy in radio medium usage, let us assume that a *transmission burst* has a fixed duration T. Such a burst,

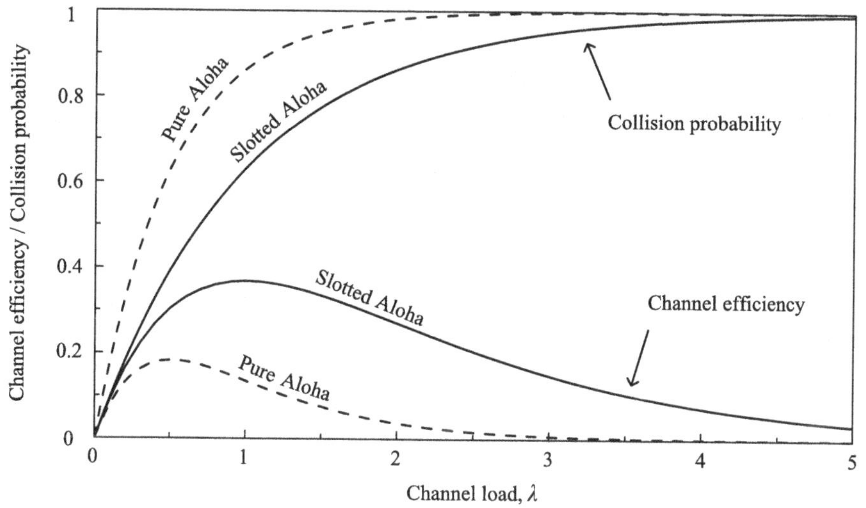

Fig. 2.2 Collision probability and channel efficiency for pure (dashed lines) and slotted (solid) random accesses

transmitted at a time t by a station, is vulnerable to any co-channel transmissions by other stations within range, happening during the period from $t - T$ to $t + T$. Clearly, no transmission from any other station should occur during this period for the burst to be successful. Further, assuming the channel load (the number of co-channel transmission attempts by stations) is *Poisson*-distributed, and the average number of transmissions per T is λ, the probability of no transmission during $[t - T, t + T]$ of length $2T$ is thus $e^{-2\lambda}$. The average number of successful transmissions can be calculated as the rate of transmission attempts multiplied by the probability of success, i.e., $\lambda e^{-2\lambda}$, which is no greater than 0.18 (which happens when the channel load is $\lambda = 0.5$). This means that the channel efficiency under the random access method is less than 18%.

It is evident that reducing the vulnerable time of a transmission burst increases the probability of success and increases channel efficiency. A straightforward method is to impose the time slot constraint on all transmissions, under which the channel time (or the time resource) is evenly divided into slots, each of length T, and any transmission burst must align with the slot boundary. This notion of time slot effectively reduces the vulnerable period from $2T$ to T, and the channel efficiency is thus increased to $\lambda e^{-\lambda}$ with a maximum of 0.37 at $\lambda = 1$.

Figure 2.2 plots the collision probability and the channel efficiency against the offered load for both random access strategies. It is seen that the slotted Aloha is superior to the original or pure Aloha. This improvement comes at the cost of the need for maintaining a common time reference at each communicating station.

2.2.1.2 Sensing-Based Medium Access

Sensing-based medium access typically involves each mobile station monitoring the transmission activities of neighboring mobile stations on a specific radio channel to ensure that only one transmitter is transmitting on this channel at any time within its transmission range to avoid collisions at the destined receiver. Thus, transmissions must wait if there is "evidence" that other transmitters are occupying the channel. Typically, a *clear channel assessment* (CCA) is used to determine whether a channel is available for transmission. CCA checks the RF energy of the channel to determine its activity before transmission. If the detected energy exceeds a threshold, the channel is considered "busy" or occupied [3].

The effectiveness of CCA depends heavily on the accuracy of the energy detection, which is prone to errors in estimating channel energy and selecting the threshold. An improvement to this energy-based detection method is the waveform-based detection method, in which a station performing CCA looks for a "signature waveform" under the assumption that this particular waveform is embedded in all transmission bursts. This a priori knowledge improves the detection sensitivity. A further gain in detection performance can be achieved by embedding additional medium access information in the burst for coordination among competing stations [4]. We will see detailed examples of this medium access method in Chap. 7.

The advantage of decentralized communication is the great flexibility that enables direct communication without the supervision of a control station; nevertheless, the disadvantage is also the lack of supervision of a control station, which gives rise to frequent collisions amongst transmitting stations in a dense traffic area, such as ports, harbors, and waterways. Frequent collisions cause not only poor overall system efficiency that limits the system capacity but also the instability of the system.

2.2.2 Infrastructure-Based Communication

The infrastructure-based wireless communication relies on a preexisting managed infrastructure consisting of control stations connected to a core network that provides services to mobile stations. All communications are through the control stations (space station or shore station) in a "star network" topology, as graphically illustrated in Fig. 2.1. A control station provides a common connection point or an access point for mobile stations to the network within its coverage or footprint, commonly referred to as a *radio cell* or simply a *cell*. The network is distributed over a geographic area covered by these interconnected cells, facilitating wide-area wireless connectivity to the service providers for the mobile stations.

2.2.2.1 Centralized Medium Access

The network-originated message is communicated from a control station over the *downlink* to the mobile station(s), and the mobile-originated message is communicated over the *uplink* to the control station. The downlink is thus a one-to-many communication link (i.e., one transmitter with multiple receivers) from a control station to the mobile stations, and the uplink is a many-to-one communication link (i.e., multiple transmitters with one intended receiver) from the mobile stations to the control station. This topology allows for *centralized medium access control* (MAC), which is characterized by the principle that any communication taking place has to go through the control stations.

The control station supervises the medium or radio channel usage by distributing the communication resources, i.e., frequency bandwidth and time of the radio channel, among the "competing" mobile stations. Specifically, the control station MAC includes a dynamic *resource scheduler* that assigns the resources to both downlink and uplink traffic. The assignments are notified to the mobile stations. For that, signaling of control messages between the control station and the mobile stations is required. A portion of the radio resource is thus set aside as the *signaling channels* for that purpose, and the remainder as the *traffic channels* for traffic (e.g., application data). We will go through the design of these channels in greater detail in the following chapters.

When a mobile station has a local uplink message (e.g., some message from a position reporting application) to send (e.g., to the application server), it signals an uplink radio resource request to the network on the uplink signaling channel; the control station resource scheduler allocates the uplink traffic channel resource per the request and announces the assignment/grant over the downlink signaling channel, indicating the granted uplink traffic channel resource that the requesting mobile station is to use for the upcoming message transmission.

On the other hand, a mobile station, expecting to receive messages from the network (e.g., some message from a navigation information service), monitors the downlink signaling channel regularly for notification of the pending downlink message. If a notification is detected (i.e., the mobile station is being paged), the mobile station notifies the control station via the uplink signaling channel that it is ready to receive the message. The pending downlink message is then transmitted over the downlink traffic channel on the resource as indicated in the downlink resource assignments announced over the downlink signaling channel.

Figure 2.3 shows these two typical centralized medium access control processes for data exchange between the mobile station and the MTC network (via the control station). Consequently, unlike distributed medium access, collisions are prevented under the centralized medium access control; however, signaling of control messages between the control station and the mobile stations requires the frequent use of the signaling channels, which incurs additional system overhead. The overhead needs to be balanced among various system design parameters, such as overall system spectral efficiency, coverage, and latency (per the service requirements).

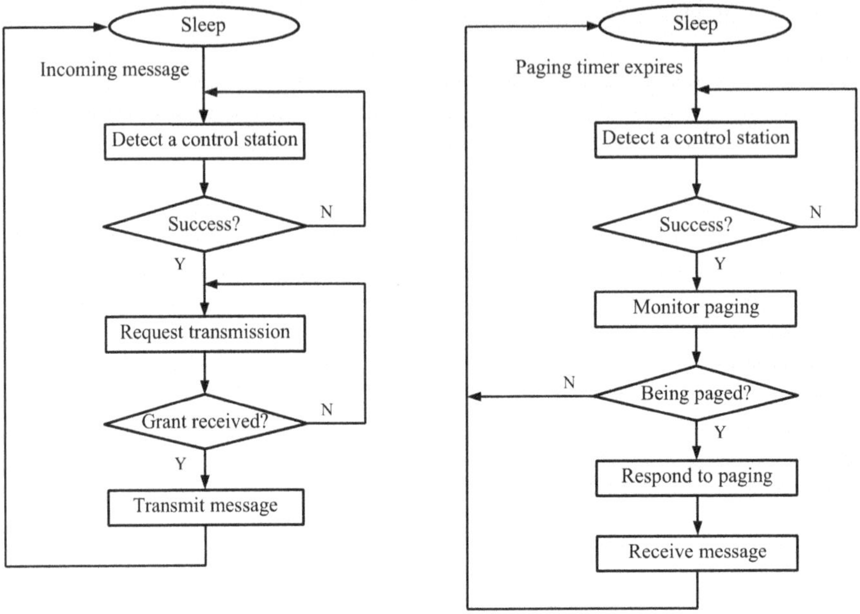

Fig. 2.3 Typical operations performed by a mobile station during "wake-up" in response to a data transfer request (incoming messge) from, e.g., position reporting, or in response to paging timer expiration, in a centralized medium access control network

Nevertheless, it is typically controlled below 30%, resulting in system resource usage efficiency higher than 70%.

However, efficiency is not the only benefit from centralized medium access. In maximizing the overall system spectral efficiency by eliminating transmission collisions among mobile stations, centralized medium access control also empowers the network to easily take into account factors like the radio channel conditions, the quality of service (QoS) requirement, and fairness when allocating available radio resources among competing mobile stations, which is not obvious for distributed medium access.

A resource scheduling strategy that targets only the spectral efficiency would always allow the mobile station with the best radio channel to transmit, maximizing the system spectral efficiency. However, this scheduling policy may starve mobile stations with unfavorable channel conditions, as is often the case for those farther away from the control station and those with limited communication capability. At the other extreme, a scheduler with only fairness in mind would simply divide radio resources evenly among competing mobile stations at the expense of system spectral efficiency and failure to meet some application's QoS requirements. That said, a scheduler only to satisfy the QoS requirement, such as latency, may harm the system's spectral efficiency. A practical scheduler is thus often a compromised solution that takes care of both QoS requirements and fairness and, at the same time, the overall system spectral efficiency or throughput.

2.2.2.2 Frequency- and Time-Division Duplexing

Two operating modes for uplink and downlink transmissions are possible: *frequency-division duplexing* (FDD) and *time-division duplexing* (TDD).

As depicted in Fig. 2.4, in FDD, one radio channel is used to communicate in one direction and the other in the opposite direction. Since a *simplex channel* only allows one-way transmission, either uplink or downlink, a pair of simplex channels provide, ideally, two orthogonal channels for uplink and downlink, where "orthogonal" in non-mathematical terms means that the two links do not interfere with each other, i.e., no cross-link interference. Each communicating end (a control station or a mobile station) uses two paired simplex channels for simultaneous transmitter and receiver operation to establish a *full-duplex* communication link.

To maintain the orthogonality, an essential issue for FDD is the protection of the receiver from the co-located transmitter at each communicating end. For an ideal transmitter, all emission power would be 100% confined within the transmit band. However, in reality, due to imperfect RF isolation, as sketched in Fig. 2.5, the receiver of a station's transceiver during full-duplex communication is continuously exposed to the unwanted out-of-band emission from its co-located transmitter. Here, a transceiver refers to a combined radio transmitter and receiver, typically sharing common circuitry and a single housing. The out-of-band emissions enter into the adjacent receive band and affect the whole receive path. Due to the huge difference in the spatial distance separating the transceiver and the intended remote transmitter at the other communicating end, the local transmitter's out-of-band leakage could be significant (relative to the intended received signal), breaching the orthogonality and hence degrading the receiver sensitivity.

A simple way to avoid the cross-link self-interference at a transceiver is to operate in the *half-duplex* mode. As depicted in Fig. 2.6, two communicating ends, i.e., a control station and a mobile station, take turns transmitting on the downlink and uplink channels, respectively, thereby negating the need for a transceiver to transmit and receive simultaneously. Also, half-duplexing allows reusing or "time-sharing" the RF components along the RF chain of a transceiver, such as the antennas, filters, mixers, frequency sources, and synthesizers. Nonetheless, the duty cycle per channel

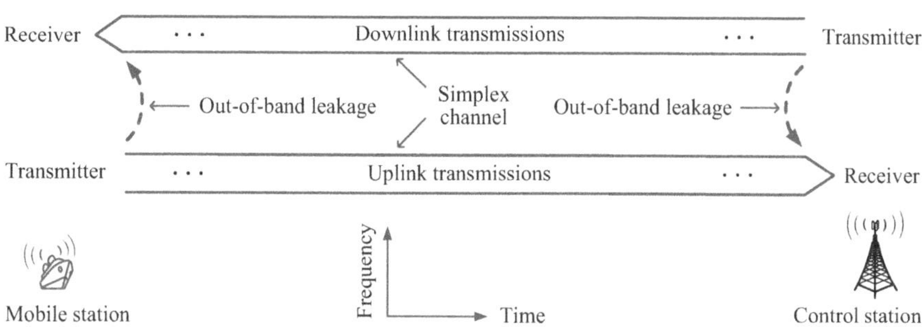

Fig.2.4 Illustration of frequency-division-duplexing of uplink and downlink transmissions

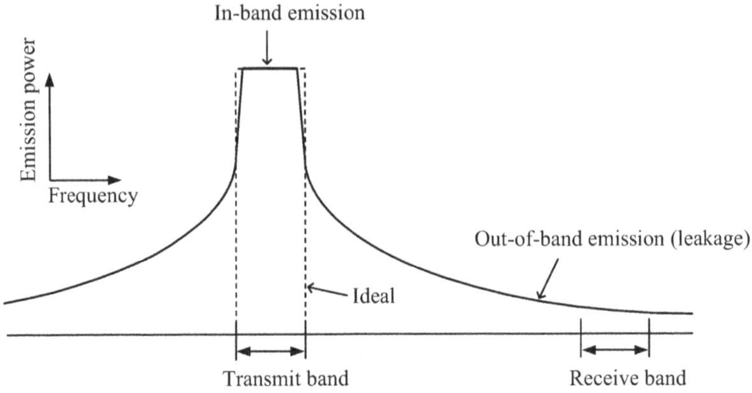

Fig. 2.5 Illustration of the emission spectrum of a transmitter and the cross-link interference at a transceiver operating under full-duplex FDD

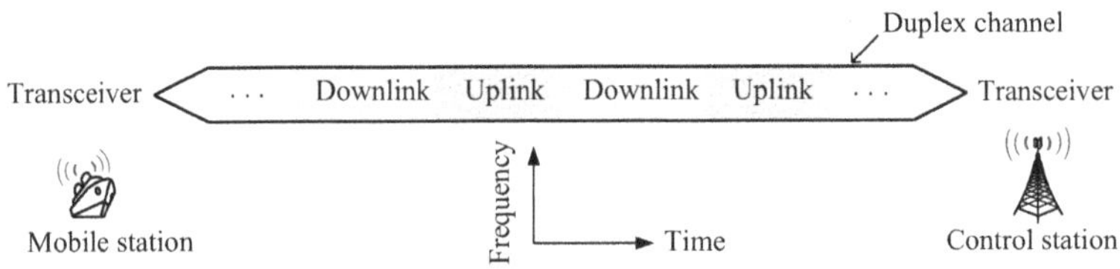

Fig. 2.6 Uplink and downlink TDD in a star-network topology. Here a frequency channel is defined as *duplex* if both uplink and downlink transmissions are allowed on this channel

is halved, so is the channel usage. From the standpoint of a control station operating in half-duplex, it means that the cell spectral efficiency is, in effect, cut down to half.

In order to retain full-duplex, an FDD transceiver relies on the RF filter coupled with "spatial separation" between the transmit and receive RF chains to bring the interference down to a manageable level. Since spatial separation is costly and oftentimes physically hard to achieve, most FDD systems mainly depend on "frequency separation" between the two paired simplex radio channels to ease the transceiver design, to the extent that the same antenna can be used for both transmit and receive with the help of an RF duplexer.

If good frequency separation is not available, a solution to circumvent this cross-link self-interference without sacrificing the spectral efficiency is to employ the TDD transmission strategy over a duplex channel shown in Fig. 2.6. Here, a radio channel is defined as a *duplex channel* if both uplink and downlink transmissions are allowed on this channel. In TDD, frequency separation is replaced by "time separation" to mitigate the cross-link interference. As such, the transceiver of a station never needs to transmit and receive at the same time—the same effect attained by half-duplexing in FDD, but still maintains full channel utilization to maximize the spectral efficiency since the same frequency is used for both directions. Another useful

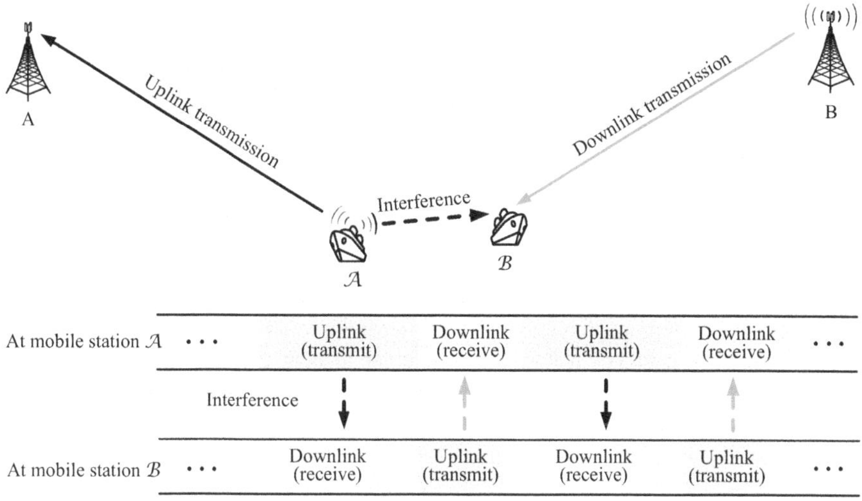

Fig. 2.7 Illustration of cross-link interference under unsynchronized TDD, where station A and station B share the same TDD channel. The dashed line denotes the co-channel interference between mobile stations at a cell edge, where the interference is much stronger than the intended signal due to the near-far effect

advantage of TDD is that the radio channel time can be partitioned in proportion to the downlink and uplink traffic—one extra degree of freedom for handling asymmetric application traffic. For instance, in an uplink-dominant application scenario, more channel time can be allocated to uplink to accommodate heavier uplink traffic, and the same works for the other direction.

However, everything comes with a price. TDD requires synchronizing all control stations so that the downlink timing and the uplink timing are aligned among control stations to avoid the potential cross-link interference at a mobile station. In this case, this cross-link interference is from the co-channel transmission from a mobile station that is transmitting on the uplink. This uplink transmission unintentionally blocks the nearby mobile station in a neighboring cell from receiving the downlink signal from its serving control station.

As exemplified in Fig. 2.7, two mobile stations are being served by two respective control stations that operate in TDD mode and share the same duplex radio channel, but their downlink and uplink are not synchronized, meaning that one station is in the uplink transmission period while the other in the downlink period and vice versa. Consequently, while receiving the downlink transmission from its serving control station B, mobile station $\mathcal{B}$ is interfered with by the uplink transmission from mobile station $\mathcal{A}$ to its serving control station A. Since mobile station $\mathcal{B}$ is much closer to mobile station $\mathcal{A}$ than to its serving control station B, the co-channel interference from mobile $\mathcal{A}$ could easily be several orders of magnitude larger than the desired downlink signal from its serving control station B, causing communication impossible at the cell edge (where the footprints of two control stations meet).

Synchronization of uplink and downlink transmissions between control stations increases network complexity and therefore cost. It also implies that all control stations are forced to use the same uplink/downlink partition pattern, including those space and shore stations that share the same spectrum, if any, offsetting the benefit of the flexibility of TDD that accommodates asymmetric traffic and time-varying uplink and downlink demands.

The cross-link interference issue in both FDD and TDD stems from the so-called *near-far effect*, a prominent phenomenon/issue in wireless communications due to the rapid propagation loss over a wireless channel, where the signal strength falls off quickly with propagation distance. Thus, a small difference in the distance leads to an accentuated difference in received signal strength. As a result, an unwanted signal from a near source makes it hard for a receiver to hear the desired signal from a farther source, although the former (FDD) is attributable to the *adjacent-channel* interference and the latter *co-channel* interference. We will have more discussions on these two types of interference in later chapters of this book.

2.3 Wireless Networks

Obviously, only wireless solutions are relevant to maritime communication. However, communication at sea differs significantly from its land counterpart. The unique maritime MTC requirements impose significant challenges to not just the system architecture but the networking schemes. The abundance of maritime services and features for specific applications and deployment environments calls for the exploitation of commonalities but easy adaptation to differences. To this end, more than one type of networking technology must be in place to address the diverse maritime IoT service requirements and environments outlined in Chap. 1. At least three types of wireless communication networks can be identified from Fig. 2.1 for a full-fledged maritime MTC system.

2.3.1 Satellite Communication Networks

Major communication system solutions would need to be satellite-based overall, considering the usually harsh and remote environment of open oceans and the impracticality and overbearing costs of constructing a terrestrial infrastructure at sea. Satellite communication is the only practical alternative for deep-sea communication out of range from any terrestrial solution. Unlike most terrestrial alternatives, a satellite communication system can be rolled out faster and more economically viable for connecting remote locations across a large area due to the broad field of view or footprint (on the surface of the earth) of a satellite transceiver, even after taking into account the convoluted processes and substantial resources of launching a satellite network, despite the most recent innovative advances in space technology [5]. Nevertheless, as noted in Chap. 1, acquiring the scarce satellite radio spectrum and precious orbits, both the extremely limited natural resources of the earth, remains the biggest challenge in launching a satellite network.

Moreover, satellite communication does suffer from several inherent shortcomings and technical difficulties, mainly, (1) inefficient to cover the high-dense traffic area, like ports, harbors, coastal areas, and waterways, by virtue of the giant footprint of a satellite space station and its limited communication capacity constrained by the payload and power supply, and (2) excessive transmission delays due to the high altitude of a satellite space station, which could be problematic for ships in close quarters in dense traffic areas.

In recent years, there have been growing interests in low-earth orbit (LEO) satellite systems. As it is nearer to the earth (e.g., 600 km), satellites launched into low-earth orbits have the least propagation delay (~10 ms) and a smaller footprint (~1400 nautical miles in radius) compared to the other higher orbits. With the help of beamforming antenna technology, the footprint can be further reduced to, e.g., a 400-nautical radius. These technologies fuel maritime IoT advancement, including the crucial international spectrum allocation.

Continuous global connectivity, including over the polar regions, is provided using a satellite network of multiple interlinked LEO satellites configured in constellations with interlocking orbits [6]. Under this infrastructure, the LEO space stations are interconnected via inter-satellite links, and the backhaul connections are through feeder links to the ground stations (or teleports) positioned around the world. The inter-satellite link is either an RF beam or optical beam for connecting space stations operating in a constellation. The feeder link is a radio link that provides the connection from the earth to the space station. Therefore, a message from a mobile station received by a LEO space station traverses the satellite network via inter-satellite links to reach the ground stations through feeder links. From there, the message is finally transferred to the maritime service cloud network and vice versa.

The transmission range of a station is ultimately limited by the radio horizon, determined by the transceiver's antenna height. As such, a satellite space station has a much larger field of view than a mobile station on earth, and mobile stations within the field of view of a satellite are likely beyond each other's radio horizon. Therefore, these mobile stations are "hidden" from each other because they are unable to detect the existence of one another, rendering the sensing-based distributed medium access control mechanism ineffectual in preventing collisions at the satellite receiver. This situation is known as the *hidden node* or *hidden terminal problem* in wireless communications. As illustrated in Fig. 2.8, Ship $\mathcal{A}$ is hidden from Ship $\mathcal{B}$ and vice versa; hence they are unaware of each other's existence. Because of this vast disparity in "visibility," there are likely a large number of such mobile stations in the field of view of a space station, resulting in a high collision probability at the receiving space station. The hidden node problem is thus accentuated in satellite communication. It is apparent that under centralized medium access control, the space station, now the control station, is more effective in managing the traffic and preventing collisions within its field of view. It is a perfect scenario for centralized resource management, providing a much higher system capacity.

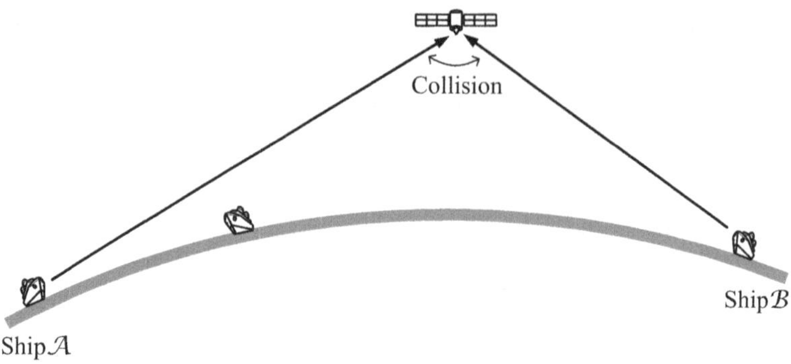

Fig. 2.8 Illustration of the hidden node problem in sensing-based distributed medium access control, where Ship 𝒜 and Ship ℬ are beyond each other's radio horizon

2.3.2 *Terrestrial Communication Networks*

Despite the reduced footprint and latency, a LEO satellite network is still not the best choice for high-traffic coastal areas. Instead, *network densification*, i.e., decreasing the footprint or cell size of a control station and increasing the density of control stations, is an effective approach for achieving capacity scaling in infrastructure-based networks. The terrestrial infrastructure is better suited for the task since a shore station is more flexible and cost-effective in adjusting its footprint (i.e., cell sizing) and reaping the spatial spectrum reuse benefit.

Simply stated, the wireless terrestrial infrastructure is such a terrestrial communication network that provides coverage, through the deployment of shore stations, mainly for the nearshore dense-traffic area, thus, best "hybridized" and integrated with the satellite network to cover areas that are unsuitable for the satellite network to serve. This hybrid model (as shown in Fig. 2.1) is an effective means to consolidate the clustered traffic and address the nonuniformity traffic issue.

Consider this hybrid model in more detail, where the terrestrial network handles high-density mobile stations clustered near shore to complement the satellite network. As graphically illustrated in Fig. 2.9, the high-dense traffic area is partitioned into multiple small areas covered by dense shore stations at scale, in which a radio frequency is fully reused. Therefore, spatial spectrum reuse is better exploited for boosting the overall system capacity to cope with the dense traffic, owing to the much smaller footprint of a shore station. The overall system capacity is thus multiplied, given the same amount of spectral resource, provided that the inter-cell interference is contained.

There exist at least two kinds of inter-cell interference under such a hybrid deployment: (1) interference between neighboring shore stations resulted from spatial spectrum reuse and (2) interference from satellite space stations due to spectrum sharing. Since the terrestrial network is mainly for limited coastal areas, its spectrum is largely underutilized, leading to inefficient and wasteful use of scarce

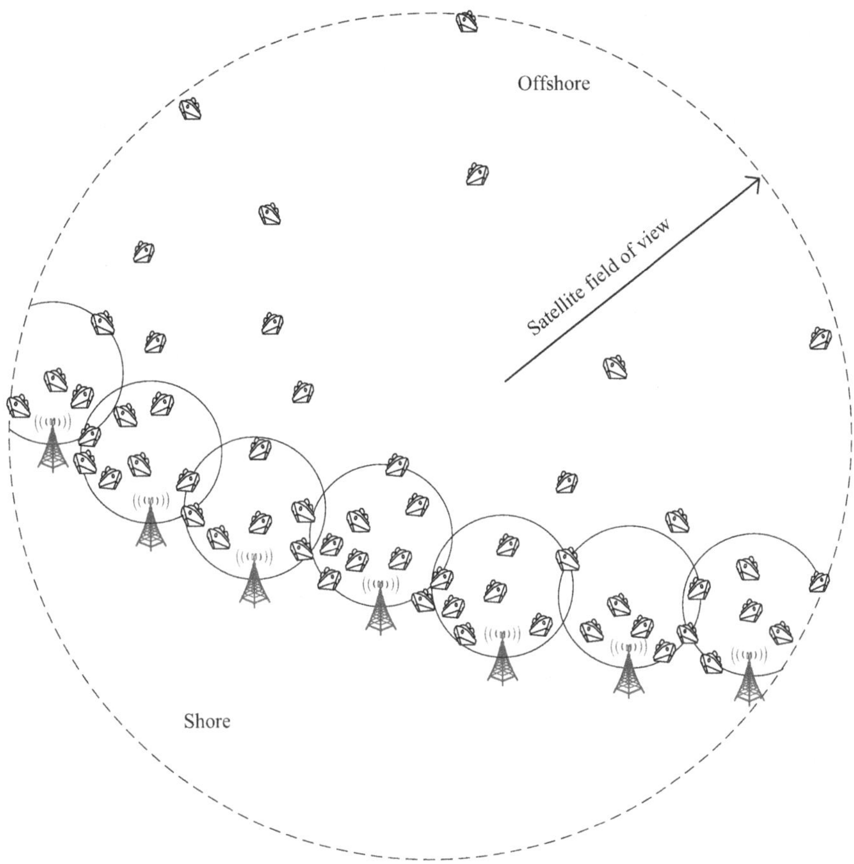

Fig. 2.9 Satellite and shore station footprints, scaled for illustration purpose

spectral resources if not shared with the satellite network. Indeed, co-channel interference due to full spectrum reuse is critical to overall system performance and more so under a hybrid deployment. Inter-cell interference mitigation is treated in later chapters, and the spectrum sharing scheme between the satellite network and the terrestrial network is detailed in Chap. 6.

Centralized medium access control is employed by the shore stations, crucial for tight resource and interference management for such an integrated system with full-spectrum reuse, leveraging the high-speed, high-bandwidth backhaul networks and the high-performance, high-capacity computing power offered by a cloud environment reaping the full benefit of the terrestrial network infrastructure. The actual capacity gain is dependent on the intra-cell co-channel interference management at the data link layer and the inter-cell and inter-network co-channel interference management at the network layer, which are treated in great detail in the following chapters.

2.3.3 *Proximity Communication Networks*

Proximity networking is a paradigm for direct interaction with the physical world, where the physical proximity of a mobile station to a network makes a local service environment immediately available to the mobile station without the presence of infrastructure. It promotes a simple integration of mobile stations into a self-organized service environment for proximity-based services via direct communication between mobile stations. It typically operates under a decentralized or distributed medium access strategy via a random or sensing-based medium access scheme, most effective for short-range self-organized ad hoc networking.

For this type of network, physical proximity plays a main role in determining whom a mobile station interacts with. Only stations in close proximity are more likely to become relevant than those farther away. As noted earlier, in wireless communication, the signal strength falls off quickly with propagation distance. The closer they are, the more likely stations can hear each other. This inherent property of wireless communication conforms with the proximity communication concept and renders it natural for proximity networking and ideal for proximity-based maritime IoT services, such as aids-to-navigation, collision avoidance, and route exchange.

These three integrated networks complement each other, and jointly, they provide a hybrid wide-area wireless access method for a truly space-earth integrated, service-centric MTC system for maritime IoT.

2.4 Applicability of Existing MTC Technologies

Many terrestrial MTC technologies are commercially available [7–12]. Among them, Narrowband IoT (NB-IoT), developed by the 3rd Generation Partnership Project (3GPP), is a mainstream MTC technology for general terrestrial IoT applications operating on the licensed terrestrial cellular bands, e.g., the refarmed 900-MHz GSM bands, deployed on the legacy LTE cellular infrastructure employing centralized medium access control.

Technically, the NB-IoT network architecture and protocols are mostly inherited from the 3GPP human-type cellular communication technology, i.e., the LTE mobile broadband technology operating on a licensed, up to 20 MHz bandwidth. LTE is optimized for broadband mobile Internet access, like media streaming and live conferencing. Hence the access protocol and interface with IP networks are not the most spectrally and energy-efficient for maritime IoT applications that are typically characterized by short burst data services, especially for power- and energy-limited applications discussed in the previous chapter.

Specifically, the NB-IoT network and data link protocols are also inherited from LTE. At the network level, NB-IoT remains a traditional network-centric rather than a service-centric network and hence is not the most efficient in dealing with IoT applications and services. Most importantly, it lacks an efficient network protocol for interworking with the IP network where most maritime IoT services nowadays reside. The LTE Packet Data Convergence Protocol may work well for wideband data streaming but is not the most efficient for short burst data services. At the data link level, NB-IoT follows a lengthy random access procedure from the legacy LTE, involving back-and-forth signaling [13], which is spectrally and energy-inefficient for short burst data services. We will have a detailed discussion on maritime MTC layering and protocols in Chaps. 3 and 4.

At the physical level, NB-IoT employs a multi-carrier waveform, i.e., the OFDM waveform, with a subcarrier interval of 3.75 kHz. A general discussion on the OFDM waveform is deferred to Chap. 5. An in-depth look at the NB-IoT waveform is referred to [14–17]. In short, it has two main drawbacks to maritime MTC. First, it is sensitive to frequency errors, whereas fast-moving satellites can have a Doppler shift of multiple kilohertz relative to a mobile station on the earth depending on the satellite's orbit. For a LEO satellite orbiting at an altitude of 600 km, the shift can be as large as ± 4 kHz off the carrier frequency at 160 MHz, for instance. Second, an OFDM waveform typically has a poor transmit power efficiency, while power efficiency is critical for transmitters aboard a LEO satellite space station and power- and energy-limited mobile stations. A unified air interface framework solution for both satellite and terrestrial domains, essential to the simplicity and low-cost requirement, is not readily available from NB-IoT. We will revisit these technical issues in greater detail in Chaps. 5, 8, and 9.

Commercially (and more importantly), maritime MTC is meant to be loyalty-free for cost-efficiency and IMO-mandatory for all vessels. NB-IoT is thus not the best candidate in terms of the potential implications of intellectual property rights. In this respect, other state-of-the-art MTC technologies, like SIGFOX [7], LoRa [8], and Weightless [9], are not under consideration since all are proprietary technologies.

Moreover, commercial technologies, like GSM and 3GPP LTE, are mostly market- and profit-driven. They come and go almost every decade and become increasingly complex and costly, leaving a significant gap in backward compatibility and uncertainty in cost and sustainability. In essence, they are not well-aligned with the core principle of a maritime MTC system, as outlined in Chap. 1.

References

1. ITU Radiocommunication Sector, Assignment and use of identities in the maritime mobile service, Rec. ITU-R M.585-7, ITU, Geneva, Switzerland, May 2015
2. N. Abramson, The aloha system: another alternative for computer communications, in Proceedings of the Fall Joint Computer Conference, pp. 281–285, 1970
3. J. Zhang et al., LTE on the license-exempt spectrum. IEEE Commun Surv Tutorials **20**(1), 647–673 (2018)
4. IEEE Std.802.11, Part 11: Wireless LAN Medium Access Control (MAC) and Physical Layer (PHY) Specifications, 2013
5. Mike Brown (2020 August). SpaceX: Elon Musk Breaks Down the cost of reusable rockets. https://www.inverse.com/innovation/spacex-elon-musk-falcon-9-economics

6. Y. Ulybyshev, Satellite constellation design for complex coverage. J. Spacecraft Rockets **45**(4), 843–849 (2008)
7. SIGFOX, SIGFOX — one network a billion dreams, whitepaper
8. N. Sornin *et al.*, LoRaWAN Specification, ver. 1.0, Jan. 2015
9. Weightless; available: http://www.weightless.org/
10. M. Wang et al., Cellular machine-type communications: physical challenges and solutions. IEEE Wireless Commun. **23**(2), 126–135 (2016)
11. 3GPP, TR 45.820, Technical Specification Group GSM/EDGE Radio Access Network Cellular System Support for Ultra-low Complexity and Low Throughput Internet of Things (CIoT), Rel. 13, v.13.1.0, Nov. 2015
12. J. Gozalvez, New 3GPP Standard for IoT. IEEE Vehic. Tech. Mag. **11**(1), 14–20 (2016)
13. M. Hua et al., Analysis of the frequency offset effect on random access signals. IEEE Trans. Commun. **61**(11), 4728–4740 (2013)
14. W. Yang, M. Wang, J. Zhang, J. Zou, M. Hua, T. Xia, X. You, Narrowband wireless access for low-power massive Internet of Things: a bandwidth perspective. IEEE Wirel. Commun. **24**(3), 138–145 (2017)
15. *3rd Generation Partnership Project Technical Speciation Group Radio Access Network Evolved Universal Terrestrial Radio Access (E-UTRA) Physical channels and modulation (Release 13) Ver. 13.2.0*, document TS 36.211, Jun. 2016
16. Y.E. Wang et al., A primer on 3GPP narrowband internet of things. IEEE Commun. Mag. **55**(3), 117–123 (2017)
17. Y.D. Beyene et al., NB-IoT technology overview and experience from cloud-RAN implementation. IEEE Wirel. Commun. **24**(3), 26–32 (2017)

Chapter 3
Protocol Structure

Abstract A large-scale heterogeneous communication system, like the maritime MTC, requires a well-designed protocol structure for interoperability and flexibility. This chapter provides a structural layout for communication under the architecture portrayed in Chap. 2. We first explain the benefit of a layered protocol structure and present a three-layer protocol stack model and the functionality of each layer under the architecture. We then describe a unified method, using the concept of layer channels and protocol and service data units, for control and service data exchange between layers and between local and remote, in a service-centric network. The concept and principle presented in this chapter serve as a framework and ground rules for the more detailed protocol designs in future chapters.

Keywords Protocol structure · Protocol stack · Protocol layer channels · Protocol data unit · Service data unit · Service-centric network · Service identifier

3.1 Protocols and Layering

In simplest terms, communication is the process of conveying information or messages from one entity to another through mutually understood rules.

A *protocol* is a set of predefined rules and formats that govern the way or process of communication between communicating peers. A single self-explained unit of data for information exchange between peer protocols is commonly referred to as the *protocol data unit* or PDU. The protocol defines the content and format of the PDU and how the PDU is exchanged. The PDU serves as a common "language" or interface such that the information exchanged between the peers is understood by peers without ambiguity. Therefore, a protocol is necessary for implementing any function or process that requires cooperation involving two or more entities.

The implementation of a communication system, like the maritime MTC system, typically involves many processes or protocols to complete the delivery of a message. Some of these protocols are not dependent on other protocols, but some are. Such dependencies between processes can be represented by a layered structure or layering. With a layered structure, what layer a protocol belongs to depends on the

M. M. Wang, J. Zhang, *Machine-Type Communication for Maritime Internet-of-Things*, Wireless Networks, https://doi.org/10.1007/978-3-030-77908-5_3

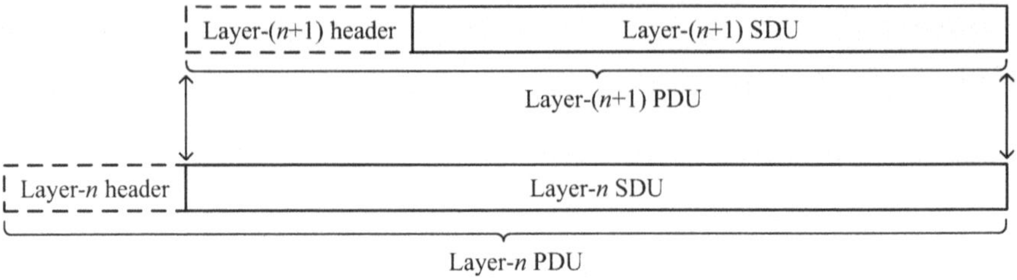

Fig. 3.1 Relationship between a PDU and an SDU, where PDU headers are denoted as dashed lines, meaning some PDUs may not have a header

service provided by the layers below it. Except for the lowest layer, a layer depends on the lower layers to get its PDUs across to its peer. The independent protocols thus reside in the lowest layer, responsible for providing services to the upper layers.

For interoperability and flexibility, as well as reduction of design and maintenance complexity, the layers are maximally decoupled to such extent that protocols are structured and implemented in a way that the protocol in a layer depends only on the layer right below it. As such, a protocol in a layer provides services to the protocols in its immediate upper layer through an interface, commonly known as the *service data unit* or SDU, which is a data unit for containing or, more precisely, encapsulating the PDU passed down from the upper layer. Here, by "encapsulating," it means that the current layer is not supposed to understand or try to understand the content of its SDU, much like a mailman is not supposed to open and read the mail—it is a federal crime for anyone except the addressee to open or destroy mail. Apparently, a Layer-n SDU is a Layer-(n+1) PDU, assuming layers are numbered from lower to upper in ascending order. The difference is that the PDU specifies the data meant to be sent to the peer protocol layer at the receiving end, and for that to happen, it needs to be passed to a local lower layer, where it is considered as an SDU and treated as the payload of the layer's PDU.

Figure 3.1 depicts this relationship, where a PDU consists of an SDU and an optional header or metadata.

The header normally contains an identifier for identifying the PDU type that the current PDU belongs to so that the layer knows which protocol it is supposed to be forwarded to once it arrives at the destined layer. Therefore, when the PDU is passed to a matching protocol layer along the path to its final destination, it helps the corresponding layer deliver its intended service, simply put, what to do with it. Nonetheless, as shown in Fig. 3.1 as dashed lines, a header may not be present. After all, a header is an overhead to the layered protocol system; it is saved for efficiency as long as the omission does not cause confusion on its journey to the destination. We will see practical examples in Chaps. 7, 8 and 9.

When taken together, the protocols organized in such a hierarchical layered structure are collectively referred to as the *protocol stack*. When two systems communicate using a layered protocol stack, they do so in a logical layer-to-layer fashion. Each layer in a protocol stack must exist on both the sending and receiving

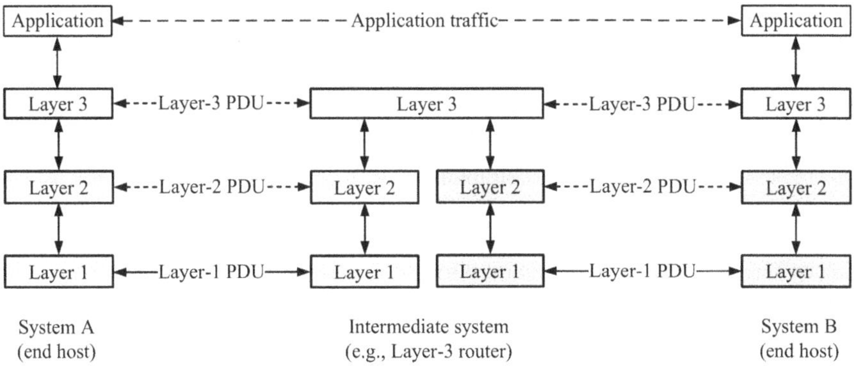

Fig. 3.2 Protocol layering and interconnection, exemplified by a four-layer protocol stack. The dashed arrow denotes a *logical* connection and a solid arrow denotes a *physical* one

hosts, and each layer on one host communicates only with its peer layer in the protocol stack on the other host. However, each layer depends on its lower layer to deliver its PDU, but the information regarding how it is delivered is shielded from its upper layer. Cross-layer information sharing is avoided, in principle.

The top layer of a communication system provides services to the ultimate clients that the system is built to serve. For the maritime MTC system, the ultimate clients are the maritime IoT applications and services. These clients reside in a layer commonly known as the *application layer* that is not intrinsic to the communication system and hence is not deemed part of the system (in this book). This detachment frees the maritime IoT applications from the MTC system, breaking the ties between the application and the communication system present in the traditional maritime communication paradigm, like AIS noted in Chap. 1. It unleashes the development of Internet-based maritime applications distributed all around the world to take the best advantage of the worldwide Internet without worrying about how the data are delivered to the end clients.

Figure 3.2 illustrates communications between application processes running in differing systems (Systems A and B, i.e., the application hosts). The model employs a protocol stack structure of three layers. Each layer performs "value-added services" at the request of the subsequent higher layer and, in turn, requests more basic services from the subsequent lower layer unless we have already hit bottom (the lowest layer). Each layer on the stack may add or provide specific functionality for its intended purpose and relies on the layer below it for the actual transfer. The higher layer protocol passes its PDU to the lower layer as the SDU of the lower layer. The lower layer encapsulates one or more SDUs in its PDU and passes it on till the lowest layer, i.e., Layer 1, where its PDU is transmitted through a transmission medium (e.g., radio/wireless channel, twisted-pair copper wire, fiber optics). At the receiving end, the process through the stack is reversed. An intermediate system, e.g., a Layer 3 router, is present on the journey of the application data in the example, although more than one intermediate system could be present. As such, the Layer 3 PDU header may contain a Layer 3 address defined by the Layer 3 protocol to help with routing. A Layer 3 protocol, i.e., routing protocol, is present in all systems across the network (including the end systems and intermediate systems). In other words, there is a common Layer 3 function in each system; nevertheless, Systems A and B do not necessarily have the same Layer 2 and Layer 1 protocols as long as the intermediate system has both. These protocols communicate with each other by exchanging Layer 3 messages via Layer 3 PDUs, and ultimately, they transfer application data between Systems A and B, i.e., the end hosts of the application. At the host, the top layer directly interacts with the application through an application programming interface, commonly known as API.

The benefits of this layering structure are obvious. First, it breaks up a complex problem into more manageable sub-problems—a typical "divide and conquer" paradigm, simplifying how protocols are designed. Second, the upper layer protocols can share the lower layer functionality with greater interoperability and flexibility, reducing the overall system complexity and increasing versatility and manageability. Finally, but importantly, it promotes abstraction or virtualization of the layer functions, hiding the implementation details. This property is of particular importance for the maritime MTC system, which establishes a framework that enables sharing the infrastructure assets among different nations or organizations without getting into the underlying physical implementation.

Nonetheless, the drawback of enforcing strict layer boundaries is that it prohibits "cross-layer optimization" that requires layer-specific information exchange across different layers, which may improve the overall performance. This is a trade-off between performance and simplicity. Therefore, the layering constraint is usually relaxed for performance optimization in most practical wireless system designs where the radio spectral efficiency is just "too big to ignore."

3.2 Maritime MTC Protocol Structure

The maritime MTC follows the generic layered structure described in Sect. 3.1. Specifically, the protocol stack consists of three native layers: *network layer*, *data link layer*, and *physical layer*, as depicted in Fig. 3.3. Layer 3, the network layer, performs the connection management for applications. The lower layers, i.e., Layer 2 (the data link layer) and Layer 1 (the physical layer), jointly handle the physical delivery over the air.

In general, the data link layer manages the network layer PDU delivery over the physical layer, between two or more communicating stations within the same local network, such as neighboring mobile stations in a proximity network or between the control station and mobile stations within a cell, satellite or terrestrial, without concern for the ultimate destination. The delivery typically does not cross the boundaries of a local area network.

The network layer is above the data link layer, mainly because the network layer has a broader "view" of the overall network (i.e., a wide area network) than the data

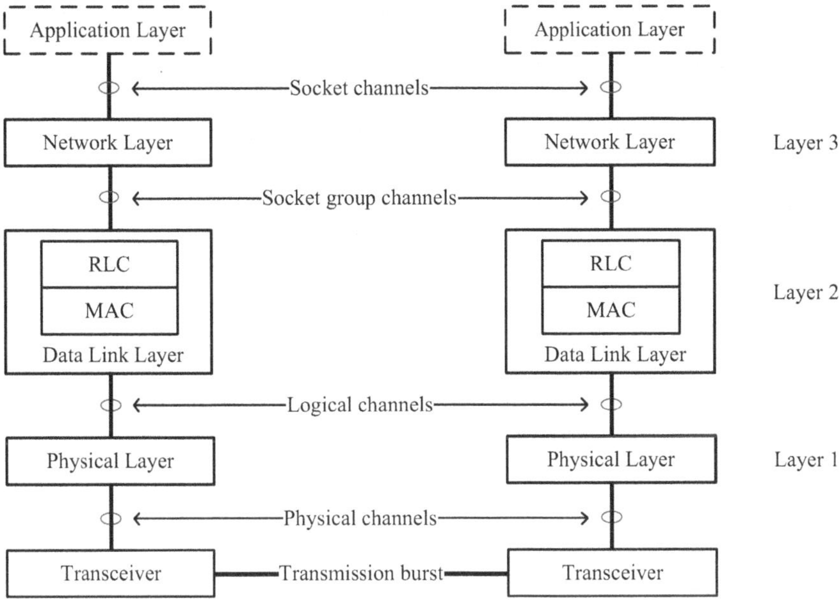

Fig. 3.3 Maritime MTC protocol stack consisting of three native layers: the network layer, the data link layer, and the physical layer

link layer. It oversees end-to-end service delivery of application traffic, possibly through intermediate systems and across multiple networks, likely heterogeneous (e.g., the maritime MTC network and the external IP networks).

3.2.1 Network Layer

Specifically, the maritime MTC network layer performs the end-to-end connection management for applications, including network resource management and interworking between the internal (the maritime MTC network) and external networks for IP-based maritime IoT applications. As a service-centric network, end-to-end means from application to application. It encompasses a suite of network protocols to perform intended network functions. Among them, the network layer's primary function is to maintain the interconnection of application hosts, with quality of service or QoS management, maritime IoT service discovery, service address resolution, and internetworking protocol adaptation and convergence if the connection involves two different networks.

For proximity communication, the connection is between the applications (possibly situated on a local private network, e.g., Ethernet, WiFi, RFID) connected to the respective mobile stations. And for terrestrial and satellite communications, it is between the application (possibly on a local private network) connected to a mobile station and the server, i.e., the application service provider, on the maritime cloud that resides on an IP network, e.g., the Internet.

For maritime MTC in particular, the network layer is also responsible for proximity, satellite, and terrestrial domain integration and radio resource management. Radio resource management encompasses a wide range of functions and procedures, including frequency band availability and assignments, spectrum sharing and interference management, coordination between control stations, radio-related measurements, and selection and reselection of serving control stations. Frequency band assignment assigns a set of frequency bands available for communication in the given geographical area to a station. In the case of centralized resource control, the time and frequency resources of the assigned bands may be coordinated in a pattern to prevent the neighboring control stations from using the same resource to avoid the inter-cell interference, either in a fixed or dynamic manner under which the amount of allocated resource may reflect, e.g., the traffic load of the cell. Radio resource management also involves interference assessment. For instance, a mobile station may be requested to measure the interference from neighboring control stations to determine if coordinated transmission and reception or switching the serving cell to a neighboring cell is warranted.

3.2.2 Data Link Layer

Layer 2 or the data link layer manages the physical delivery of network PDUs between the maritime MTC stations (i.e., between a mobile station and a control station or between mobile stations) over the physical medium, air or space. The network PDUs from the network layer, containing both network control traffic and application traffic, are processed by the data link layer before being handed over to the physical layer for over-the-air/space transmission.

Specifically, the data link layer is responsible for (1) the adaptation between the upper network layer and the lower physical layer and (2) mobile station transmission medium access, QoS and reliability control. As such, the data link layer is further divided into the radio link control (RLC) sublayer and the medium access control or MAC sublayer, referring back to Fig. 3.3.

3.2.2.1 Layer Adaptation

In particular, the adaptation function of the data link layer includes:

1. *Segmentation*. The network layer PDU may exceed the maximum size of the SDU of the physical layer, dependent on the transceiver's capability of a mobile station. When this is the case, the network layer PDU must be split into multiple fragments suitable for transmission or processing by the physical layer – a process known as segmentation (also known as *fragmentation*).

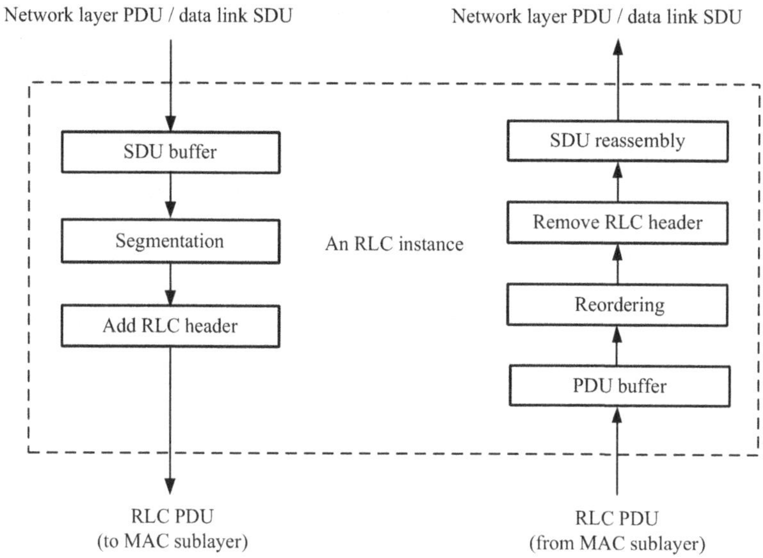

Fig. 3.4 Network layer PDU segmentation and reassembly by an RLC instance

In a practical communication system, the size of data per transmission (i.e., the burst payload) is upper-limited, mainly by the transceiver complexity (such as buffers) and other factors like system latency—a larger data block takes longer transmission time and potentially introduces excess medium occupancy time and blocking other users from using the medium. Hence, the data link layer makes sure that its PDU delivered to the subsequent layer, i.e., the physical layer, does not exceed the SDU limit set by the physical layer. Therefore, the RLC sublayer of the data link layer performs segmentation by breaking down a large data link SDU (i.e., the network layer PDU) into several smaller SDUs if necessary. The peer RLC protocol is responsible for reassembling the original SDU before forwarding it to the layer above (i.e., the network layer).

This process is diagrammed in Fig. 3.4, where an RLC instance is created for each socket group to differentiate network PDUs from different socket groups, which will be explained in Sect. 3.3.

2. *Multiplexing*. In the opposite case, a PDU can be much smaller than the minimum size of the SDU of the subsequent layer (i.e., the physical layer)—a constraint often imposed by the transmission granularity of a practical system, determined by the radio resource granularity and minimum channel code block length. For instance, the code block size of a turbo code is typically greater than 100 bits for a good coding gain (discussed in Chap. 5). To maximize the resource efficiency and the coding gain, the MAC sublayer of the data link layer may pack multiple small MAC PDUs into the payload of a single data link PDU.

3.2.2.2 Medium Access and Reliability Control

The control function of the data link layer, specifically the MAC sublayer, includes the following:

1. *Medium access control.* Before a data link PDU is passed on to the physical layer for transmission, the data link layer must perform transmission resource allocation and determine the corresponding modulation and coding scheme (detailed in Chap. 5). As mentioned in Chap. 2, for distributed resource allocation, random medium access and sensing-based medium access strategies are used for contention-based medium access control. For centralized resource allocation, the medium access control protocol of the data link layer schedules the radio resources for both downlink and uplink traffic by taking into account the radio channel conditions, as well as the traffic volume, the QoS requirement, and possibly fairness when sharing resources among mobile stations. This process also includes the random access used by a mobile station to send a short message, e.g., an uplink scheduling request, to the control station without the need for resource scheduling. Also of primary concern for battery-powered IoT applications are the mobile station transmission power constraint and energy conservation, which strongly affect the way resources are allocated and scheduled.
2. *Reliability control.* Data integrity is probably the most critical aspect of the design and implementation of a communication system. Any unintended change to data as the result of transfer and processing is a failure of data integrity. The unintended change includes 1) missing data from the original data and 2) altered data, i.e., bit errors. A communication system must ensure that information or message is transferred intact from its transmitting peer to its receiving peer. A common method to correct the error is to request retransmission once a missing piece of data or a data error is detected by the receiving end. This technique is generally known as the *automatic retransmission request* or ARQ, but its implementation can be quite different in wired and wireless channels.

   Since bit errors are rare in a wired communication link, ARQ is normally not provided in lower layers, meaning that incorrect SDUs are detected but only canceled, not retransmitted. Instead, data integrity via retransmissions relies on higher-layer protocols, i.e., the transport protocols of the network layer or the application itself in the application layer. The reason is that the common cause of the loss of data integrity in wired communication is the loss of data that happens in the network layer due to, e.g., IP datagram loss during network congestion when the buffer of a Layer 3 router overflows.

   Nonetheless, radio channels are arguably the most erroneous and unreliable communication media due to interference, weather, and fading, which impairs the RF waveform and causes a high probability of data corruption or bit errors. The constant retransmissions invoked by the higher layer protocols entail excessive delays, such as the Transport Control Protocol (TCP) of the Internet Protocol or IP. Moreover, in addition to retransmissions in response, TCP also treats frequent loss of datagrams as an indication of network congestion (for a reason mentioned above) and drastically slows down the transmission—most likely unnecessary in a wireless system. Consequently, in a radio communication system, ARQ is necessary at the lower layer (i.e., the physical layer controlled by the MAC sublayer) closer to where the error happens and thus can be handled more efficiently. More details regarding ARQ and its implementation are treated in Chaps. 5, 7 and 8.

3.2.3 Physical Layer

Layer 1 or the physical layer executes the physical delivery of data link layer PDUs over the air or space between remote stations within the maritime MTC network, per the supervision of the data link layer, the MAC sublayer in particular. It handles the transmission and reception of its SDU, i.e., the data link PDU, through the protocol's defined physical transmission medium, i.e., the radio channel.

The physical layer's job is to move the raw bits of the data link PDU from one station to another over the radio channel. To this end, the bit sequence is transformed into a sequence of constellation symbols on the transmission resource (prescribed by the MAC sublayer) to be transmitted over the radio channel by the radio transceiver. This transformation process is rather involved and directly impacts the system performance and ergo warrants a chapter of its own (Chap. 5).

It is also the physical layer's responsibility to measure and report the radio channel conditions to the MAC sublayer to aid the MAC in selecting the best radio resources and waveform for over-the-air/space transmissions. This responsibility extends to other radio measurements, including inter-cell interference, per the instruction from the radio resource management protocol of the network layer.

3.2.4 Cross-Layer Optimization

Strict boundaries between layers are enforced in the original layering model for maximum decoupling of layers by keeping protocol information strictly within its layer. This layering scheme works well for traditional wired networks, like the Internet layering model. As aforementioned, layering violations are sometimes tolerated to allow access to the information and control of the lower layer. In this cross-layer optimization model, each layer is characterized by a few key parameters and control inputs. The parameters and controls increase the observability and controllability at all levels of the network to enable fast adaptation and highest spectral efficiency for best overall performance, which is especially beneficial to wireless networks due to the scarcity of radio spectrum, high mobility of end-users, and the dynamic and uncertain nature inherent to wireless radio channels. For that, we are willing to pay the penalty of increased complexity. We will see detailed design examples in Chaps. 7, 8, and 9.

3.2.5 Application Traffic and Control Traffic

The traffic generated by each layer includes that from the upper layer and may also include that originated from the protocols of the current layer, i.e., the control traffic between the protocol peers.

For example, besides carrying the application traffic (data exchange between the application layer peers), the network layer PDU may carry network control messages, like the network resource update information for network management, generated by a network layer protocol and consumed by the protocol peer at a mobile station or vice versa. We refer to the network layer PDU that carries the application traffic as the *network layer traffic PDU* and the network layer PDU that carries the network control traffic as the *network layer control PDU*.

The same happens in the lower layers. The data link layer's primary responsibility is to deliver the network layer PDUs to the peer protocol, which includes both traffic and control PDUs. They may have different QoS requirements, but both belong to the traffic PDUs from the eyes of the data link layer. To carry out this responsibility and meet the QoS requirements, MAC control protocols, like medium access control and ARQ protocols, execute a suite of MAC control functions via the so-called data link layer signaling PDUs or MAC Signaling PDUs in particular. Likewise, both data link traffic and signaling PDUs belong to "traffic" as far as the physical layer is concerned. Physical signals (e.g., preamble, pilots) for signaling synchronization and channel information are often embedded in a transmission burst for a receiver to detect the transmitted burst and recover the physical layer PDU (the traffic) over the physical medium.

Apparently, only the application traffic throughput is relevant from the system's perspective – after all, this is where the money is. On the other hand, the control or signaling traffic is simply the system overhead or cost and hence should be minimized, provided that the requirements are met. As will become clear in future chapters, a high QoS requirement requires tight control at every level of the protocol stack and incurs high control and signaling overhead.

3.3 Protocol Layer Channels

As already seen, the maritime MTC protocol stack consists of three protocol layers; information flowing between these layers is through the concept of "channels," referring back to Fig. 3.3. These *layer channels* are used to distinguish different protocol data, enabling orderly and defined segregation. They provide interfaces to each layer within the protocol stack and allow protocol data to be transported between layers with specific value-added treatments at each layer. A layer channel is thus a logical construct for ease of design and implementation, which is the main topic of this section.

3.3.1 Socket Channels

The *socket channel* or simply *socket* is the communication channel between a maritime application and the maritime MTC network. As depicted in Figs. 3.5 and 3.6, the network layer receives application data (either from the local application layer or from a premises network) through the socket channel.

In particular, a socket channel is a structure that serves as an endpoint for sending and receiving application data between the application layer and the network layer. The structure and properties of a socket are defined by an application programming interface or API. The socket channel concept is mainly adopted from the Internet socket for the maritime MTC network layer to interwork with the application. A socket is usually created and managed by the embedded operating system and used by the network layer for identification and interaction with applications in a practical system.

An application running on a host on an IP network (the default) is identified by a combination of an *IP address* and a *port number*, and a socket channel is bound to a port number so that the network layer can identify the application to which data is destined to be sent. On the client-side, the maritime application client is assumed to know the server's network location (IP address and port number). Once a socket is successfully created, the client can use the socket channel to communicate with the application on the other end through the maritime MTC network by writing to or reading from the socket channel. The same goes for the client-to-client scenario.

Since maritime MTC is a service-oriented network, an application is uniquely identified by the service name or service identifier within the network, i.e., a Layer 3 identifier. It is more efficient (and natural) for an application to present the service identifier to the network when engaging in communication with its peer of interest, even if the host is on an IP network. It is the network layer's responsibility to identify the best host and convert the service identifier to the peer's best IP network location.

3.3.2 Socket Group Channels

The application data, now the network layer SDUs, are grouped and passed on to the data link layer in the form of network layer Traffic PDUs through the *socket group channel*, created and maintained by the network layer PDUs control protocol. Each is associated with a specific QoS of the maritime IoT application, represented by, e.g., some predefined priority index. Socket groups thus define what type of data is to be transferred. The data link layer makes use of this information to manage the transmission of various types of network layer traffic (application or control) and strives to meet the corresponding service requirement using the communication resources specified by the network layer. The RLC instances are created just for this purpose to handle individual socket groups. The top m socket groups, Socket Groups 0 to $m-1$, are the default for the control traffic, typically of the highest priority, and the rest for various types of application traffic.

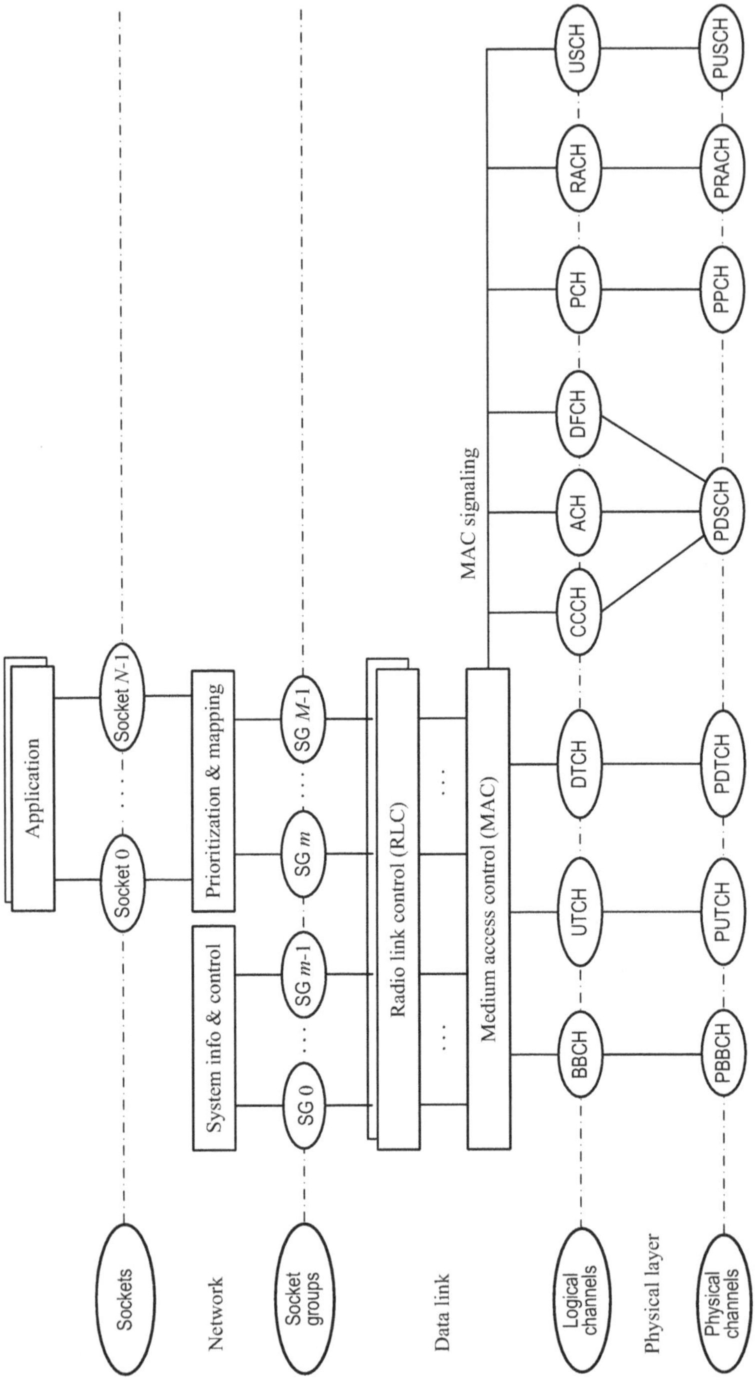

Fig. 3.5 Layer channels and mapping for centralized medium access control with a star topology

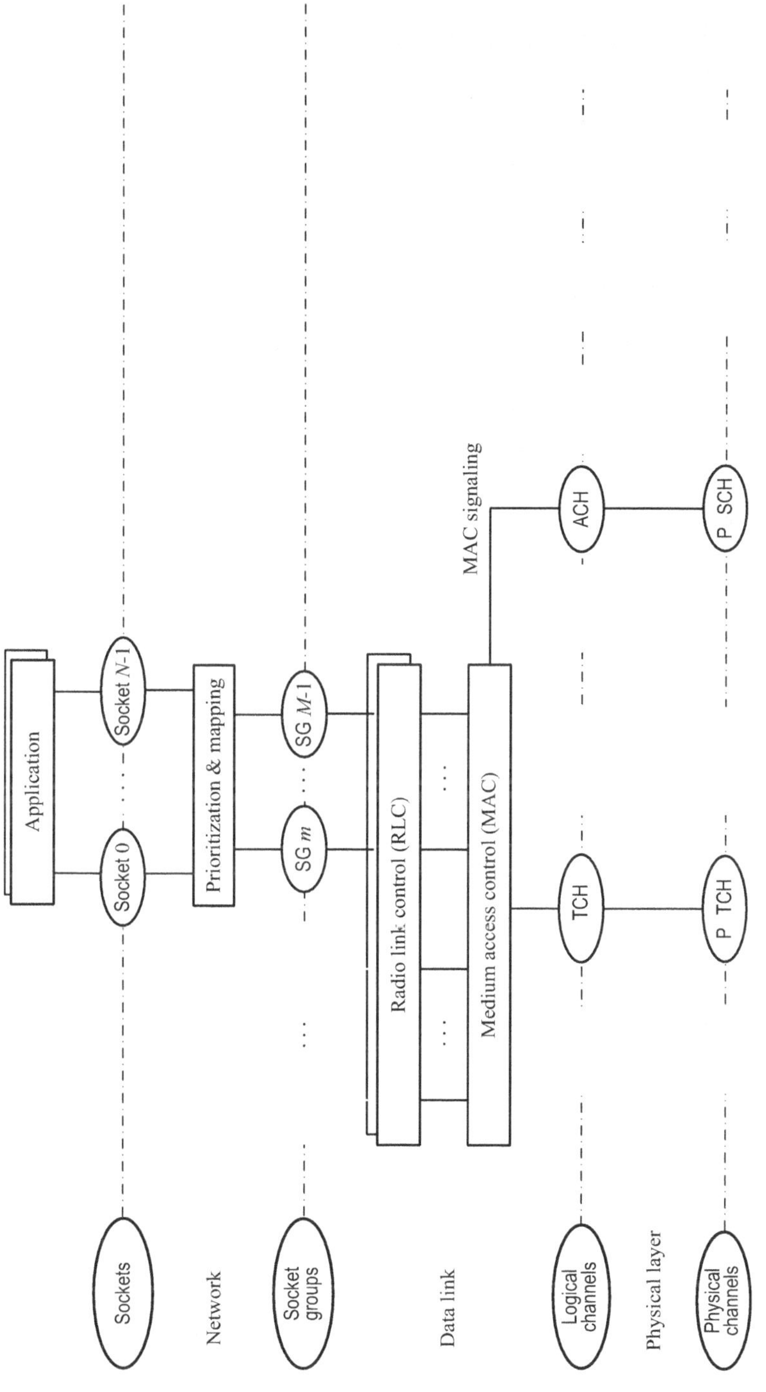

Fig. 3.6 Layer channels and mapping for air interface with distributed medium access control

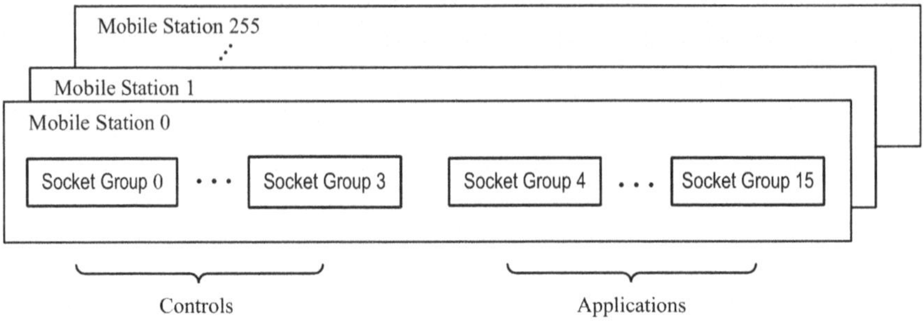

Fig. 3.7 Socket group indexing example at a control station, where $M = 16$ and $m = 4$ (cf. Fig. 3.5), and the mobile station is identified by its MMSI (a Layer 2 identifier)

Since a control station deals with more than one mobile station, their socket groups need to be locally managed separately. Figure 3.7 shows an example of socket group indexing, where a total of four bits are used to index the socket group: four for control and 12 for traffic.

3.3.3 Logical Channels

A *logical channel* is for data exchange between the data link layer and the physical layer. Different types of logical channels are utilized by the data link layer, particularly the MAC sublayer, to carry out specific tasks, mainly to transport MAC traffic (i.e., MAC Traffic PDUs) and signal MAC control messages (i.e., MAC Signaling PDUs).

For ad hoc communication (Fig. 3.6), the Traffic Channel (TCH) is mainly for transporting MAC Traffic PDUs. The Assignment Channel (ACH) is for signaling medium access control-related messages, including resource assignment and modulation and coding scheme for the physical channel that carries TCH. All other MAC Signaling PDUs are through TCH, either alone or piggybacked with MAC Traffic PDUs, in a data link layer PDU.

In infrastructure-based communication (Fig. 3.5), e.g., satellite and terrestrial MTC, there is a notion of "uplink" (from mobile station to control station) and "downlink" (from control to mobile). Hence depending on its uplink traffic and downlink traffic nature, a traffic channel, i.e., TCH, is further distinguished by a directional prefix, uplink or downlink, to read Uplink TCH (UTCH) or Downlink TCH (DTCH), respectively.

Pertinent to the centralized medium access control, the Bulletin Board Channel (BBCH) is designated for the network layer of a control station to broadcast the top-level system information like system configuration and operational parameters necessary for accessing the network. It is an express highway between the network layer peers of a control station and a mobile station.

The Common Control Channel (CCCH) is for broadcasting the resource configuration message from the control station to all mobile stations, containing (and modifying) a subset of the system parameters broadcast by BBCH but at a faster update rate to increase the system's responsiveness at a lower overhead.

The Paging Channel (PCH) is for the MAC sublayer to page a set of mobile stations with pending downlink data. PCH is typically transmitted at a low duty cycle so that the mobile station can enjoy a long period of sleep to preserve energy. It is designed for mobile stations with prioritization of energy efficiency over latency, suitable for services less sensitive to delay. For latency-sensitive services, a mobile station should monitor the ACH at all times for incoming data at the expense of energy efficiency. We revisit this issue in Chap. 5 on mobile categorization. The ACH is for the control station to announce the physical channel (UTCH or DTCH) resource assignment to the (layer 2) addressed mobile station.

The Downlink Feedback Channel (DFCH) is for signaling MAC feedback messages, e.g., the MAC Feedback PDU from a control station to a mobile station regarding the state of UTCH.

Also pertinent to the centralized medium access control is the Random Access Channel (RACH) used by a mobile station to send short messages to the control station without requesting uplink resources first. Such messages include Uplink Resource Request for mobile-originated uplink traffic and Response to Paging for retrieving pending downlink traffic.

In contrast, an Uplink Signaling Channel (USCH) is reserved for a mobile station currently using the traffic channels to send related MAC control messages, such as the MAC Feedback PDU regarding the state of DTCH and the downlink channel status.

3.3.4 Physical Channels

A physical channel is for physical layer PDU exchange between remote physical layer peers. For that, a physical layer PDU (a bit sequence) is transformed into a constellation symbol sequence, on a specific resource block, via a modulation and coding scheme prescribed by the medium access control protocol of the data link layer (the MAC sublayer in particular). It is transmitted over the air in the form of a transmission burst for physically carrying the physical channel PDU over a specific set of radio resources.

For ad hoc communication with distributed medium access control, the Physical Traffic Channel (PTCH) is defined for carrying TCH. The Physical Signaling Channel (PSCH) is used to carry ACH, as depicted in Fig. 3.6.

Referring to Fig. 3.5 for infrastructure-based communication under centralized medium access control, the Physical Uplink Traffic Channel (PUTCH) and the Physical Downlink Traffic Channel (PDTCH) are mainly for carrying UTCH and DTCH, respectively. The physical resources for PUTCH and PDTCH are assigned by the medium access control function of the MAC sublayer via ACH.

The Physical Downlink Signaling Channel (PDSCH) carries DFCHs, along with ACHs and the CCCH for better channel coding efficiency or a higher coding gain (treated in Chap. 5).

The PCH is transmitted on the dedicated Physical Paging Channel (PPCH), and the BBCH is broadcast over the dedicated Physical Bulletin Board Channel (PBBCH).

The Physical Uplink Signaling Channel (PUSCH) is dedicated to the carriage of USCH. The resource configuration for PUSCHs is by BBCH and CCCH.

RACH is transmitted on a Physical Random Access Channel (PRACH) selected by the accessing mobile station from a pool of shared PRACHs among mobile stations in a contention-based manner (hence subject to interference). Techniques to mitigate the interference are described in Chap. 8.

3.4 Protocol Layer Data Units

Layer channels are defined for data exchange between two adjacent protocol layers of a station. As noted at the beginning of this chapter, for proper interworking between peer protocol layers of two communicating ends, the protocol data unit or PDU must also be defined. Therefore, a PDU is typically composed of protocol-specific control information for use by the layer protocol, i.e., a *header* or *trailer*, and the payload for carrying upper layer data or the current protocol-generated data. This section describes a generic maritime MTC layer PDU framework. Specific PDU examples associated with various layer protocols are deferred to future chapters as they become relevant.

Figure 3.8 on the next page demonstrates the entire process happening at each protocol layer of encapsulating application traffic into a physical layer PDU, including local protocol generated traffic at each layer.

3.4.1 Network Layer PDU

There are two types of traffic that the network layer needs to process: the *application traffic* (*traffic* for short) generated by maritime applications and services from the application layer and the *network control traffic* generated by various network protocols. The header of the network layer PDU that contains the application traffic, i.e., the network application Traffic PDU, commonly includes a `destination` field and a `source` field. For truly service-centric networking, routing and forwarding are based on service identifiers rather than host identifiers or network locations in a traditional host-centric networking environment like the Internet. Service requests are sent by the client to the network using a name that identifies the service; the `destination` field thus carries the service identifier.

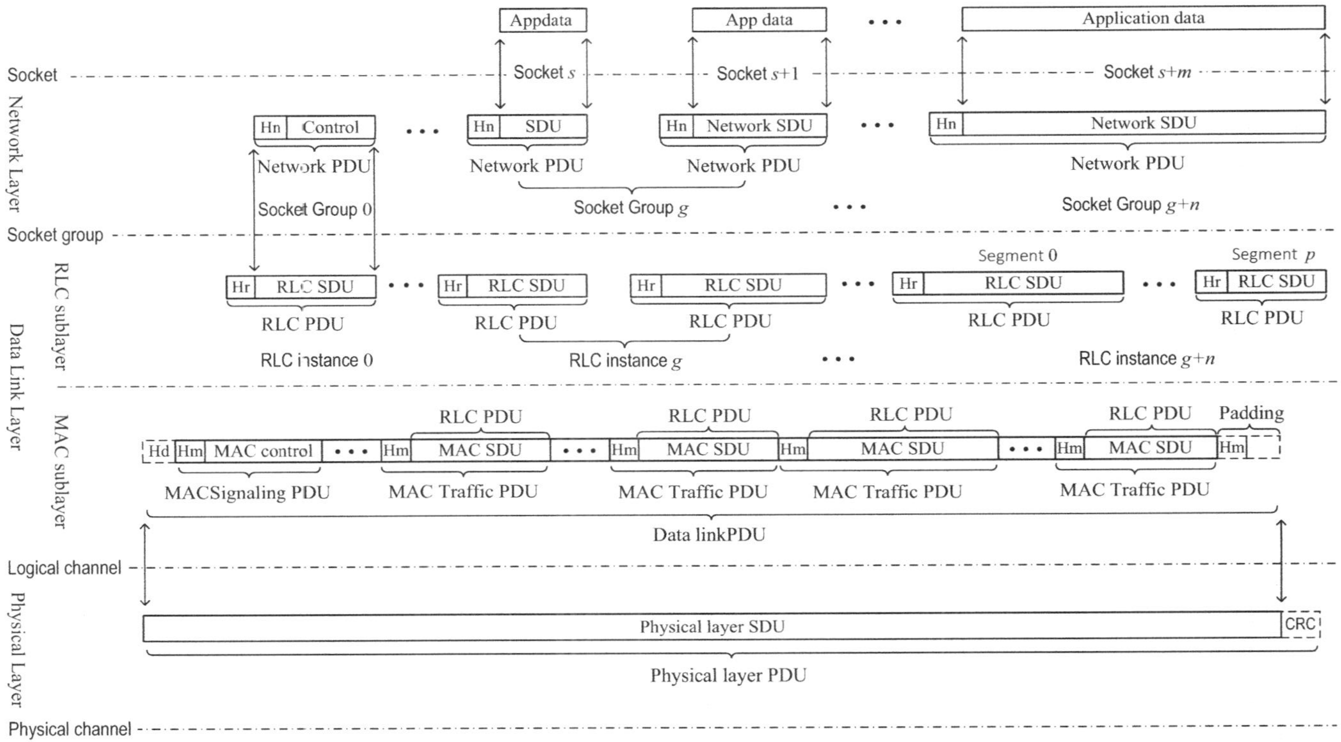

Fig. 3.8 Illustration of protocol layer PDU transformation, where "Hn" stands for the network layer header, "Hr" the RLC header, "Hm" the MAC header, and "Hd" (optional) the data link header. CRC (optional) denotes cyclic redundancy check

The header of a Network Layer Control PDU for carrying network control traffic includes a *control function* field that identifies various network layer control functions or protocols. Since a network layer PDU that carries the Network Layer Control PDU and the Network Layer Traffic PDU is transported through different socket groups and taken care of by corresponding RLC instances, there is no confusion between these two types of PDUs at the receiving end.

The detailed specification of the service identifier for maritime MTC involves interworking with the external networks as well as spectral and energy efficiency constraints. It is deferred to Chap. 4, where an in-depth discussion of MTC networking is provided.

3.4.2 Data Link Layer PDU

A suite of control or signaling protocols must be in place for the transport of the network layer PDU to its peer protocol safely and timely per the requirement, between mobile stations or mobile and control stations. These protocols reside in the data link layer and communicate with their peers via the data link layer PDU.

3.4.2.1 RLC PDU

An RLC instance is associated with a socket group, characterized by a specific QoS and other parameters inherited from the socket group, and identified by the same socket group number.

When necessary, the network layer PDU (i.e., the data link SDU) is segmented into multiple segments in the form of RLC PDUs. For reassembly, PDUs from the same RLC instance are numbered sequentially with a `sequence index` included in the RLC PDU header listed in Table 3.1. Each RLC PDU is tagged with a `segmentation status` to indicate the segment's position in the original RLC SDU after segmentation. With the RLC PDU header, the RLC peer on the receiving end can reassemble these RLC SDUs to recover the original network layer PDU once all the fragments are received.

Table 3.1 RLC PDU header

`Segmentation Status`	`Sequence Index`
`00`: No segmentation	6 bits (0 – 63)
`01`: The first segment	
`10`: Intermediate segment	
`11`: The last segment	

Table 3.2 MAC PDU header

PDU Type	Field	Bit width	Value
Traffic PDU	`type`	1	`0`
	`RLC instance`	4	0 – 15
	`SDU size`	8	0 – 255
	`reserved`	3	
Signaling PDU	`type`	1	`1`
	`control function identifier`	7	`0000000 - 1111110`
			`11111111` (padding)

3.4.2.2 MAC PDU

These RLC PDUs are then individually packed into MAC PDUs, henceforth referred to as the MAC Traffic PDUs. Each is tagged with an `RLC instance` for identification and QoS control. Apart from carrying these protocol data from the upper layer, the MAC protocol may use its PDU to carry its own protocol data for, e.g., medium access, QoS, and reliability controls, as noted earlier, which requires message exchanges or signaling between MAC protocol peers. These PDUs are called MAC Signaling PDUs. Depending on their priorities or QoS of these MAC PDUs and the payload of the data link PDU as well as the scheduling policy, MAC Traffic PDUs are multiplexed into a single data link PDU and transported via DTCH, as illustrated in Fig. 3.8. As such, an `SDU size` field is included in the header for demultiplexing at the receiver.

The MAC Signaling PDUs that carry certain frequently used critical MAC signaling information are conveyed through dedicated logical channels. For instance, for the downlink, the MAC Resource Assignment PDU is transported through ACH. Under centralized medium access control, the MAC Common Medium Access Control PDU is through CCCH, the MAC Paging PDU through PCH, and the downlink MAC Feedback PDU through DFCH. For the uplink, the MAC Feedback PDU is through USCH and the MAC Random Access PDU through RACH. Furthermore, since the format of these MAC Signaling PDUs is not as dynamic as the MAC Traffic PDU, the header for these PDUs is often omitted for efficiency. Any dynamic for these MAC Signaling PDUs is handled by the BBCH that is broadcast at a much lower duty cycle for reduced overhead. The design of these special MAC Signaling PDUs has a direct impact on the system performance. It hence is demonstrated through detailed examples provided in Chap. 7 for proximity communication with distributed medium access control and Chaps. 8 and 9 for terrestrial and satellite communications with centralized medium access control.

MAC Signaling PDUs that do not have dedicated logical channels all go through the logical traffic channel (TCH or DTCH/UTCH). For efficiency, they can be piggybacked with the MAC Traffic PDUs, encapsulated in a data link PDU (see Fig. 3.8). Thus, any MAC PDU using the traffic channel must adopt the MAC PDU header enumerated in Table 3.2, with a `type` field distinguishing the MAC

Signaling PDUs from the MAC Traffic PDUs. A MAC Signaling PDU contains a `control function identifier` field to indicate various MAC functions.

For instance, a MAC Traffic PDU header may look like

$$\underbrace{\texttt{0}}_{\texttt{type}}\ \underbrace{\texttt{1010}}_{\texttt{RLC instance}}\ \underbrace{\texttt{00000110}}_{\texttt{SDU size}},$$

where a `type` value of `0` makes the PDU a MAC Traffic PDU, the `RLC instance` field, `1010`, implies that the payload of this MAC traffic PDU (i.e., a MAC SDU) contains an RLC PDU from RLC instance 10 (i.e., Socket Group 10), and a value of `00000110` of the `SDU size` field indexes to the 6th entry of a predefined MAC SDU size table.

A `type` field value of `1` means that the PDU is a MAC Signaling PDU. The MAC Timing Control PDU is an example of such a MAC Signaling PDU. It is generated by the MAC timing control function or protocol of a control station to apprise a mobile station of its uplink transmission timing error. In this case, the MAC PDU header takes this form,

$$\underbrace{\texttt{1}}_{\texttt{type}}\ \underbrace{\texttt{0001010}}_{\texttt{function}},$$

where the `type` value of `1` indicates that it is a Signaling PDU and the `control function identifier` field of `0001010` implies that the control message is the MAC Timing Control PDU from the MAC timing control function. Uplink timing adjustment is an essential MAC function, especially in satellite communication. More details on this topic are given in Chap. 9. The `size` field is absent in the MAC Signaling PDU header since the length of the MAC Signaling PDU is predefined and hence self-explained.

Also included in the `function` field is a special MAC Signaling PDU used as a delimiter for filling the space unfilled by MAC PDUs at the end of a data link PDU. It is necessary because the physical layer SDU may have discontiguous sizes (in byte), and the data link PDU itself does not have a header for indicating its actual length. There would be ambiguity as to the number of MAC PDUs in the data link PDU. As such, a dedicated header with eight bits (one byte) of consecutive `1`s at the beginning of the MAC control PDU header is used to indicate `padding` for filling the unused byte(s) in a data link PDU.

A data link layer PDU may contain multiple MAC PDUs from the same mobile station. In the case where the data link layer deals with multiple mobile stations, the MAC PDUs belonging to different mobile stations are packed into separate data link PDUs. These data link PDUs may be loaded into the same physical layer PDU through separate logical traffic channels and transmitted on the same physical traffic channel. This situation happens when the physical channel is too big for a single mobile station.

3.4.2.3 MAC PDU Scheduler

The MAC protocol in charge of the resource assignment, i.e., the MAC resource assignment protocol, employs a "MAC PDU scheduler" (i.e., the resource scheduler) to multiplex the MAC PDUs from the logical traffic channel, TCH or DTCH/UTCH, onto a data link layer PDU and then the physical traffic channel, PTCH or PDTCH/PUTCH, under distributed or centralized medium access control, respectively (see Fig. 3.8). In general, a scheduler must consider multiple factors regarding the specific service requirements (such as latency) and network operation goals (such as spectral efficiency or system throughput and fairness) within the physical constraints. These constraints include the available resources for the traffic channels, radio link conditions, and mobile station communication capabilities. In this sense, the design of a scheduling strategy often lies in finding a good trade-off among factors of relevance. The factors and trade-offs are often represented by a numerical indicator known as the *scheduling metric*.

For example, the scheduling metric for the simplest "first come, first served" policy can be implemented as

$$\partial_i = t - t_i, \tag{3.1}$$

where t is the current time and t_i the time when the service i PDU enters the system. The value of this metric determines the priority of a MAC PDU, represented by a *scheduling queue*, as diagrammed in Fig. 3.9. Obviously, retransmission naturally raises the priority and helps push the failed MAC PDU ahead in the queue. A delay-sensitive scheduling metric can be expressed as

$$\partial_i = (T_i - \tau_i)^{-1}, \tag{3.2}$$

where T_i is the delay threshold for service i and τ_i is the current delay of the MAC PDU from this service.

For distributed medium access control, the scheduler resides at the mobile station. The burst configuration, including the PTCH MCS and the assigned resources, is communicated to the receiving peer via ACH over PSCH.

For centralized medium access control, the scheduler is naturally located at the control station, in charge of both uplink and downlink MAC PDU scheduling (i.e., uplink and downlink schedulers) for all active mobile stations. The scheduling metric (and queue) is updated at every scheduling interval before the scheduler makes the assignment decision. The assignment information is broadcast in the MAC Resource Assignment PDU to mobile stations via ACH over the PDSCH.

Since MAC PDU scheduling under centralized medium access control involves distributing resources among multiple mobile stations, the *multi-user diversity gain* can be exploited. Such a scheduler takes advantage of the channel conditions experienced by different mobile stations and assigns priority to the mobile stations

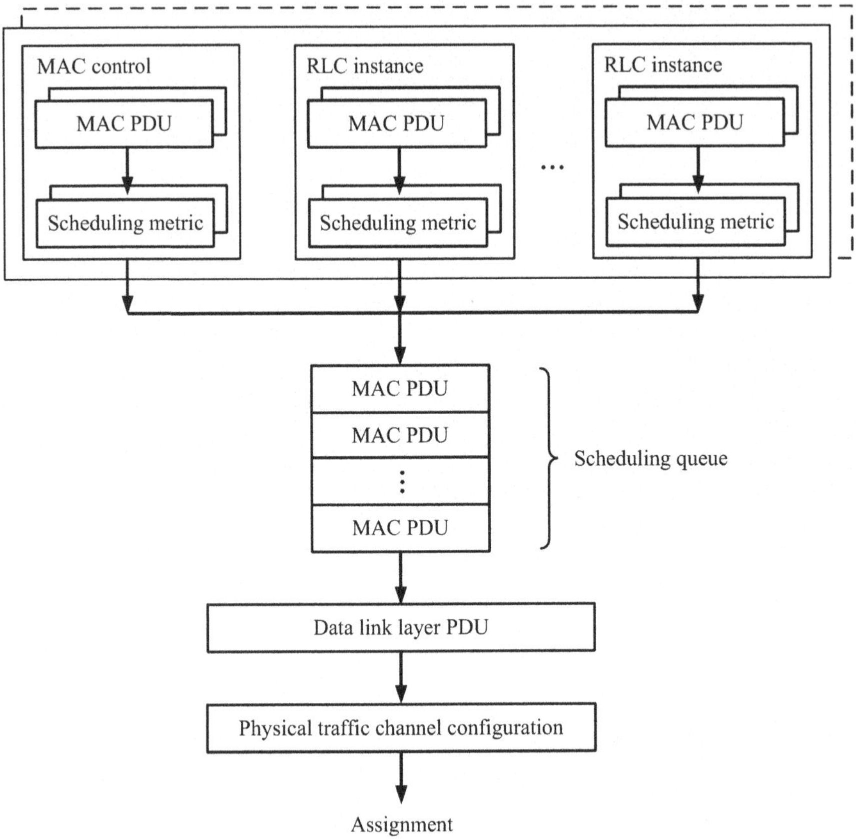

Fig. 3.9 A high-level view of a MAC PDU scheduler. The stacked dashed plane denotes multiple mobile stations requesting PDTCHs or PUTCHs under centralized medium access control

that can achieve the maximum data rate on the current resource, hence maximized system throughput. Such scheduling metric can be expressed as

$$\partial_j = r_j, \tag{3.3}$$

where r_j is the expected data rate on the resource of mobile station j, a function of a mobile station's channel condition and communication capability. The metric can be modified to reflect the balance between system throughput and fairness,

$$\partial_j = r_j^{\alpha}/\bar{r}_j^{\beta}, \tag{3.4}$$

where $\bar{r}_j$ is the average data rate of mobile station j, and α and β are weighting parameters for fine-tuning the trade-off between spectral efficiency and fairness.

A practical scheduler often incorporates multiple factors of relevance into the scheduling metric to achieve a balanced outcome.

3.4.3 Physical Layer PDU

The physical layer PDU is for carrying the data link PDU through the physical layer SDU, as shown in Fig. 3.8. The data link PDU that the SDU receives may contain MAC Signaling PDUs or MAC Traffic PDUs, or a combination of both. As mentioned above, a physical layer PDU may contain multiple physical layer SDUs that carry multiple data link PDUs (although only one physical layer SDU is shown in the example of Fig. 3.8) but always occupies one and only one physical channel.

The integrity of the physical layer PDU is normally protected by an error detection code, like the cyclic redundancy check (CRC). Chapter 5 provides greater depth into this topic on channel coding.

Chapter 4
Network Structure

Abstract This chapter introduces a general design of maritime MTC from the network perspective under the protocol structure presented in the previous chapter to address the maritime IoT service requirements outlined in Chap. 1. A central solution is a service-centric network structure that leverages three functional network components and corresponding virtualized network functions to provide service differentiation and adaptation according to per-service requirements, which is not available in traditional maritime communication systems, leading to a cloud-based, highly adaptive, service-centric, and spectral- and energy-efficient maritime MTC network. Under this networking framework, each maritime IoT application owns a virtually isolated networking solution customized dynamically to their physical environments and device capabilities while sharing the underlying infrastructure assets and network resources with other IoT applications that may possess vastly different service characteristics and policies and deployment scenarios.

Keywords Service-centric networking · Service profile · Service instance · Spectrum and energy-efficiency · Network controller · Maritime application server · Control plane · Application plane · Virtual network functions · Cloud service delivery

4.1 Functional Components

As noted in Chap. 3, the ultimate role of the network layer is the end-to-end connection management for a broad range of IoT applications, including (1) network resource management to establish and maintain an appropriate connection and (2) interworking between the internal network (i.e., the maritime MTC network) and external networks (i.e., the IP-based maritime information and service networks or clouds where maritime service providers reside).

Most, if not all, of today's maritime communication systems are *physically* isolated systems to deliver different services in the sense that the supporting communication systems and networks are custom-made only for a specific application class with solutions that only address certain requirements, thereby producing many

M. M. Wang, J. Zhang, *Machine-Type Communication for Maritime Internet-of-Things*, Wireless Networks, https://doi.org/10.1007/978-3-030-77908-5_4

systems with intrinsically coupled hardware, software, and networks. The efficiency, adaptability, and scalability of such a service delivery model are inherently limited. Another fundamental problem with today's maritime communication systems is the lack of interworking capability with IP networks where most applications and services exist. We all know too well that it is rather difficult and inefficient to reuse these legacy systems, e.g., AIS, for new applications. In some terrestrial scenarios, we have seen the IoT applications employ the existing communication networks such as the cellular network (e.g., the GSM network built for human-type communication) to avoid the high cost of building a brand new infrastructure. Work-around is often needed and messy, and inefficiency is inevitable. Applications are constrained by the underlying supporting communication network. This "best effort, one-size-fits-all" approach to maritime IoT is just no longer viable. Therefore, instead of adopting such traditional service-specific and network-centric structures, a highly adaptive and service-centric network is essential to meet the maritime IoT requirements. Towards this endeavor, the current network design leverages three network components: *network controller*, *maritime application server*, and *control station*, interconnected through a transport backbone, as depicted in Fig. 4.1 and elaborated in the following.

4.1.1 Network Controller

The maritime MTC network controller is a logically centralized component responsible for the network infrastructure supervision and management from a global perspective to provide network-related control functionalities essential for MTC. In particular, it aims to leverage the pliable and agile integration of the terrestrial and satellite segments to provide network capabilities for mobile stations to access a wide variety of maritime applications and services. It performs resource management that utilizes the maritime MTC network infrastructure to classify data from different applications according to priority and select appropriate network resources to provide quality assurance for maritime applications and services. Here "resource," in general, includes the radio communication resources, storage resources, computational resources, middleware services, network functions, as well as information resources.

Bringing *network function virtualization* and *chaining* [1] into both domains enables establishing an end-to-end virtualized network. It allows the network layer to exploit the virtualized network function-based maritime resources management to facilitate automated provisioning of maritime applications that may have diverse characteristics and ensure that specific applications are delivered with the proper network resources or characteristics over highly dynamic and heterogeneous network environments. Virtualization of maritime MTC network function provides the separation of network layer functions (see Chap. 3) from the underlying infrastructure to create a virtualized global network that can run on disparate classes of network infrastructures and, more importantly, allowing interoperability and cloud services. For maritime MTC, this concept has another layer of significance: allowing

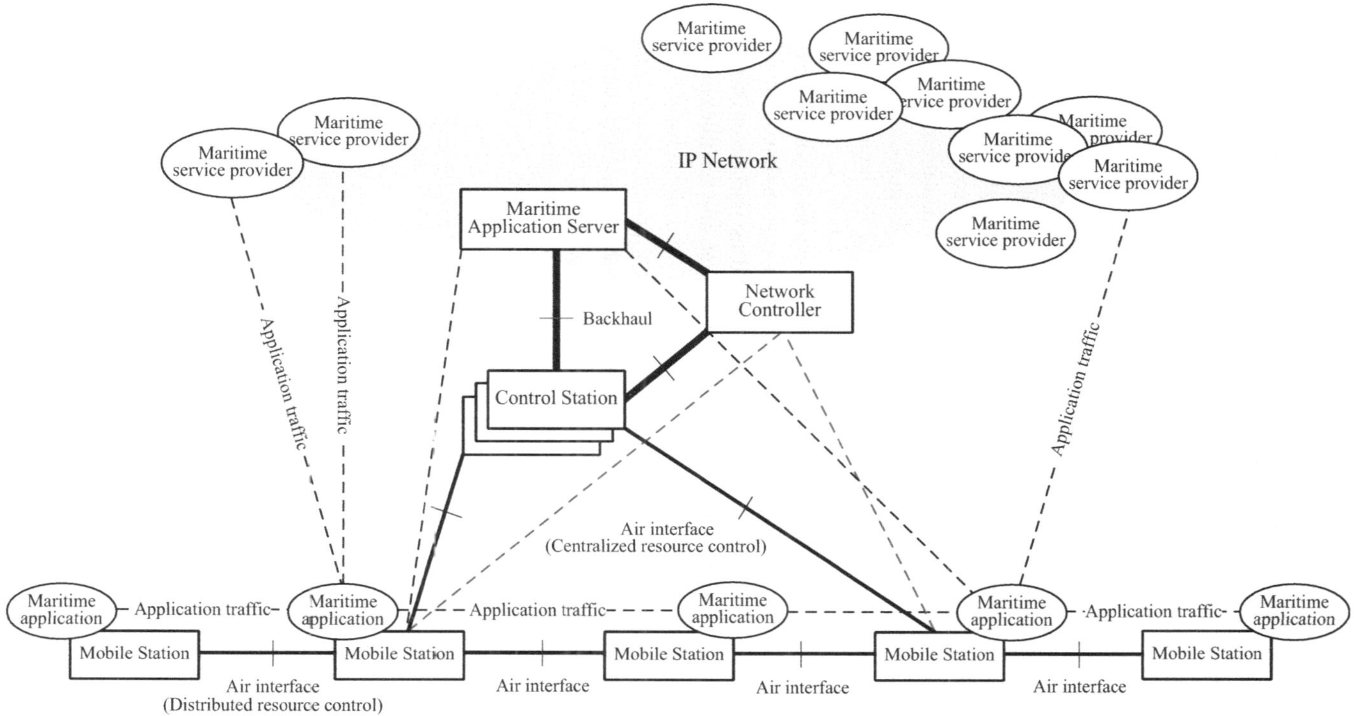

Fig. 4.1 Maritime MTC network functional view and topology for service-centric maritime MTC networking, where solid lines denote physical links, and dashed lines logical links. The transport options for the backhaul include a combination of fiber-optic network and microwave links

sharing the underlying network resources among nations and organizations and supporting diverse IoT applications, each of which may have a different set of services, policies, parameters, and deployment scenarios.

With the benefit of virtualized network functions, the maritime MTC network becomes centrally manageable, enabling the network controller to properly configure all the resources and infrastructure components necessary for a specific maritime application or service. The network controller executes the resource control function via a *network resource control* (NRC) entity responsible for application requirement translation, configuration, provisioning, optimization, remediation, and terrestrial and satellite system integration, among many other things, based on changing traffic scenarios and deployment characteristics.

Specifically, NRC is in charge of proximity, satellite, and terrestrial network management, including (1) network configuration per the service requirement, (2) information updating over the network, (3) radio resource management, such as frequency band assignment and sharing, radio-related measurements and interference management, and (4) control station coordination. As introduced in Chap. 2, control stations are physically distributed deployments of the logically centralized network component, directly interacting with mobile stations to provide network access to the mobile stations. A control station maintains a local NRC, i.e., the *edge NRC*, whose peer, the NRC agent, is embedded in a mobile station.

Frequency band assignment by NRC identifies and assigns a set of frequency bands available for communication (proximity, terrestrial, and satellite) in the given geographical area. In the case of centralized resource control, the network resource is configured under the supervision of NRC to provide the proper connection quality for requested services while maximizing frequency reuse. For instance, the time and frequency resources of the assigned bands may be configured by NRC in a pattern to prevent the neighboring control stations from using the same resource to avoid heavy inter-cell interference in certain areas of the cell, e.g., cell edges, either in a fixed or dynamic fashion. Inter-cell interference measurements may be ordered by NRC to determine the best serving and coordinating control stations for a mobile station. When a high-quality connection is required for a high-QoS service, NRC may configure such a connection by committing more network resources. Under this configuration, a mobile station may be served by multiple control stations, orchestrated by NRC, engaging in coordinated transmission and reception to mitigate inter-cell interference and enhance signal quality at cell edges. Here, coordinated transmission means that the downlink data is transmitted to a mobile station jointly from multiple control stations, thereby not only eliminating the inter-cell interference but increasing the received signal strength at the mobile station. Coordinated reception means that the signals received at these associated control stations are jointly processed for enhanced reception performance. Chapter 8 provides more detailed descriptions of network-supervised interference coordination. The setup and maintenance of the connection are through virtualized functions so that NRC does not need to concern itself with implementation details but to focus itself on connection solutions tailored to the physical environments and device capabilities while sharing the underlying network resources with other IoT applications that may possess completely different service requirements and deployment scenarios.

4.1.2 Maritime Application Server

The maritime application server is another logically centralized network component whose primary function is to enable service-centric networking, which automates the *end-to-end* provisioning (the setup, management, and maintenance) of the maritime services and applications. It maintains the interconnection of an application running on different hosts (which may reside on different networks) with QoS management, maritime IoT service discovery, Layer 3 address resolution, and inter-network protocol adaptation and convergence if the connection involves two different networks. Therefore, the maritime application server is oriented more towards maritime services and lets the network controller worry about overcoming service deficiencies, e.g., providing reliable services over unreliable lower layers. Therefore, from the application's perspective, the maritime application server is the "transport network."

To that end, the maritime application server maintains a *federated* repository of provisioned maritime services, i.e., the *maritime service registry* (MSR) entity, which provides a runtime registry for service discovery that contains the most current information as to where and how to reach a running maritime service. Notably, the operational content of MSR contains, but is not limited to, the following:

1. Identification, authentication, and abstraction of maritime services in the form of the *service profile*, including network locations and settings, pre-shared keys, traffic patterns, QoS, and other service characteristics.
2. Presentation of the service profiles to the network controller, which creates a transport network configuration to match the service attributes and policies prescribed in the service profile. Specifically, the MSR utilizes the socket group construct (described in Chap. 3) to implement transport level QoS marking and enforcement.
3. Convergence and adaptation between the internal service-centric MTC network and the external IP-based host-oriented network to provide efficient and seamless interoperability with the service providers, including interface adaptation and service name and host address resolution emphasizing service by making it directly addressable and routable. A virtual connection construct is instantiated, termed *service instance*, and recorded in the service portfolio. The service instance abstracts complete means and attributes to deliver an end-to-end service (yet transparent to the end hosts), including the transport network configuration and the internetworking information. In general, there can be multiple instances of the same service, all under the same service identifier and portfolio. As a service request is received, the proper service instance is activated on the relevant transport assets and network resources.
4. Maintaining the service portfolio and populating and updating service instances across the network. Service portfolios may change, and new services may emerge. It is the central MSR's responsibility to update the edge and agent MSRs on the changes.

In a traditional small-scale network, a service runs on a well-known network location and is delivered through a well-determined network environment. The service instance is thus relatively static or singular. Nevertheless, it may not be the case in a colossal, highly dynamic, cloud-based, heterogeneous maritime MTC environment. The service instances registered on the service portfolio may not be equally the best choice in a given environment, and some may even be temporarily unavailable in a particular geographical area. Thus, when a service request is sent by the client from a mobile station to the network using a service name or identifier, the request is routed to the most appropriate service location somewhere in the maritime cloud network that is load-balanced and close to the control station currently serving the client that invokes the service. Therefore, an appropriate service instance can be started through the underlying maritime MTC infrastructure.

The maritime application server may also contain a Maritime Identity Registry (MIR) responsible for identifying and authenticating mobile stations, a federated authority for identities of nationalities, organizations, or ships using the network. Authentication is a process by which the network verifies the identity of a station that wishes to access the network. Since access control is normally based on the identity of the mobile station that requests access to a resource and service, authentication is essential to security. Contrasting to identification that refers to the act of stating a station's identity, authentication is the process of actually confirming the stated identity. Such handshakes are carried out between two MIR peers, i.e., the MIR agent at the mobile station and the central or edge MIR.

Furthermore, the location of a mobile station may constantly change, and the situation at sea where the vessel is encountered is continuously varying, which necessitates maritime services that support safe and efficient operation worldwide. Therefore, another function of the network controller is mobility management that tracks the location of mobile stations worldwide that enables the maritime service to use the location information of the mobile station. As such, the network controller may maintain a Maritime Messaging Service (MMS) that supports store-and-forward functionality for reliable delivery of messages and maps geographic location-related information to appropriate target areas for geo-casting of messages. Messages can then be timely and reliably exchanged between mobile stations and a maritime service provider in a target area using unicast or broadcast.

Compared to MSR, MIR or MMS does not have strong enough relevance to service-centricity to make it to the essential kernel of maritime MTC. In the current design, they are not deemed native functions of the maritime application server; instead, they are treated as services in the application layer.

The network controller and the maritime application server represent a logical and functional separation of the application and control functions but do not imply any particular physical implementation. Together, they enable holistic visibility and operational consistency for end-to-end service orchestration of how the services cross the domains and cloud environments that make up the maritime MTC network to support existing and emerging services and business models. However, no amount of this grand goal will be fulfilled without control stations.

4.1.3 Control Station

A control station, either a shore station or a space station, resides at the "edge" of the network. In addition to serving as a wireless access point to the MTC network for mobile stations, the wide geographical distribution of the physical control stations extends the centralized control and services by fostering an "edge deployment" of a part of the centralized functionalities. Relevant system information is cached by the control station. Relying on the physical proximity to the mobile stations or end-users, ergo more awareness of geolocation and context, they help determine the best way to perform the functions and enforce them among individually managed mobile stations. It interplays with its central counterpart (i.e., the network controller and maritime application server) to perform functionalities that require up-to-date centralized information and management, thereby maintaining a logically centralized but physically distributed vision of the network. This structure segregates the network infrastructure management from application traffic flows and provides the centralization of these functions with distributed control, allowing automation of the entire network through the centralized virtual control functions without stressing the network or limiting scalability. It enhances the real-time responsiveness to service needs and the adaptability to workload and changing environments (e.g., radio channel conditions and service instances).

4.2 Control Plane and Application Plane

As we recall from Chap. 3, the maritime MTC system follows a layered structure to address characteristics such as interoperability, scalability, reliability, modularity, and simplicity. As shown in Fig. 4.2 [2], the protocol structure adopts the four-layer model, where the layer is assigned protocols to perform well-defined tasks. Simply stated, Layer 3 or the network layer is designed for end-to-end connection management, and the lower layers (Layers 1 and 2) perform the physical delivery within the local network. Together they provide end-to-end connectivity for maritime IoT applications and services. From the network perspective, the functionalities are executed through virtualized network functions, over two separate virtual planes on the network layer, referred to as the *control plane* and the *application plane*, under a distributed architecture that includes the maritime MTC infrastructure edge nodes, i.e., the control stations. This logical construct enables monitoring network statistics that provide observability to not only the overall network topology but specific application performance. The information is then used to calculate and establish communication paths over either the terrestrial or satellite domain and push down rules to the network edge on a per-user and per-service basis. Through these two virtual planes, the network controller and the application server supervise the network and the mobile station behavior at the desired application traffic granularity. As such, application traffic can be engineered pliably and precisely to

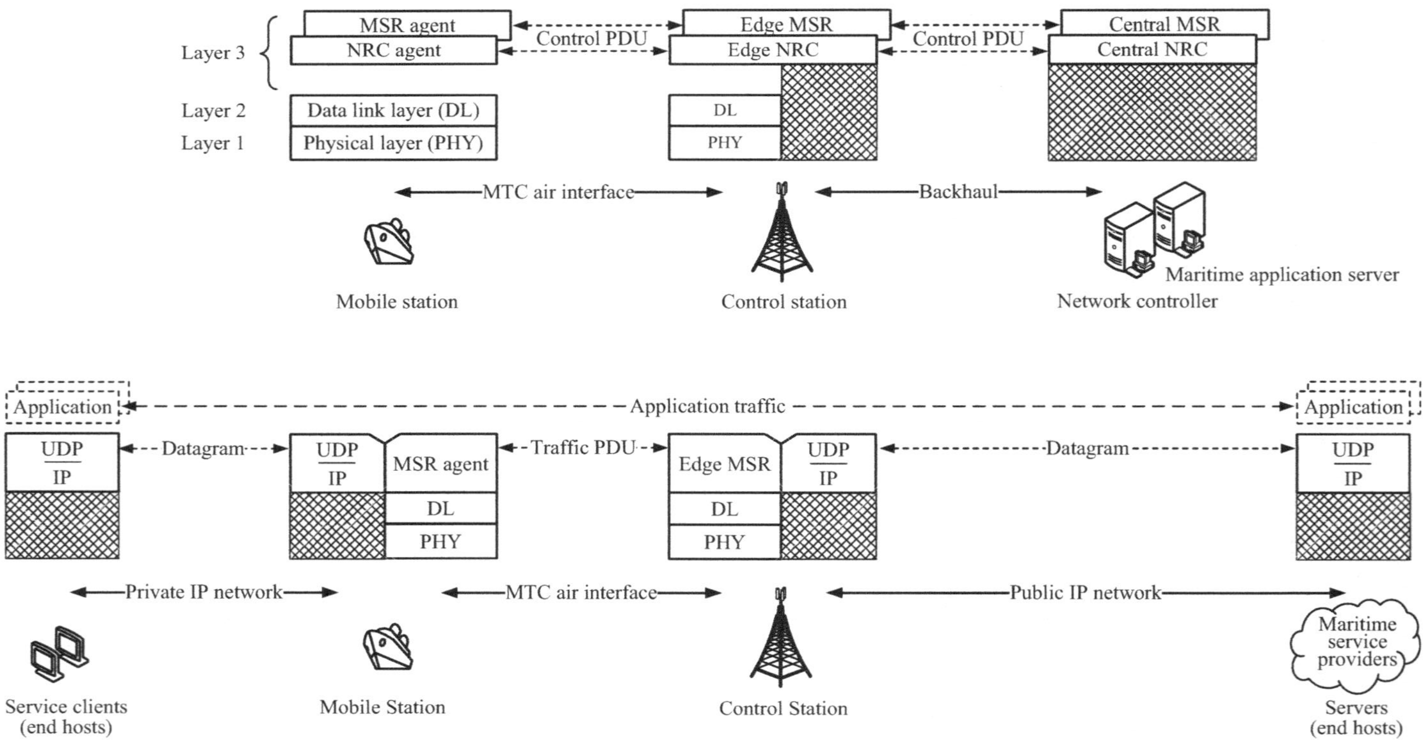

Fig. 4.2 Maritime MTC control plane (top) and application plane (bottom). The shaded blocks denote the native layer services of the backhaul networks. The MSR agent at the mobile station may serve as a gateway from the maritime MTC network to a private IP network, e.g., a premises network aboard a ship

one or a combination of goals, such as maximizing aggregate network utilization, optimizing load balancing, minimizing power consumption, lowering latency, and other generic network and traffic optimization techniques, and most importantly, providing service differentiation and adaptation.

4.2.1 Control Plane

The network controller and maritime application server naturally reside in the network layer. They abstract a common set of virtualized control functions or modules and execute them over the control plane, as depicted in Fig. 4.2 (top). The generated traffic over the control plane is referred to as the network "control traffic" in the form of network layer Control PDU, as discussed in Chap. 3.

The NRC of the network controller, under the virtual connection construct, configures all the resources and infrastructure components necessary for a specific maritime application or service through the network layer Control PDUs. All stations participating in the communication exchange control information, such as communication capability and environment, to make the real data exchange (i.e., the application data over the application plane) possible.

NRC defines a generic information model for the operation and control of radio link network elements, intended to facilitate the integration of air interfaces and frequency spectrum solutions under a common and single control framework, regardless of satellite or terrestrial. These configurations change dynamically at the time of communication in light of changing environments, particularly the application traffic characteristics observed by the maritime application server. NRC configures the air interface and communicates it to the NRC agent over the control plane, which includes system synchronization information, available communication resources, and their configuration parameters, physically broadcast on PBBCH, necessary for mobile stations to access the network. The NRC agent at the mobile station discloses its identity, communication capability, radio channel conditions at the time of access, and requests for the network resource and the service from its peer NRC, i.e., the edge NRC of the network controller. NRC also monitors the interference status fed back by the NRC agent through the control plane for coordinated transmission and reception among control stations, all through the control plane.

Similarly, the maritime application server uses the control plane to populate and update maritime service-oriented information (new services and service portfolio updates, for instance) to enable service-centric networking. MSR keeps the service instances of the maritime IoT application service portfolio up-to-date at every tier (edge MSR and MSR agent) via the control plane, which is essential for providing end-to-end services over the application plane. Since the service instances are relatively static and hence do not invoke frequent updates, the control traffic that it generates over the control plane of the MTC network is minimal.

4.2.2 *Application Plane*

While the MTC connection or path for a maritime IoT service is established and supervised over the control plane, the application plane that is logically separated from the control plane is for execution and accomplishment of the services by the maritime application server at the network layer, depicted in Fig. 4.2 (bottom).

When it comes to today's data communication, it would be hard to argue against the fact that the IP network is the largest data communication network on earth. Almost everything nowadays travels over the IP network, which, not surprisingly, includes maritime IoT applications and services, just as the name "maritime IoT" indicates. Typically, the application server resides on the public IP network (e.g., Internet), whereas the client application is on an on-premises network – a private IP network (e.g., Ethernet, Wi-Fi). The message generated by the maritime IoT application and received by the network layer of a mobile station is expected to be IP-based datagrams such as the Transport Control Protocol (TCP) datagrams and User Datagram Protocol (UDP) datagrams. It is, therefore, essential for maritime MTC to interwork efficiently with the IP network.

TCP provides reliable, ordered, and error-checked end-to-end delivery between applications via an IP network. The function of TCP is to control the transfer of IP packets so that it is reliable through connection management and reliability control, which simplifies the implementation of an application and hence is the default transport protocol for most applications. Connection management includes connection initialization (a three-way handshake) and its termination. TCP opens the connection and completes all the handshaking formalities before transferring the datagram. Hence, even a short datagram needs a minimum of seven message exchanges, thereby introducing excessive overhead and delays. Reliability is achieved by a process known as automatic retransmission request or ARQ, in which the sender detects lost segments and transmits them. A retransmission of the TCP segment occurs after the segment is lost, i.e., when the sender does not receive the acknowledgment after a timeout. TCP is reliable in the sense that the protocol itself checks to see if everything that has been transmitted is delivered to the receiving end. Another issue with TCP is that it includes a network congestion control mechanism that treats unacknowledged or lost datagrams as an indication of network congestion. The protocol mitigates the congestion by slowing down the transmission after it occurs, as indicated by packet losses. It makes perfect sense for the wired IP network, where the loss of data is caused by routers dropping IP packets due to buffer overflow at the network layer as a result of network congestion. However, in a wireless network like maritime MTC, data loss is normal due to errors caused by the notoriously unreliable wireless channels. Therefore, the handling of lost data by the network layer, whereas caused by the wireless channel at the physical layer, is no longer appropriate or efficient.

On the other hand, UDP provides a connectionless datagram service that emphasizes reduced latency and overhead over reliability. UDP does not require creating a connection; a datagram is transferred without handshaking. It is not reliable in the sense that the protocol itself does not ensure the delivery of the IP datagram to the destination. For this reason, UDP is a transport protocol more suitable for maritime MTC, in which the reliability is taken care of by the lower layer, as noted briefly in Chap. 3. More details on this topic will be covered in Chap. 5, where we see that reliability for an erroneous channel can be handled with greater efficiency through channel coding at the physical layer where the error most likely occurs. Nevertheless, it does not prevent the application from adding additional ARQ for reliability (at the application level) if deemed necessary.

Regardless, both UDP and TCP rely on the IP. The problem with the IP is that it is itself "costly" in that a 16-byte source and destination address included in an IPv6 packet, plus 2-byte TCP or UDP port number, mean that a minimum size of a packet is 36 bytes before any other overhead, i.e., the metadata required for routing and delivery. This IP overhead is a significant burden for maritime MTC since most maritime IoT applications are characterized by short burst data services, in which case the conventional IP header compression [3, 4], as often seen in a human-type wireless communication system, where the traffic is dominated by streaming media services, is no longer effective. A more aggressive compression approach is desirable to keep the overhead more manageable, especially for a practical wireless system with a limited bandwidth (as we will see in Chap. 6) and serving devices with limited energy.

As shown in Fig. 4.2, the air interface over the narrowband radio channel is surely the weakest link of the entire data exchange chain; its limited capacity is the bottleneck of the MTC system and hence is inherently unsuited for IP traffic. The adaptation and coordination between the MTC network and the IP network is thus another key function of the maritime application server to ensure, at the network level, a minimum efficiency loss during the data exchange between these two distinct networks.

Therefore, the maritime application server at the network layer serves as a "gateway" that sits between the internal maritime MTC network and the external IP network, aggregating traffic, performing inter-network adaptation, and routing it to the destination across the boundary of two very different networks: the service-centric maritime MTC network, where communication is based on service identifiers, and the IP network, where message forwarding is based on host identifiers, i.e., IP addresses. The goal is to hide the topology and complexity of the IP network from the maritime MTC network (and vice versa), which is in perfect accord with the maritime MTC service-centric principle by providing a resource- and energy-efficient means for mobile stations to interact with the service providers without being overburdened by the resource and energy-hungry wired protocols that drive the IP network.

Based on these understandings, Fig. 4.3 sketches how service data (application traffic) is exchanged between applications on the end hosts across three networks: a private on-premises IP network, the maritime MTC network, and the public IP network.

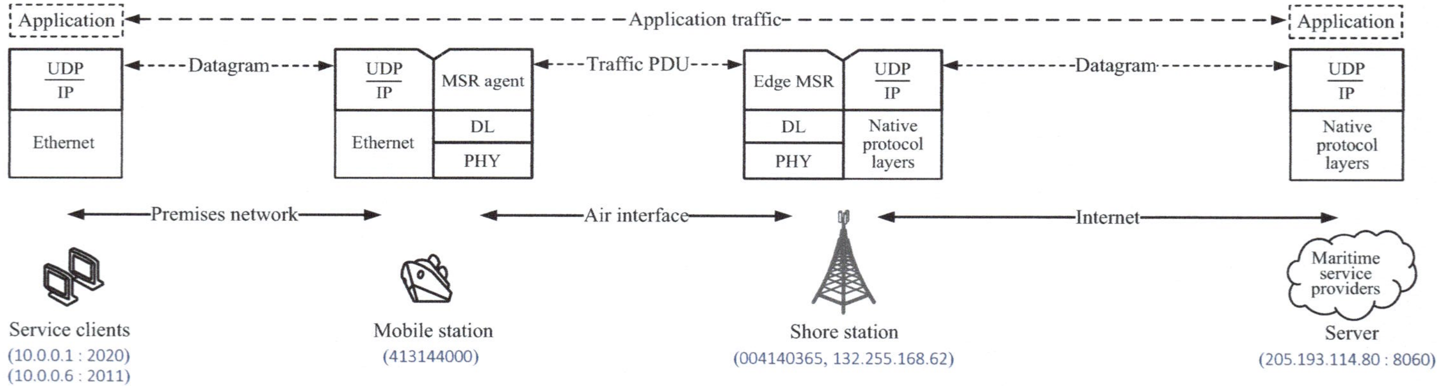

Inter-Network Adaptation Tables
(maintained by the mobile station)

Premises network (Private IP : Port)	Service client (Client ID)	Service provider		Priority
		Service ID	Public IP : Port	
10.0.0.1 : 2020	23	10	120.198.203.28 : 8080	1
10.0.0.6 : 2011	19	6800	219.137.32.74 : 8060	3

Default server IP

Inter-Network Adaptation Tables
(maintained by the shore station / MSR)

Service client (MMSI : Client ID)	Shore station (Public IP : Port)	Service provider		Priority
		Service ID	Public IP : Port	
413144000 : 23	132.255.168.62 : 1025	10	120.198.208.28 : 8080	1
413144000 : 19	132.255.168.62 : 1026	6800	205.193.114.80 : 8060	3

Selected service instance

Fig. 4.3 Illustration of the interactions between the service clients reside on a premises network (e.g., Ethernet) and the service provider host on the public IP network via the maritime MTC network. In this example, a navigational warning service client (`Client 19`) aboard a ship receives the service from the service instance at `205.193.114.80 : 8060`

4.2.2.1 Server and Client Address Resolution

Without the IP network location (IP address and port number), the service data inside the maritime MTC network cannot be routed to the external maritime service providers on the public IP networks and vice versa. As such, the MSR of the maritime application server maintains a service profile for the authenticated maritime application and service, along with a universal maritime service identifier or `Service ID` (e.g., two bytes). As aforementioned, this information is provisioned to the mobile station and control station over the control plane as part of the control traffic in the form of the network layer Control PDU (e.g., Service Connectivity Update PDU). A mobile station MSR (i.e., the MSR agent) may also request such information over the control plane through a network layer Control PDU (e.g., the Service Connectivity Request PDU). Nonetheless, specifying the network layer Control PDUs for all virtualized network functions is beyond the scope of this book. Within the MTC network, the service provider is addressed by a `Service ID`, a unique Layer 3 address of the MTC network maintained by the MSR of the maritime application server. A two-byte `Service ID` identifies up to 65,536 services. At the other end (i.e., the mobile station) of the maritime MTC network layer, the application/service on the client end is represented by a `Client ID`, a Layer 3 identifier for the service client; for example, a one-byte `Client ID` supports up to 256 clients. The network layer traffic PDU header then consists of a destination field for a `Service ID` or a `Client ID` and a source field for a `Client ID` or a `Service ID`, depending on the communication direction, as tabulated in Table 4.1.

The control station MSR in Fig. 4.3 keeps a record for mapping between the service client (`Client ID` and `MMSI`) for use inside the maritime MTC network and its public IP location (IP address and port) on the Internet for service data transactions between these two networks. Similarly, the mobile station keeps a record of the mapping between the service client (`Client ID`) for use inside the maritime MTC network and their private network location (IP address and port) on the local on-premises network for service data transactions between these two networks.

4.2.2.2 Service Delivery

For outbound application traffic (i.e., the service data from a service client on the maritime MTC network to the service provider on the public IP network), the MSR agent at the mobile station receives the UDP/IP datagram from the service client through one of the socket channels, destined to the service provider addressed by the server's public IP location (destination `IP` and `port`). The MSR agent extracts the payload (service data) from the datagram and places it in a maritime MTC network

Table 4.1 Network layer Traffic PDU header

Destination identifier	Source identifier
Client ID / Service ID	Service ID / Client ID

layer Traffic PDU, replacing the destination IP location with the corresponding `Service ID` and the service client's private IP location (source `IP` and `port`) with the `Client ID`. It then pushes down the traffic PDU to the lower layers through one of the socket group channels (SG_m to SG_{M-1} as depicted in Fig. 3.5) for transport over the air interface. The `Client ID` is used to identify the service client running on an end host on the premises network, as shown in Fig. 4.2 (bottom). The network layer Traffic PDU header then consists of a two-byte `destination` field for `Service ID` and a one-byte `source` field for `Client ID`, as tabulated in Table 4.1.

Once received by the control station, the `Service ID` is converted back to the service location (public `IP` and `port`) as the destination. The `Client ID`, associated with the mobile station's identification or `MMSI`, is mapped to a port number associated with the control station's public IP location (as the source `IP` and `port` of an IP datagram). A new UDP/IP datagram is then constructed and routed to the destined service location over the public IP network. Note that one IP address supports up to 65,536 port numbers. If *ever* necessary, the control station may be provided with multiple IP addresses to scale with the population of mobile stations.

For inbound application traffic (i.e., when the service provider responds to the service client), the control station MSR takes the incoming datagram destined to it that carries the service data from a service provider and maps the destination port of the datagram back to the `MMSI` of the mobile station and the `Client ID`. By `MMSI`, the service data associated with the `Client ID` is dispatched to the mobile station over the maritime MTC network. Once received, the MSR agent at the mobile station looks up the private IP address and the port number of the service client by the `Client ID` and forwards the service data over the premises network. Now, the `destination identifier` field of the network traffic PDU contains the `Client ID`, whereas the `source identifier` field contains the `Service ID`.

It is thus apparent that during the message exchange between a mobile station (in the maritime MTC network) and a service provider (on the IP network), the over-the-air overhead incurred by the IP traffic is kept to a minimum. This strategy improves the system spectral efficiency and reduces energy consumption for those energy-limited maritime devices. Furthermore, this ensures that only the qualified or authenticated services are "visible" to the mobile station and vice versa.

More importantly, as pointed out in Chap. 3, for truly service-centric networking, network addressing should be based on service identifiers rather than host identifiers or network locations in a traditional host-centric networking environment like the IP network. The use of `Service ID` at the network layer as the Layer 3 identifier replacing the IP header is in perfect agreement with the service-centric principle, in addition to the above-mentioned benefits. As a final note regarding the Layer 3 identifier, although both are Layer 3 identifiers under the current addressing scheme, unlike `Service ID`, the `Client ID` is used to identify a service client hosted by a mobile station or an on-premises network attached to the mobile station. It is thus not universal at the MTC network level and hence only makes sense when associated with a mobile station or MMSI – a Layer 2 identifier.

References

1. K. Gray, T. D. Nadeau, *Network Function Virtualization*, 2016
2. M.M. Wang, J. Zhang, X. You, Machine-type communication for maritime Internet of Things: a design. IEEE Commun. Surv Tutorials **22**(4), 2550–2585 (2020)
3. E. Ertekin, C.A. Christou, Internet protocol header compression, robust header compression, and their applicability in the global information grid. IEEE Commun. Mag. **42**(11), 106–116 (Nov. 2004)
4. M. Tömösközi, P. Seeling, P. Ekler, F.H.P. Fitzek, Performance evaluation and implementation of IP and robust header compression schemes for TCP and UDP traffic in the wireless context, in 2015 4th Eastern European Regional Conference on the Engineering of Computer-Based Systems, Brno, Czech Republic, 2015, pp. 45–50

Chapter 5
Air Interface

Abstract There are various ways to incorporate "wireless" functionality into a communication system. One of these is the use of electromagnetic radiation, which is the basis for radio frequency communication or simply radio communication. The air interface is the radio communication link between two communicating stations via a transceiver, which involves both the physical and data link layers (i.e., Layers 1 and 2) in the layering model presented in Chap. 3. Various types of applications and use cases demand different design strategies in communication protocol and radio transceiver. Their specifications and operating frequency spectrum impose limitations or offer freedoms on the transceiver characteristics and impact the complexity, cost, power consumption, and even the form factor of a mobile station. This chapter serves as a repository of air interface fundamental components focusing on the physical layer aspect and emphasizing mobile stations. It reviews the main components and primary characteristics of a radio transceiver and techniques commonly used in the MTC air interface that will become relevant and helpful in the remaining chapters of the book. Specifically, it explains the basic concepts of radio spectrum and bandwidth, antennas and propagation, carrier and modulation, passband and baseband signals, and synchronization and reception to provide the reader with a technical foundation on the radio communication aspect of MTC, the tutorial value of which will soon become evident. The materials in this chapter are presented in a descriptive way and thus do not rely on the reader's full grasp of the wireless communication theory.

Keywords Radio transceiver · Radio frequency carrier · Radio wave · Radio frequency spectrum · Carrier modulation · Passband signal · Baseband signal · Channel coding · Timing and carrier synchronization · Circular retransmission · Rate matching · Processing gain

5.1 Transceiver Structure

Information transfer is the process of moving messages from a source to a destination through a communication medium. In wireless communication, it involves the use of a wireless transceiver, i.e., a combination of a transmitter and a receiver, that

M. M. Wang, J. Zhang, *Machine-Type Communication for Maritime Internet-of-Things*, Wireless Networks, https://doi.org/10.1007/978-3-030-77908-5_5

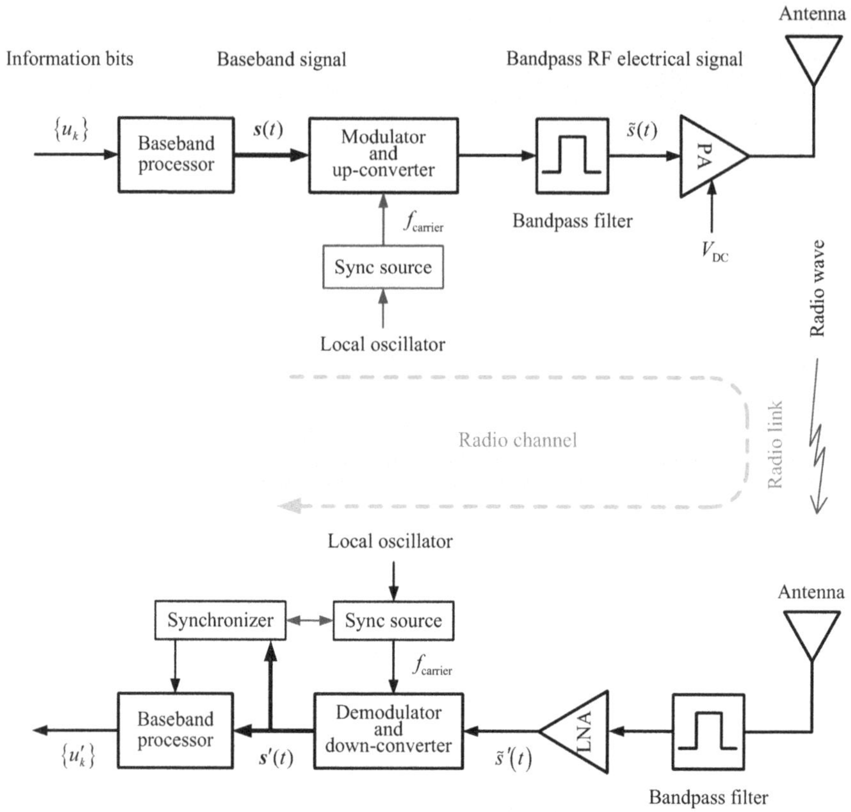

Fig. 5.1 Simplified transceiver block diagram, where PA stands for RF power amplifier and LNA for low-noise RF amplifier. In this book, the term "radio channel" is, in general, interchangeable with "radio link," except in the context of "channel gain" and "channel estimation," where the "channel" explicitly includes the RF paths of the transmitter and the receiver

gives a mobile station its wireless communication capability through a wireless medium. For maritime MTC, the transceiver is a radio transceiver that communicates via *radio waves* at a specified *radio frequency* (RF) confined within a given bandwidth, henceforth referred to as the *radio link* or *radio channel.*

Practical transceivers may look quite different, but they can still be traced back to a common fundamental structure. A functional block diagram of a radio transceiver should include the integrated elements depicted in Fig. 5.1. In this section, we briefly review the functional properties of each element in the diagram.

5.1.1 Radio Waves and Spectrum

In wireless communications, a *carrier* is an electromagnetic wave at a specific frequency that carries information across space from the transmitter to the distant receiver at the speed of light. This section looks at the phenomenon—the propagation of radio waves that lets us transfer information from one point to another with or without any physical medium.

The electromagnetic spectrum is divided into radio waves, infrared, visible light, ultraviolet, X-rays, and gamma rays, as shown in Fig. 5.2. For radio communication, the carrier frequency ranges from 3 kHz to 3 GHz. This range has now extended to 300 GHz, corresponding to a wavelength roughly from 100 km to 1 mm. The corresponding electromagnetic wave is called a *radio wave*.

Motivated by the recent advancements in light-emitting diodes (LEDs), the radio spectrum has recently extended to even higher frequencies in certain specific applications, e.g., infrared communications and visible light communications.

Under the ITU designation, the radio frequency spectrum is further divided into Very Low-Frequency (VLF), Low-Frequency (LF), Medium-Frequency (MF), High-Frequency (HF), Very High-Frequency (VHF), Ultra High-frequency (UHF), Super High-frequency (SHF), and Extremely High-Frequency (EHF) bands. As a form of electromagnetic radiation, the radio waves in these frequency bands have distinct propagation behaviors.

The overall behavior of radio waves is described by Maxwell's equations. Besides *line-of-sight*, it is affected by *reflection*, *refraction*, and *diffraction*. Depending on which phenomenon is dominant, radio waves can propagate from transmitter to receiver as *ground waves*, *sky waves*, or *space waves*.

Radio waves at low and medium frequencies in the VLF, LF, and MF bands propagate over the earth's surface and travel beyond the horizon as "ground waves" or "surface waves." These long and medium waves can bend over obstacles like hills and the earth's curvature due to diffraction, hence capable of traveling far beyond the horizon, following the earth's contour.

MF and HF band radio waves can be refracted or reflected from a layer of charged particles (ions) high in the atmosphere (i.e., the ionosphere), ergo the name "sky waves." This means that these medium and short waves can be reflected back to the earth beyond the horizon, useful for transmission over great distances, even transcontinental distances.

Radio waves in VHF and above are pronounced by the line-of-sight transmission path from the transmit antenna to the receive antenna. They are comprised of a direct or line-of-sight wave and reflected waves from the earth's stratosphere. Line-of-sight is the only possible wave at microwave frequencies and above, capable of penetrating the earth's atmosphere and reaching orbiting satellites, hence the name "space waves." VHF offers a much larger transmission range but grants less bandwidth occupation than the higher frequency space waves.

The radio frequency range on the electromagnetic spectrum is upper-bounded by the propagation range and lower-limited by the antenna dimension and the available bandwidth. As we will see next, both the efficiency of an antenna and the propagation range of the electromagnetic wave are dependent on the frequency.

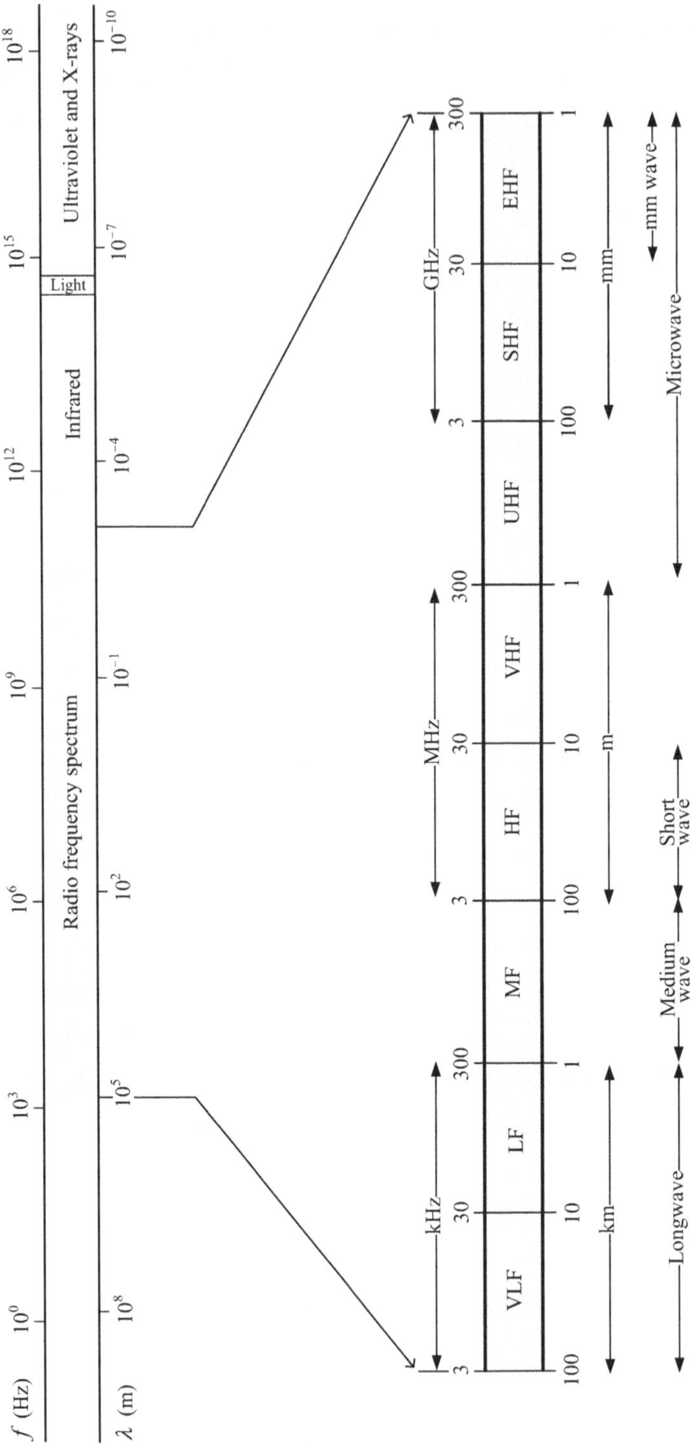

Fig. 5.2 Electromagnetic spectrum and breakdown of radio-frequency spectrum, where f denotes the frequency and $\lambda = c/f$ the wavelength (c the speed of light)

5.1.2 Antenna

Antennas are used at the transceiver for radiation and interception of radio waves. When an electrical current flows in a conductor, it tends to hang out on the conductor's surface, namely, the skin effect. The electrical energy begins to escape as electromagnetic waves when the size of the conductor becomes similar to the wavelength of the electrical current. In fact, every circuit that draws current radiates electromagnetic energy to some extent, either intentionally or unintentionally, according to Maxwell's equations. In wireless communication, an antenna is a *transducer* made purposely for converting the electrical energy to electromagnetic energy at a transmitter or the other way around at a receiver.

Specifically, an antenna receives an electrical signal from the transmitter circuit, e.g., the RF *power amplifier*, and converts it into electromagnetic waves. The efficiency of the antenna in performing this conversion, known as *antenna radiation efficiency*, is defined as the ratio of the electromagnetic power dissipated into space relative to the electrical power delivered to the antenna by an RF power amplifier. During the reverse operation, an antenna converts an electromagnetic wave into an electrical signal. The efficiency of this reverse operation is the same as the antenna radiation efficiency and thereby collectively called *antenna efficiency*.

An ideal isotropic antenna is a lossless antenna, i.e., with 0-dB efficiency, and radiates radio energy uniformly in all directions. In free space (i.e., no medium), the *power flux density* (PFD), defined as electromagnetic power per unit area, at a displacement of **d** away from an isotropic radiation source with power P, is

$$\rho = \frac{1}{4\pi d^2} \cdot P, \tag{5.1}$$

where $d = |\mathbf{d}|$, and $4\pi d^2$ is the area of the sphere at radius d, as illustrated in Fig. 5.3. (5.1) is also known as the *Friis' free-space equation*.

Unlike an isotropic antenna, a directional antenna concentrates the radiation in a particular direction and can be considered to have a gain relative to the isotropic antenna, called *directivity gain*. By this definition, an ideal isotropic antenna has a unit *antenna gain*. An antenna gain combines directivity with efficiency to describe how well the antenna converts input electrical power into radio waves headed in a specified direction, $\boldsymbol{\varphi} \triangleq (\vartheta, \psi)$, and vice versa, where $-\pi/2 \leq \vartheta \leq \pi/2$, and $-\pi \leq \psi \leq \pi$. Under this definition, (5.1) can be rewritten for a general antenna as

$$\begin{aligned} \rho(\mathbf{d}) &= \frac{1}{4\pi d^2} \cdot G_{\text{tx}}(\boldsymbol{\varphi}_{\text{tx}}) \cdot P \\ &= \frac{1}{4\pi d^2} \cdot \Lambda(\boldsymbol{\varphi}_{\text{tx}}), \end{aligned} \tag{5.2}$$

where $G_{\text{tx}}(\boldsymbol{\varphi}_{\text{tx}})$ is the transmit antenna gain and

$$\Lambda(\boldsymbol{\varphi}_{\text{tx}}) \triangleq G_{\text{tx}}(\boldsymbol{\varphi}_{\text{tx}}) \cdot P \tag{5.3}$$

is commonly known as the *equivalent isotropic radiated power* or EIRP.

Fig. 5.3 Illustration of power flux density of an isotropic antenna

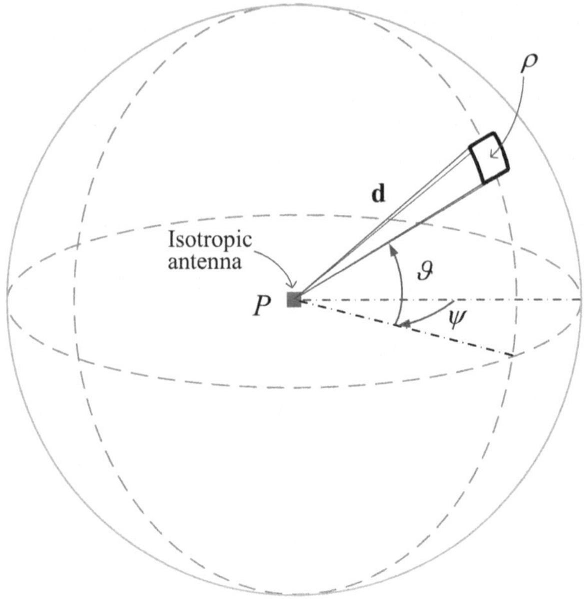

The received power of an antenna at the distance of **d** is

$$p = \rho(\mathbf{d}) \cdot A(\boldsymbol{\varphi}_{\mathrm{rx}}), \tag{5.4}$$

where A is the *effective area* of the antenna, also known as the *antenna aperture*. Since an ideal isotropic antenna has an aperture of

$$A^{\mathrm{iso}} = \frac{\lambda^2}{4\pi}, \tag{5.5}$$

and a unit gain $G^{\mathrm{iso}} = 1$, the relationship between the aperture, A, and the gain, G, of a general antenna is

$$\begin{aligned} A &= \frac{G}{G^{\mathrm{iso}}} \cdot A^{\mathrm{iso}} \\ &= \frac{\lambda^2}{4\pi} \cdot G, \end{aligned} \tag{5.6}$$

where λ is the wavelength. Substituting (5.6) into (5.4) gives

$$p = \ell^{-1} \cdot \Lambda(\boldsymbol{\varphi}_{\mathrm{tx}}) \cdot G_{\mathrm{rx}}(\boldsymbol{\varphi}_{\mathrm{rx}}), \tag{5.7}$$

where

$$\ell \triangleq \left(4\pi \frac{d}{\lambda}\right)^2 \tag{5.8}$$

is the free-space propagation loss.

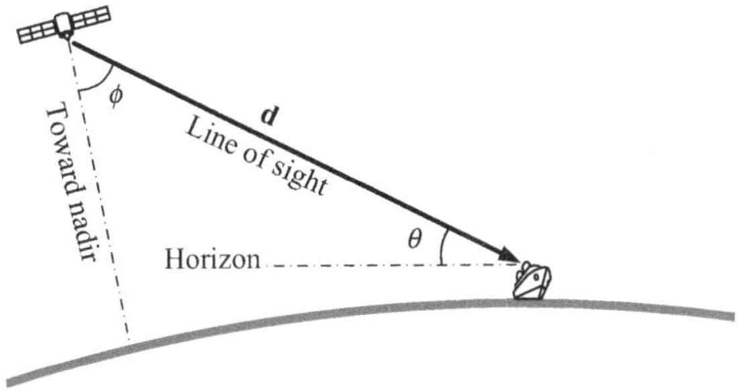

Fig. 5.4 Illustration of the off-nadir angle, ϕ, and elevation angle, θ. Illustrated is the satellite downlink

For ease of illustration and explanation, the following discussions will be confined within a two-dimensional vertical plane. The direction is then represented by a one-dimensional angle. For example, in the case of the downlink as illustrated in Fig. 5.4, $\boldsymbol{\varphi}_{\text{tx}}$ is the *off-nadir angle* of the control station, denoted as ϕ and $\boldsymbol{\varphi}_{\text{rx}}$ the *elevation angle* of the mobile station, denoted as θ, whereas for the uplink, $\boldsymbol{\varphi}_{\text{tx}}$ becomes θ and $\boldsymbol{\varphi}_{\text{rx}}$ is ϕ. When no direction is specified, the antenna gain is understood to refer to the peak value of the gain, i.e., the gain in the direction of the main lobe of the antenna. A control station typically employs a directional antenna, and the footprint of the main lobe defines the coverage, i.e., the cell.

5.1.3 Power Amplifier

An RF power amplifier is a type of active device that converts a low-power RF signal into a higher-power signal to drive the load, i.e., the antenna of a transmitter. It is the final stage of amplification prior to spaceborne. This active device is used as a controlled current source, i.e.,

$$i_{\text{out}}(t) = f(u_{\text{in}}(t)), \tag{5.9}$$

where $u_{\text{in}}(t)$ is the voltage of the input RF signal, f the transfer function, and $i_{\text{out}}(t)$ the output current flowing in the load, which is illustrated in Fig. 5.5.

A wireless transmitter relies on the RF power amplifier to deliver power to the antenna, high enough for reaching the required transmission distance. It is one of the major power consumption areas of a transceiver. Accordingly, it should be made as efficient as possible to help preserve battery life. For a given link budget, we wish a maximum power gain or efficiency, η_{PA}, from a power amplifier since efficiency means low power consumption for desired power, which is particularly important for power- and energy-limited mobile stations.

A commonly used *power amplifier efficiency* metric, known as the *drain efficiency*, is defined as the ratio of the output power, p_{out}, to the consumed power by the amplifier, i.e., P_{DC} the direct current (DC) power supply to the amplifier,

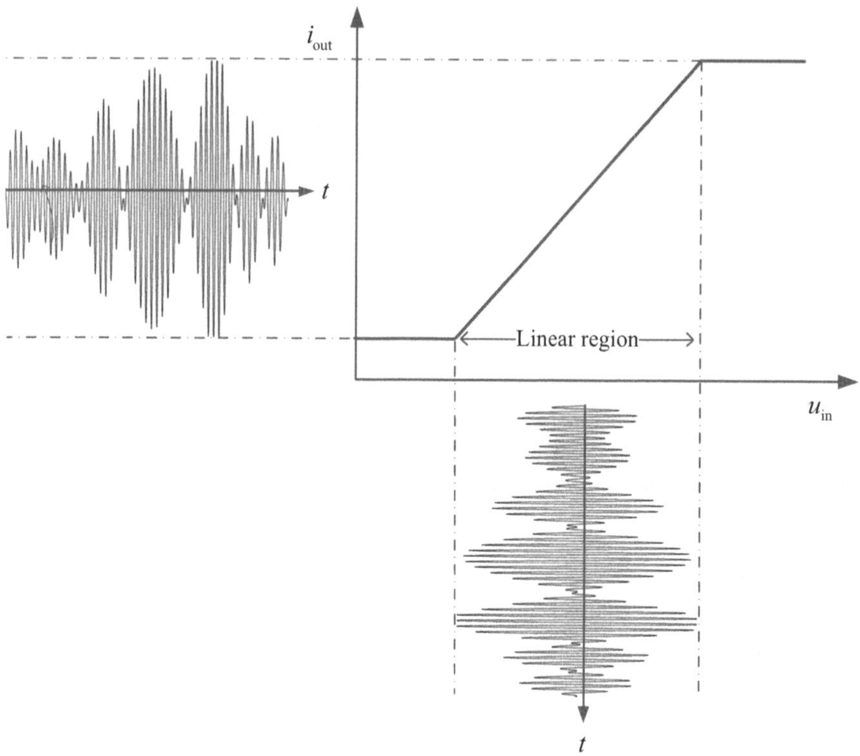

Fig. 5.5 Illustration of the current-voltage characteristic of a power amplifier

$$\eta_{PA} = \frac{p_{out}}{P_{DC}}. \tag{5.10}$$

Since P_{DC}, a constant, is independent of p_{out}, a greater p_{out} means higher efficiency.

For an amplifier, the natural next question is: What about the gain? How is the gain related to the efficiency? Defining the *power gain* as

$$g_{PA} = \frac{p_{out}}{p_{in}}, \tag{5.11}$$

or

$$p_{out} = g_{PA} \cdot p_{in}, \tag{5.12}$$

yields

$$\eta_{PA} = g_{PA} \cdot \frac{p_{in}}{P_{DC}}, \tag{5.13}$$

where p_{in} is the input power to the amplifier.

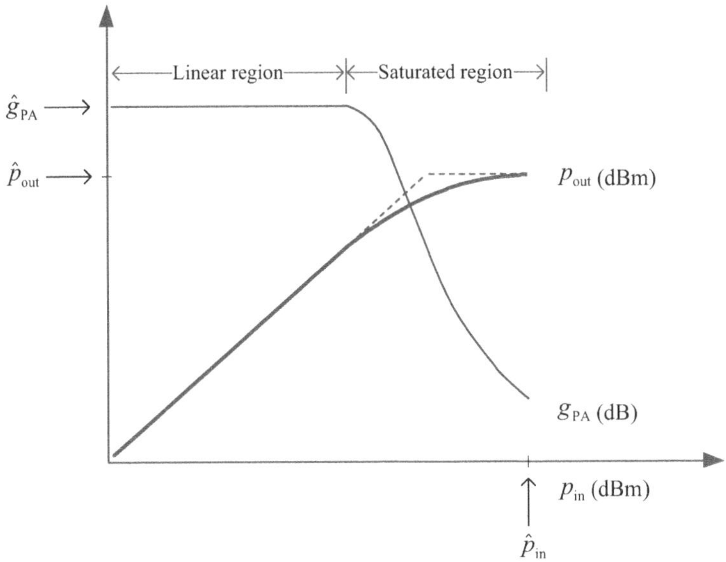

Fig. 5.6 Illustration of the relation between the input and output of a power amplifier, where $\widehat{p}_{in}$ is the peak efficiency operating point and $\widehat{p}_{out}$ the peak output power. In a realistic power amplifier, the change from linear power to saturated power is a smoother transition, as shown by the solid curve

A close look at Fig. 5.5 shows that f in (5.9) is a nonlinear function. The linearity of a power amplifier is thus compromised. This intrinsic nonlinearity of a realistic power amplifier has two major effects: 1) reduced power amplifier efficiency and 2) spectral regrowth in the output signal.

Note that the power gain, g_{PA}, holds constant at a peak level of $\widehat{g}_{PA}$, only when p_{in} is confined in the linear region, within which p_{out} scales linearly with p_{in}, as graphed in Fig. 5.6. Consequently, higher p_{in} leads to higher power efficiency, η_{PA}, as shown in (5.13). In the nonlinear region or saturated region, the gain is being compressed more and more as the input power increases since increasing input power does not produce a corresponding increase in output power. Nevertheless, the power efficiency continues to increase with input signal power as prescribed by (5.13). In this regard, an amplifier should operate in the saturated region for maximum power efficiency.

However, everything comes with a price. Running an amplifier into the saturated region results in waveform distortion, which not only creates the *in-band noise* degrading the output signal quality but causes unwanted *out-of-band emissions* to appear at the output, interfering with adjacent radio channels. The effect of the in-band noise on the output signal quality is commonly measured by the error vector magnitude metric or EVM, which will be treated shortly. The interference with the transmissions in adjacent radio channels created by the out-of-band emission is measured by the *adjacent channel leakage ratio* (ACLR), defined as the ratio of

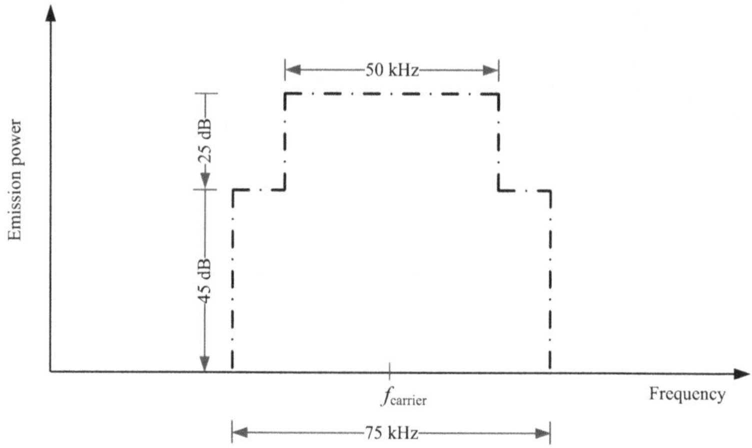

Fig. 5.7 A spectrum emission mask example, where f_{carrier} denotes the carrier frequency

the transmitted power on the currently assigned channel to the power leaked into the adjacent radio channels.

To limit the adjacent channel leakage, typical radio communication standards stipulate the maximum permitted out-of-band emission levels with respect to the carrier, usually in the form of a transmit spectrum "mask," as exemplified in Fig. 5.7. These constraints restrict the dynamic range of the input RF signal, $u_{\text{in}}(t)$. To maximize the mean input power, $\overline{p}_{\text{in}} = E\{p_{\text{in}}(t)\}$ (so as to maximize the mean output power, $\overline{p}_{\text{out}}$, yet with the out-of-band emission contained within the mask), the dynamic range of the input signal power must be minimized. It is equivalent to say that a highly efficient amplifier favors signals with low power variations.

Extension of the linear region reduces the out-of-band emission, nevertheless incurs higher costs. A low-cost power amplifier has much less than 50% efficiency, whereas the peak efficiency can be higher than 50% for a high-end power amplifier, usually found in the transmitter of a control station. A 50-percent power efficiency means that one-half of the DC power is transformed to the RF signal power for exciting the antenna, while the other half is dissipated in the power amplifier, heating up the device.

The amount of power variation of a waveform is usually captured by the *peak-to-average-power ratio* or PAPR, defined as the ratio of the peak power with respect to the average or mean power of a waveform, i.e.,

$$\gamma_{\text{PAPR}} = \frac{\max\{p_{\text{in}}(t)\}}{E\{p_{\text{in}}(t)\}} \quad \text{or} \quad \frac{\widehat{p}_{\text{in}}}{\overline{p}_{\text{in}}}, \tag{5.14}$$

where $\widehat{p}_{\text{in}} \triangleq \max\{p_{\text{in}}(t)\}$. Clearly, $\gamma_{\text{PAPR}} \geq 0$ dB.

For a sinusoidal waveform (such as an unmodulated RF waveform) of amplitude $\mathcal{A}$, the PAPR is

$$\gamma_{\text{PAPR}} = \frac{\mathcal{A}^2}{\mathcal{A}^2/2} = 2 \text{ or } 3 \ \text{dB}. \tag{5.15}$$

As a rule of thumb, a PAPR of 3 dB means that, if we wish to operate the power amplifier at 10-dBm (average) output power, a power amplifier of saturation power 13 dBm should be used to keep 3-dB headroom from peak power. That is to say, $\overline{p}_{\text{out}}$ =10 dBm, P_{DC}=13 dBm, and hence

$$\eta_{\text{PA}} = \overline{p}_{\text{out}} - P_{\text{DC}} = -3\text{dB}. \tag{5.16}$$

The power efficiency is thus less than 50%.

We will use this metric in the waveform design later in this chapter.

In some communication scenarios, e.g., a shared transmission resource scenario, like PUSCH and PRACH, the output power is also constrained by other factors. In this case, the output power, $\overline{p}_{\text{out}}$, is purposely adjusted to a desired level, called *power control*, just enough for the intended receiver to recover the message but at the same time causing minimum interference to other unintended receivers that share the resource. For a power amplifier working in the linear region, where the power gain is a constant, $\widehat{g}_{\text{PA}}$, power control is therefore most effective through controlling the input power, $\overline{p}_{\text{in}}$, per (5.12). We will have more discussion on PUSCH and PRACH power control in Chap. 8.

5.1.4 Power Supply

It is typically difficult to run a mains supply to the mobile station, although there are certain occasions, e.g., large ships, where mains power might be available. MTC mobile stations are usually resource-constrained, small-form-factor devices designed to run on battery power. In the case where there is no mains power supply, energy is stored either in batteries or capacitors. Batteries, both rechargeable and non-rechargeable, consist of one or more cells (nickel-cadmium, nickel-zinc, nickel-metal hydride, and lithium-ion), in which chemical energy is converted into electricity. They combine good energy density with the energy supply at precisely the standard voltage levels, negating the need for intermediate power conditioners. Capacitors, on the other hand, store energy in the form of electrical charges. They have significantly higher power density than batteries, nevertheless, much lower energy density. Today's energy supplies are even able to renew their energy through energy harvesting from solar sources, radio waves, temperature differences, or vibrations like ocean waves. Nonetheless, the harvested energy is typically low, from 100s of microwatts to a few milliwatts.

A mobile station consumes energy mainly for communicating and processing. More energy is required during communication than in any other state. A mobile station can be said to remain in one of the three states: *active state*, *idle state*, and *sleep state*. The energy consumed in any state depends on the power dissipated and the time spent in that state. In the active state, the mobile station is turned on and communicating, either transmitting or receiving. During transmit, the transmitter spends a considerable amount of power radiating the electromagnetic power to overcome the typically large propagation loss when traveling over long distances to the intended receiver. During receiving, the transmitter is not radiating, and most power is consumed on processing the received signal, e.g., filtering, demodulation, and channel-decoding. In the idle state, the transceiver is on but not receiving or transmitting. In the sleep state, almost all parts of the mobile station are turned off. The only power consumed during this state is from the leakage in the circuit. Hence, the sleep state is the least power-consuming state.

Defining

$$\overline{\eta}_{\mathrm{PA}} = \frac{\overline{p}_{\mathrm{out}}}{P_{\mathrm{DC}}} \approx \frac{P}{P_{\mathrm{DC}}} \tag{5.17}$$

the mean power amplifier efficiency, the total current drawn from the DC power supply during transmission is roughly

$$I_{\mathrm{tx}} \approx I_0 + (P/\overline{\eta}_{\mathrm{PA}}/V_{\mathrm{DC}}), \tag{5.18}$$

where P is the transmitted or radiated power (from the antenna), I_0 the current drawn by baseband processing, e.g., channel-encoding, and V_{DC} the DC voltage to the power amplifier. The total current drawn during the reception is $I_{\mathrm{rx}} \approx I_0$. Assuming I_0=100 mA, P=20 dBm, V_{DC}=3 V and $\overline{\eta}_{\mathrm{PA}} = 30\%$, the total current drawn is 211 mA during the transmission and 100 mA during the reception. The battery consumption during transmission and reception can be approximated as $T_{\mathrm{tx}}I_{\mathrm{tx}}$ and $T_{\mathrm{rx}}I_{\mathrm{rx}}$, respectively, where T_{tx} and T_{rx} are the transmit and receive time, respectively.

Depending on the type of device, the typical capacity of a battery ranges from 500 mAh (e.g., button cell batteries) to 5000 mAh (e.g., two AA batteries). The required battery life ranges from days to years. Nevertheless, it is clear that the "air time" of a transceiver has a profound effect on the battery life, and its optimization is vital to most IoT applications.

Besides air time, energy can be wasted in other scenarios:

1. Transmission collisions among mobile stations: if transmissions collide, all data involved are garbled, which ensues another round of transmission attempts.
2. Control overhead: from an MTC system's point of view, energy spent in transferring application traffic is energy usefully spent, whereas the energy spent on control plane activities, i.e., the control traffic, is overhead and should be kept to a minimum.
3. Transceiver efficiency: transceiver efficiency, mainly the power amplifier efficiency, dominates the total energy consumption during air time.

Indeed, energy efficiency is a key issue at every level of an MTC system. We will revisit and elaborate on these topics throughout the remainder of the book.

5.1.5 Baseband and Passband Signals

A *baseband signal* refers to the information-bearing bandwidth-limited electrical waveform in its original frequency range (typically in the low-frequency range around DC). Therefore, a baseband signal itself cannot propagate across the space with a practical antenna. If, instead, the information-bearing baseband signal is used to modulate an RF sinusoidal electrical waveform, i.e., altering some aspects of the time-domain waveform such as amplitude, phase, or frequency, to impress the information on the waveform, it then can be radiated into the space via an antenna. This process of imposing a baseband signal onto an RF waveform, i.e., the carrier, is called carrier *modulation,* and the information-bearing baseband signal is named the *modulation signal.* Transmission of baseband signals via an RF carrier also enables multiple baseband signals to be transmitted at different frequencies simultaneously, known as frequency-division multiplexing. FDD of the uplink and downlink signals (Chap. 2) is an example.

Mathematically, the modulated RF signal (by the modulation signal) can be represented as

$$\begin{aligned}\tilde{s}(t) &= \mathcal{Re}\left\{\mathcal{A}(t)\cdot e^{\,j(2\pi f_{\text{carrier}}t+\varphi(t))}\right\}\\ &= \mathcal{Re}\left\{s(t)\cdot e^{\,j2\pi f_{\text{carrier}}t}\right\},\end{aligned} \tag{5.19}$$

where $e^{\,j2\pi f_{\text{carrier}}t}$ is the unmodulated complex RF waveform and

$$s(t)\triangleq\mathcal{A}(t)\cdot e^{j\varphi(t)} \tag{5.20}$$

is a complex-valued representation of the baseband modulation signal, a function of time, t.

Apparently, the modulation process entails *up-conversion* that translates a baseband signal to a higher carrier frequency, f_{carrier}. The passband signal, $\tilde{s}(t)$, is simply a translation of the baseband signal, $s(t)$, to a real-valued RF signal. We can dispense with the shift to RF in most discussions for simplicity without loss of generality.

The modulated RF electrical waveform is passband-filtered to remove the unwanted spectral components (if any) to meet the emission spectrum mask, such as the one in Fig. 5.7, hence the name *passband signal.* Rewriting (5.19) to have the following form,

$$\tilde{s}(t) = \underbrace{\mathcal{Re}\{s(t)\}\cdot\cos\left(2\pi f_{\text{carrier}}t\right)}_{\text{I-channel}} - \underbrace{\mathcal{Im}\{s(t)\}\cdot\sin\left(2\pi f_{\text{carrier}}t\right)}_{\text{Q-channel}}, \tag{5.21}$$

indicates that the real part of the modulation symbol, $\mathcal{Re}\{s(t)\}$, is used to modulate the RF sinusoidal waveform, $\cos(2\pi f_{\text{carrier}}t)$, and the imaginary part, $\mathcal{Im}\{s(t)\}$, to modulate $\sin(2\pi f_{\text{carrier}}t)$, where the former is commonly known as the *in-phase*

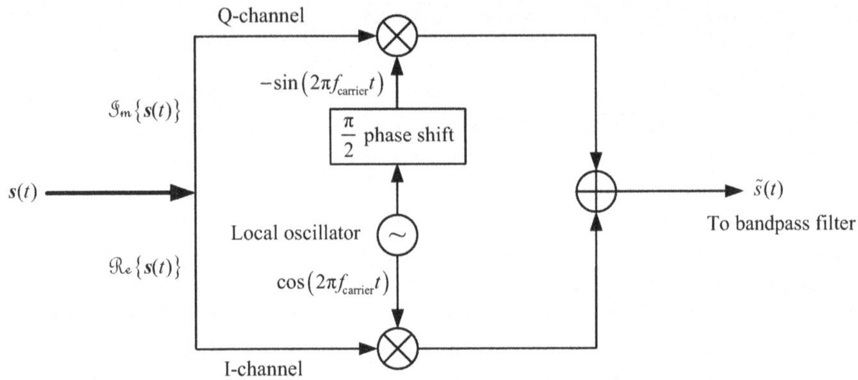

Fig. 5.8 Implementation of a quadrature modulator for carrier modulation and up-conversion from a baseband modulation signal (complex in general), $s(t)$, to a quadrature-modulated RF carrier (real), $\tilde{s}(t)$. Demodulation or down-conversion is the reverse operation

(I) *carrier*, and the latter the *quadrature* (Q) *carrier* since it is phase-shifted by $\pi/2$. The I and Q carriers are thus orthogonal to each other and serve as the two bases in a two-dimensional signal space. Similarly, $\mathcal{R}e\{s(t)\}$ and $Im\{s(t)\}$ are referred to as the I and Q components of the modulation signal, $s(t)$, respectively.

Figure 5.8 depicts a practical implementation of a quadrature modulator. A *quadrature modulator* is thus a modulator that has two channels, with one being phase-offset by 90 degrees or one-quarter of a cycle. It can be used to create amplitude, phase, and frequency-modulated RF signals, as we will see in Section 5.2.

Figure 5.9 shows the representation of the baseband modulation signal in the I-Q plane, i.e., the space spanned by the in-phase and quadrature sinusoids. The demodulation of the passband signal is the reverse operation of Fig. 5.8, resulting in the I and Q components of the modulation signal. Under this representation, a radio channel can be considered as complex for ease of mathematical treatments.

Since $\cos(2\pi f_{carrier}t)$ and $\sin(2\pi f_{carrier}t)$ are orthogonal, $s(t)$ can be fully recovered from $\tilde{s}(t)$, as long as the oscillators at the transmitter and receiver are frequency and phase synchronized.

For spaceborne, the passband signal is amplified to boost the power level (e.g., P_{tx}=20 dBm) to excite the antenna, resulting in an information-bearing radio wave radiated into the space and travels at the speed of light. If the radio wave has sufficient power to overcome the propagation loss, it will eventually make its way to the destination at a power greater than the receiver sensitivity, where the operation is reversed. When the radio wave is incident on the receive antenna, it induces a weak RF electrical current in it, at a power level of, e.g., $P_{rx}=-100$dBm. It is immediately bandpass-filtered and amplified using a low-noise amplifier (LNA) with a noise figure typically 1–2 dB and then fed to the demodulator. If the signal strength

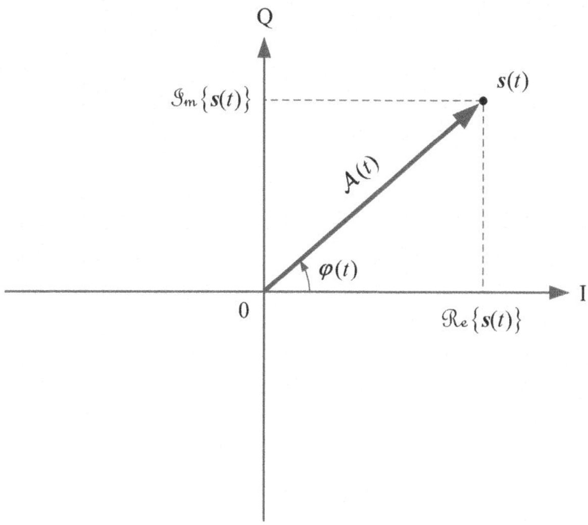

Fig. 5.9 Representation of modulation symbol, $s(t)$, in the space spanned by cos $(2\pi f_{\text{carrier}}t)$ and sin $(2\pi f_{\text{carrier}}t)$, corresponding to the I- and Q-channel of a radio channel, respectively

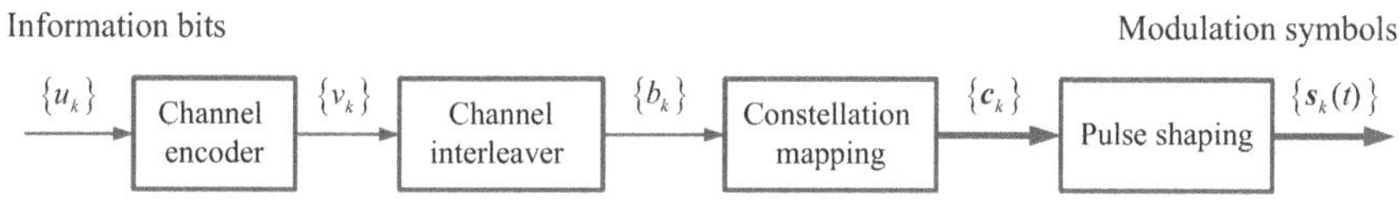

Fig. 5.10 Baseband processing block diagram

relative to the noise, i.e., the signal to noise ratio (SNR), at the demodulator's input is sufficiently large, the demodulator can recover the original information-bearing modulation signal with a high success probability. With channel error-correction coding, the original information bits can be truthfully recovered even in the presence of demodulation errors.

If a radio wave were the shipping carrier, the baseband signal would be the parcel or container for packaging the "goods," i.e., the information. Obviously, not all shipments are created equal; the destination, size, weight, and priority (speed and reliability) will affect the final shipping price. In wireless communication, such factors determine the output power from the RF power amplifier needed to drive the antenna and the amount of radio resources (i.e., bandwidth and time) to transport the information to the destination. Given the transmit power, the role of a baseband processor is to create the proper baseband waveform for the given bandwidth and time to represent the information for efficient (low cost) and reliable (low probability of package loss) transportation of the information. This process typically involves channel coding, channel interleaving, constellation symbol mapping, and pulse shaping, as shown in Fig. 5.10. It is the primary focus of this chapter.

5.1.6 Synchronization Source

As previously noted in Sect. 5.1.5, the recovery of the information-bearing modulating signal from the RF carrier requires the synchronization between the transmitter and the receiver. As shown in Fig. 5.1, the frequency offset between the transmitter and the receiver of two communicating stations is determined by the accuracy and stability of the local "free-running" crystal oscillator of a mobile station. The offset is the deviation of the oscillator frequency from the specified target frequency (i.e., the network carrier frequency). Factors such as temperature, crystal aging, humidity, and vibration affect the frequency accuracy and stability of the crystal. Among these, the temperature is the primary factor that affects the accuracy. The offset (without compensation) is typically greater than ±5ppm, depending on the quality of the crystal. For a low-cost MTC device, the initial frequency error could be well above ±5ppm, e.g., ±20ppm, especially for the first initial system acquisition when the local crystal oscillator is never synchronized to the network (i.e., no compensation/correction has ever been made).

Assume that the frequency error of a local oscillator of a mobile station transceiver is ζ_{mobile} ppm and a control station transceiver ζ_{control} ppm. From (5.19), the received bandpass signal is

$$\begin{aligned}\tilde{s}'(t) &= \mathcal{Re}\left\{s(t)\cdot e^{j2\pi(1+\zeta_{\text{mobile}}-\zeta_{\text{control}})f_{\text{carrier}}t}\right\}\\ &= \mathcal{Re}\left\{s'(t)\cdot e^{j2\pi f_{\text{carrier}}t}\right\},\end{aligned} \tag{5.22}$$

where

$$s'(t) \triangleq s(t)\cdot e^{j2\pi(\zeta_{\text{mobile}}-\zeta_{\text{control}})f_{\text{carrier}}t} \tag{5.23}$$

is the received baseband signal. It demonstrates that the frequency errors, in effect, introduce a linear *phase ramp* in the received baseband signal. Since $\zeta_{\text{mobile}} \gg \zeta_{\text{control}}$, (5.23) can be rewritten as

$$s'(t) \approx s(t)\cdot e^{j2\pi\zeta_{\text{mobile}}f_{\text{carrier}}t}. \tag{5.24}$$

For a mobile station operating at f_{carrier}=160 MHz with ζ_{mobile}=20 ppm, the phase of the baseband signal drifts by more than 6π per ms. This impact on the baseband signal caused by the RF frontend is often equivalently considered part of the radio channel impairment and needs to be taken into consideration in baseband signal design and baseband processor design.

Moreover, the local oscillator frequency error may cause receiver timing to drift by $\zeta_{\text{mobile}}t$, potentially causing the receiver to lose its lock on the incoming signal timing after a prolonged period. For a mobile station with ζ_{mobile}=20 ppm, the timing drifts by 1.2 ms per minute. Since the frequency accuracy of a crystal depends heavily on temperature, a high-end transceiver is thus typically equipped with a temperature-compensated crystal oscillator (TCXO). A TCXO reduces the environmental changes to produce a stable output frequency. Nonetheless, not only is a TCXO costly, but it is also power-consuming to maintain, a luxury that a low-end MTC device cannot afford. Moreover, even for a TCXO, the smallest errors can significantly add up over a long period.

Global Positioning System or GPS (and now other global navigation systems) receivers have superior long-term stability, yet the GPS signal is available worldwide. Using the GPS signal to regulate a local oscillator is a common practice in communication systems to control frequency and timing drifting, providing atomic clock accuracy without a local atomic clock. This GPS-disciplined oscillator works by "steering" a local oscillator. It locks the frequency of the oscillator to a GPS signal via a tracking loop that works in a similar way to a phase-locked loop [1]. It may even go further to compensate for not only the phase and frequency changes of the local oscillator but the "learned" effects of aging, temperature, and other environmental parameters, which becomes useful when the GPS signal is not available.

Indeed, GPS reception may not always be stable. Various problems can occur due to factors in the environment, such as signal blockage, a weather-related outage, interference, and other appreciable impairments. Since the timing is sustained by the accuracy of the oscillator during GPS outages, the oscillator output does not go away when the GPS signal does. It just starts to drift as time elapses, gradually deteriorating the accuracy.

A regular GPS receiver module draws about 50 mA of current and costs less than 50 dollars. Nonetheless, it is too high a cost for some energy- and cost-sensitive deployment scenarios. For mobile stations in these types of deployments, synchronization is a challenge, which is treated in great detail in the following chapters of this book.

5.2 Modulation

As described in the previous section, in radio communication, the information (in the form of bits) is carried over the space by the radio wave at a specified radio frequency, confined within a bandwidth defined by a spectrum emission mask. The information bits must be properly "packaged" in a form suitable for being carried by the radio wave. To that end, a baseband waveform that falls into the specified bandwidth is used to represent the information bits as a modulation signal to modulate an RF carrier at the specified frequency. Therefore, a practical baseband waveform has the following required characteristics: (1) a waveform spectrum that meets the specified emission mask, (2) easy detection or information recovery from the waveform at a receiver, (3) transceiver specifications that comply the MTC requirements, and (4) maximized information transfer rate. In this section, we discuss the common modulation methods. The focal point is placed on the modulation schemes and underlying principles that are most relevant to maritime MTC.

5.2.1 *Phase and Amplitude Modulation*

Phase and amplitude modulation, generally termed *quadrature amplitude modulation* (QAM), is the most spectrally-efficient modulation scheme. When combined with channel error correction coding, it can be one of the most energy-efficient modulation and coding schemes as well.

5.2.1.1 Modulation Symbol

For amplitude and phase modulation or QAM, a baseband modulation signal in (5.20) can be represented as a superposition of a sequence of modulation symbols,

$$\begin{aligned} s(t) &= \sum_{k=0}^{N_{\mathrm{sym}}-1} s_k(t) \\ &= \sum_{k=0}^{N_{\mathrm{sym}}-1} c_k \cdot p_k(t), \end{aligned} \tag{5.25}$$

hence the name *linear modulation*, where N_{sym} is the number of modulation symbols in the signal, and

$$\begin{aligned} s_k(t) &\triangleq c_k \cdot p_k(t) \\ &= c_k \cdot p\left(t - kT_{\mathrm{sym}}\right) \end{aligned} \tag{5.26}$$

is the modulation symbol, $p_k(t) \triangleq p(t - kT_{\mathrm{sym}})$ is T_{sym}-spaced time shifts of a pulse function, $p(t)$– a real-valued pulse with unit norm, and $\boldsymbol{c}_k \in \mathbb{C}$ is a function of the information bits that it represents.

Note that, depending on the pulse shape, T_{sym} is not necessarily the actual length or duration of the pulse; rather, it is the time spacing between two consecutive pulses. Hence, T_{sym}^{-1} is the symbol transmission rate or simply *channel symbol rate* and T_{sym} is naturally referred to as the *channel symbol time* or *interval.*

In particular, $\boldsymbol{c}_k \in \boldsymbol{\Omega}^m$, where

$$\boldsymbol{\Omega}^m \triangleq \{\omega_i \in \mathbb{C}, i = 0, 1, \cdots, 2^m - 1\} \tag{5.27}$$

is a set of complex values, commonly referred to as the *modulation constellation* or *constellation symbol* that maps a complex value or a *constellation point* to an m-bit sequence: $b_0, b_1, \cdots, b_{m-1}$. m is referred to as the *modulation order*. It is convenient to assume $\boldsymbol{\Omega}^m$ has an average unit norm: $\|\boldsymbol{\Omega}^m\|_2 = 1$.

The term "symbol" is an overused nomenclature and, by no means, has a universally accepted definition. In this book, the term "symbol" is reserved to refer

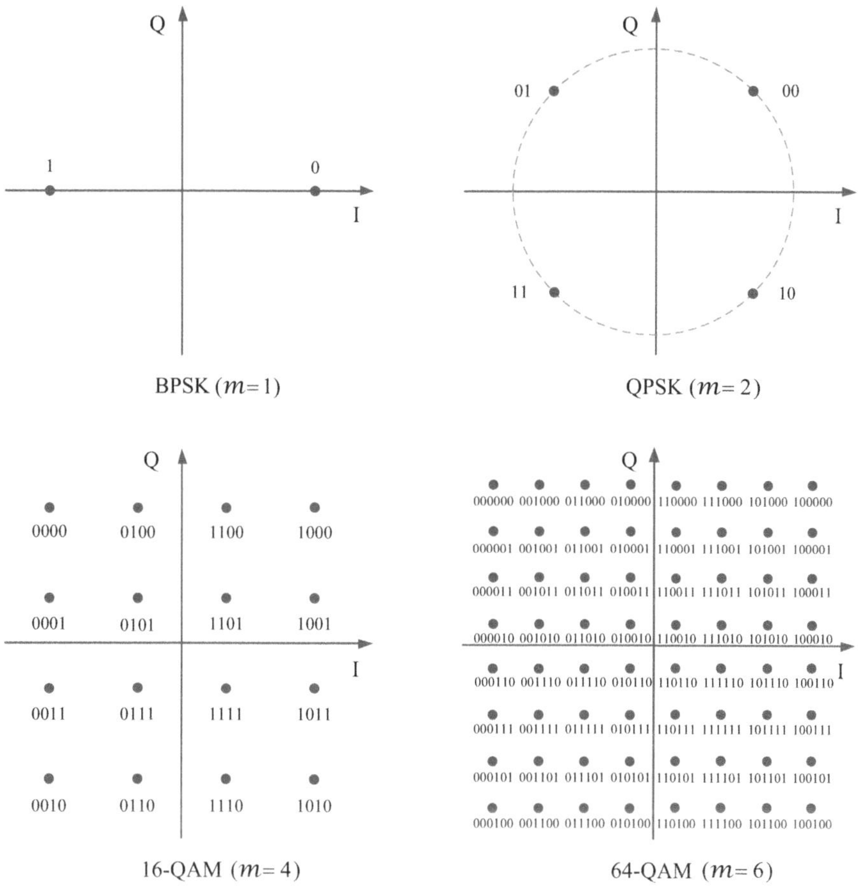

Fig. 5.11 Constellations for BPSK, and Grey-coded QPSK, 16-QAM, and 64-QAM, denoted as $\boldsymbol{\Omega}^1$, $\boldsymbol{\Omega}^2$, $\boldsymbol{\Omega}^4$,and $\boldsymbol{\Omega}^6$, respectively; each set has an average unit norm by default

to either the constellation symbol, $\boldsymbol{c}_k$, or the modulation symbol, $s_k(t)$, or the channel symbol.

Following the convention explained in the previous section, the real and imaginary parts of $\boldsymbol{c}_k$ represent the I and Q components, respectively, corresponding to the coordinate of the constellation point in the I-Q plane. A modulation order of m consists of a constellation of size $|\boldsymbol{\Omega}^m| = 2^m$, which encodes an m-bit information sequence into one of the 2^m constellation points. Figure 5.11 shows the most commonly used constellations: *binary phase shift keying* (BPSK), *quadrature phase shift keying* (QPSK), *16 quadrature amplitude modulation* (16-QAM), and *64 quadrature amplitude modulation* (64-QAM).

The representation of a baseband modulation signal in (5.25) is thus a linear combination of the basis functions, $p_k(t)$, with coefficients, $\boldsymbol{c}_k \in \boldsymbol{\Omega}^m$, k=0, 1,$\cdots$,

$N_{\text{sym}}-1$. This representation makes it possible to retrieve the information embedded in the interpolation at the receiver. Specifically, the information-bearing coefficients, $\{\boldsymbol{c}_k\}$, in (5.25) can be recovered by a receiver, i.e.,

$$\begin{aligned}\langle s(t), p_k(t)\rangle &= \int_t \left(\sum_{i=0}^{N_{\text{sym}}-1} \boldsymbol{c}_i \cdot p_i(t)\right) \cdot p_k^*(t) \cdot dt \\ &= \sum_{i=0}^{N_{\text{sym}}-1} \boldsymbol{c}_i \cdot \langle p_i(t), p_k(t)\rangle \\ &= \boldsymbol{c}_k, \quad k = 0, 1, \cdots, N_{\text{sym}} - 1,\end{aligned} \tag{5.28}$$

provided $p_i(t)$ and $p_j(t)$ are *orthogonal*, i.e.,

$$\langle p_i(t), p_k(t)\rangle = \delta(i-k), \quad i,k = 0, 1, \cdots, N_{\text{sym}} - 1, \tag{5.29}$$

or

$$\left\langle p(t - iT_{\text{sym}}), p(t - kT_{\text{sym}})\right\rangle = \delta(i-k), \quad i,k = 0, 1, \cdots, N_{\text{sym}} - 1, \tag{5.30}$$

where $\delta(\cdot)$ denotes the *Kronecker delta function*, $\langle \cdot, \cdot \rangle$ represents the dot product operator, and $(\cdot)^*$ is the complex conjugate operator. Since $p(t)$ is real, $p^*(t) = p(t)$. (5.30) implies that, for (5.28) to hold, $p(t)$ needs to have an autocorrelation function that is zero-valued at the T_{sym}-spaced lags except at the zeroth lag, namely,

$$\begin{aligned}\gamma_p(\tau) &= \langle p(t), p(t+\tau)\rangle \\ &= \delta(k), \quad \tau = kT_{\text{sym}}, k = 0, 1, \cdots, N_{\text{sym}} - 1,\end{aligned} \tag{5.31}$$

and the values in between are irrelevant (in theory). A function is said to be *Nyquist* if it possesses periodic nulls as in (5.31). A pulse whose autocorrelation is Nyquist is called a *Nyquist pulse*. An immediate significance of being Nyquist is that a correlator or *matched filter* can recover the information from the pulses.

The simplest pulse that satisfies (5.31) is the rectangular pulse,

$$p(t) = \text{rect}\left(tT_{\text{sym}}^{-1}\right), \tag{5.32}$$

where

$$\text{rect}(t) \triangleq \begin{cases} 1, & -1/2 \le t < 1/2, \\ 0, & \text{otherwise}. \end{cases} \tag{5.33}$$

The autocorrelation of (5.32), i.e., the *effective pulse* seen by a receiver at the output of a correlator or a matched filter to$p(t)$, is

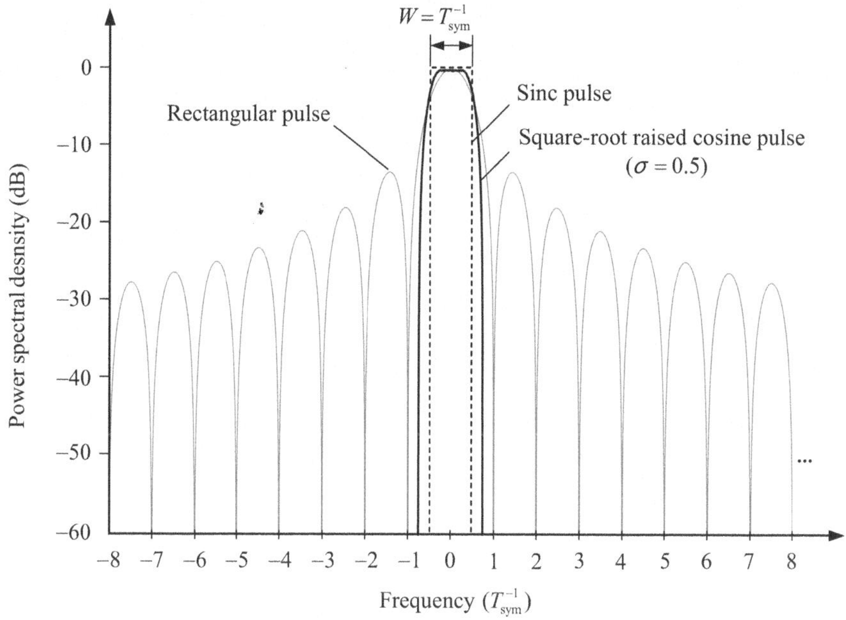

Fig. 5.12 Power spectral densities for a rectangular pulse and a square-root cosine pulse s with σ=0 (i.e., a sinc pulse) and 0.5. Note that, BPSK, QPSK, 16-QAM and 64-QAM have the same power spectral density

$$\gamma_p(t) = \begin{cases} 1 - \left|\frac{t}{T_{\text{sym}}}\right|, & -T_{\text{sym}} \le t < T_{\text{sym}}, \\ 0, & \text{otherwise}. \end{cases} \tag{5.34}$$

Since (5.34) is Nyquist, a rectangular pulse is a Nyquist pulse. However, the power localization of a rectangular pulse in the frequency domain is not only unbounded but dispersive, as can be seen from its power spectral density

$$\mathcal{P}(f) = \text{sinc}^2(\pi f T_{\text{sym}}), \tag{5.35}$$

plotted in Fig. 5.12, which likely fails to satisfy most practical spectrum emission masks. Here, sinc(·) is the sinc function defined as

$$\text{sinc}(x) \triangleq \frac{\sin(x)}{x}. \tag{5.36}$$

Another well-known Nyquist pulse is the *square-root raised cosine* pulse,

$$p(t) = \sqrt{T_{\text{sym}}^{-1}} \cdot \frac{\sin\left(\pi t T_{\text{sym}}^{-1}(1-\sigma)\right) + 4\sigma t T_{\text{sym}}^{-1} \cos\left(\pi t T_{\text{sym}}^{-1}(1+\sigma)\right)}{\pi t T_{\text{sym}}^{-1}\left(1 - \left(4\sigma t T_{\text{sym}}^{-1}\right)^2\right)}, \tag{5.37}$$

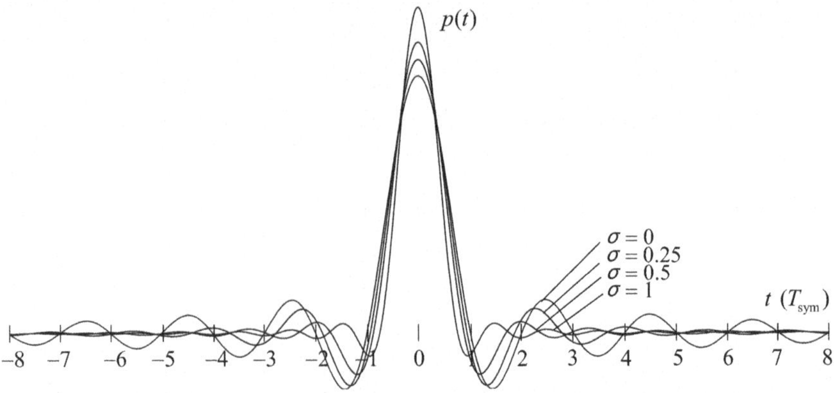

Fig. 5.13 Square-root raised cosine pulses with various roll-off factors

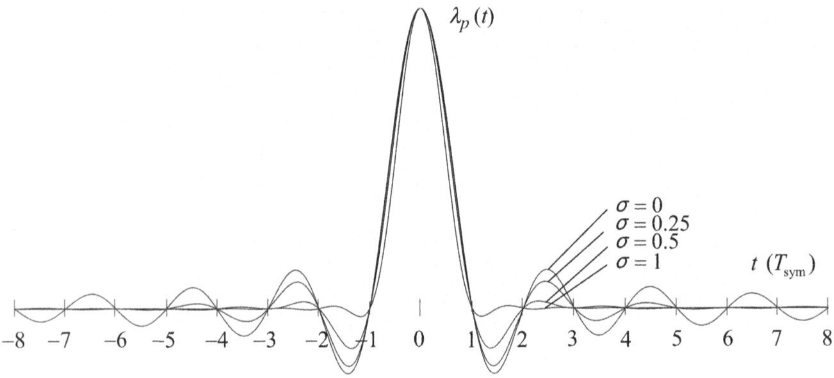

Fig. 5.14 Autocorrelation function of square-root raised cosine pulses

where σ is the *roll-off factor*, as shown in Fig. 5.13.

Substituting (5.37) into (5.31) directly results in the autocorrelation function or the effective pulse waveform of the square-root raised cosine pulse,

$$\gamma_p(t) = \frac{\cos(\pi\sigma t T_{\text{sym}}^{-1})}{1 - (2\sigma t T_{\text{sym}}^{-1})^2} \cdot \operatorname{sinc}(\pi t T_{\text{sym}}^{-1}), \tag{5.38}$$

which satisfies $\gamma_p(kT_{\text{sym}}) = \delta(k)$, as shown in Fig. 5.14. (5.38) is called a *raised cosine* function.

Make no mistake about it; the raised cosine function does not make a Nyquist pulse because its autocorrelation function does not possess the property prescribed by (5.31). After all, what makes a pulse a Nyquist pulse is that its autocorrelation is

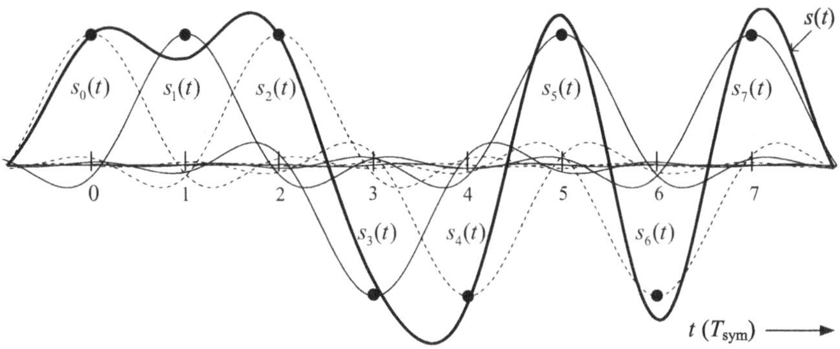

Fig. 5.15 The baseband modulating signal at the transmitter, composed of eight BPSK modulation symbols (N_{sym}=8), based on the square-root raised cosine pulse with σ=0.35

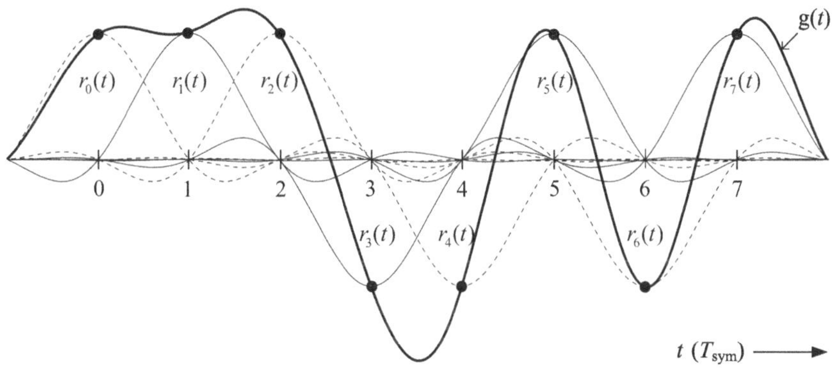

Fig. 5.16 The output and the sampled output (solid dots) of the matched filter at the receiver

Nyquist rather than the pulse itself since what matters in terms of information recovery is what the receiver observes at the output of the matched filter (i.e., the effective pulses), not what the transmitter sees (the original pulse).

Figure 5.15 is an example of the baseband modulation signal waveform, $s(t)$, where $s_k(t)$ is the square-root raised cosine pulse modulation symbol with σ=0.35. Figure 5.16 is the matched filter (to $s_k(t)$) output at the receiver, where $r_k(t)$ is given in (5.44) with an effective raised cosine pulse in (5.38). The sampled outputs are clearly inter-symbol interference-free, assuming perfect timing. The sample timing is part of the synchronization process between the transmitter and the receiver, which is touched upon in the previous section, and treated in more detail in Section 5.6. The power spectral density of a square-root raised cosine pulse can be shown to be

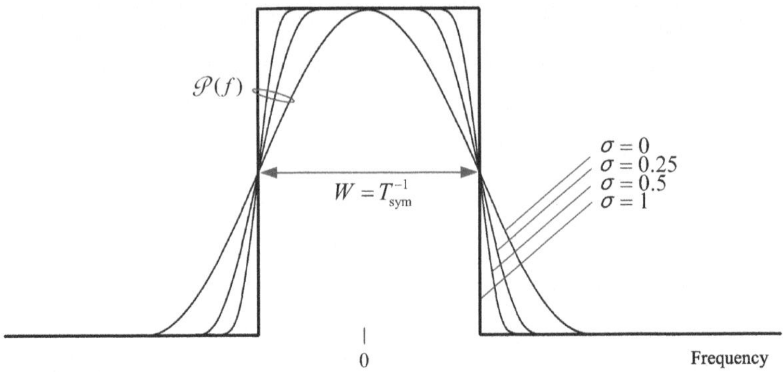

Fig. 5.17 Power spectral density of square-root raised cosine pulses from (5.39)

$$\mathcal{P}(f) = \begin{cases} W^{-1}, & 0 \leq |f| \leq \frac{1-\sigma}{2}W, \\ W^{-1}\cos^2\left(\frac{\pi}{2\sigma}\left(|f|W^{-1} - \frac{1-\sigma}{2}\right)\right), & \frac{1-\sigma}{2}W < |f| \leq \frac{1+\sigma}{2}W, \\ 0, & |f| > \frac{1+\sigma}{2}W, \end{cases} \tag{5.39}$$

where $W \triangleq T_{\text{sym}}^{-1}$. Figure 5.17 shows plots of (5.39) with various values of σ.

A close look at (5.37) shows that when $\sigma = 0$, a square-root raised cosine pulse becomes the "ideal" Nyquist pulse—the sinc pulse. It is ideal because its autocorrelation is another sinc function (hence Nyquist), as shown in (5.38), and it has an ideal "brick-wall" spectrum (Fig. 5.12), with spectral power strictly confined within the bandwidth, $W = T_{\text{sym}}^{-1}$. It is generally true that $T_{\text{sym}} = W^{-1}$ is the shortest symbol interval that a transmitter can possibly squeeze into the radio channel of bandwidth W without introducing inter-symbol interference at the receiving end; hence W sinc pulses per second is the highest symbol rate that the radio channel can support. This maximum rate is commonly known as the *Nyquist symbol rate or pulse rate.*

However, although a sinc pulse has ideal frequency localization, in the time domain, it is time-unbounded, rendering it impractical for implementation since a practical pulse must be time-limited for causality. Nonetheless, time truncation (to limit the pulse duration) gives rise to spectrum regrowth beyond the bandwidth, destroying the ideal brick-wall spectral property. Moreover, another unfavorable property of the sinc pulse observed from Figure 5.13 is that it suffers from excess time dispersion manifested by the side-lobes. Consequently, in practical implementation, any timing error in sampling leads to a series of inter-symbol interference that may diverge, caused by the side lobes of the neighboring symbols, as inferred from (5.31).

As σ becomes greater than zero (while T_{sym} remains the same), the spectral power begins to extend beyond W but is *strictly* confined within $(1 + \sigma)W$, according to

(5.39) and shown in Fig. 5.17. However, although still unbounded in the time domain, the power retracts significantly faster than $\sigma = 0$, leading to (1) less inter-symbol interference in the presence of timing error and (2) less spectral regrowth from truncation.

5.2.1.2 Matched Filter

In a practical receiver, (5.28) is commonly implemented via a matched-filter structure [2] with an impulse response of

$$h(t) = p^*(-t). \tag{5.40}$$

The output of the filter is

$$\begin{aligned} \boldsymbol{g}(t) &= \boldsymbol{s}(t) * h(t) \\ &= \int_\tau \boldsymbol{s}(\tau) \cdot h(t-\tau) \cdot d\tau \\ &= \sum_{k=0}^{N_{\text{sym}}-1} \boldsymbol{c}_k \cdot \gamma_p\big(t - kT_{\text{sym}}\big). \end{aligned} \tag{5.41}$$

It is then *sampled* at $t_k = kT_{\text{sym}}$, $k = 0, 1, \cdots, N_{\text{sym}} - 1$, to give

$$\boldsymbol{g}(t_k) = \boldsymbol{c}_k, \;\; k = 0, 1, \cdots, N_{\text{sym}} - 1, \tag{5.42}$$

based on (5.31).

We rewrite (5.41) as

$$\boldsymbol{g}(t) = \sum_{k=0}^{N_{\text{sym}}-1} \boldsymbol{r}_k(t), \tag{5.43}$$

whcrc

$$\boldsymbol{r}_k(t) \triangleq \boldsymbol{c}_k \cdot \gamma_p\big(t - kT_{\text{sym}}\big), \tag{5.44}$$

Compared with (5.26), $\gamma_p(t)$ can be deemed the pulse that the receiver actually sees at the output of the matched filter (and is what matters); this is why it is called the effective pulse.

Nevertheless, in a realistic communication environment, the radio channel has a significant effect on the waveform at the input of the matched filter and must be taken into account.

5.2.1.3 Effects of a Radio Channel

In the above discussion, the effect of the radio channel on the received signal is not taken into account. Here, the "channel" is collectively considered as the signal path from the output of the baseband signal at the transmitter all the way to the input of the matched-filter at the receiver, as pointed out earlier in this chapter (see Fig. 5.1). What the matched-filter sees at its input is the transmitted pulse, $s_k(t) = \boldsymbol{c}_k \cdot p_k(t)$, convolved with the channel impulse response $\boldsymbol{\alpha}(t) \in \mathbb{C}$, namely,

$$\begin{aligned} s'_k(t) &\triangleq s_k(t) * \boldsymbol{\alpha}(t) \\ &= \boldsymbol{c}_k \cdot \boldsymbol{p}'_k(t), \end{aligned} \tag{5.45}$$

where

$$\boldsymbol{p}'_k(t) \triangleq \int_{-\infty}^{+\infty} p_k(\tau) \cdot \boldsymbol{\alpha}(t-\tau) d\tau. \tag{5.46}$$

When the radio channel is a *frequency-flat* channel with a *single-tap impulse response*, i.e.,

$$\boldsymbol{\alpha}(t) = \boldsymbol{\alpha} \cdot \delta(t), \tag{5.47}$$

typical for line-of-sight satellite channels and narrowband channels, where $\boldsymbol{\alpha} \in \mathbb{C}$ is the complex *channel gain*, and $\delta(t)$ is a *Dirac delta function*. Hence,

$$\boldsymbol{p}'_k(t) = \boldsymbol{\alpha} \cdot p_k(t) \tag{5.48}$$

is still a Nyquist pulse and (5.42) becomes

$$g(kT_{\text{sym}}) = \boldsymbol{\alpha} \cdot \boldsymbol{c}_k, k = 0, 1, \cdots, N_{\text{sym}} - 1. \tag{5.49}$$

Here, a frequency-flat channel is a radio channel whose bandwidth is less than its coherent bandwidth, and from the receiver's perspective, it is a single-path channel that can be represented as a single-tap filter, like the one in (5.47), with a lump-sum gain, $\boldsymbol{\alpha}$.

In reality, a radio channel is a noisy channel, attributable to the thermal noise ever-present in the receiver RF chain, potential co-channel and adjacent channel interference, and the imperfections of the radio transceiver. Such channel noise is commonly modeled collectively as a complex zero-mean *additive white Gaussian noise* (AWGN) process, per the central limit theorem. A frequency-flat channel with a constant gain and AWGN is commonly referred to as the *AWGN channel*. After taking into account the effect of an AWGN channel with a *power spectral density* of N_0, (5.49), i.e., the output sample of the matched filter, becomes

$$g(kT_{\text{sym}}) = \boldsymbol{\alpha} \cdot \boldsymbol{c}_k + \boldsymbol{n}_k, k = 0, 1, \cdots, N_{\text{sym}} - 1. \tag{5.50}$$

For a matched filter (with normalized energy), it can be shown that $\boldsymbol{n}_k$ is a zero-mean complex Gaussian random variable with variance of N_0, independent and identically distributed for different values of k[3]. It is an optimal filter or receiver in the sense that the output sample SNR, i.e.,

$$\begin{aligned} \eta =& E\left(\int_{-\infty}^{+\infty} \boldsymbol{\alpha} \cdot \boldsymbol{s}_k(t) dt\right)^2 / N_0 \\ =& |\boldsymbol{\alpha}|^2 / N_0, \end{aligned} \tag{5.51}$$

is maximal, assuming ideal sample timing.

Obviously, the channel gain, typically unknown, needs to be estimated at the receiver in order to obtain $\boldsymbol{c}_k$. The estimation of $\boldsymbol{\alpha}$ is called *channel estimation* and is typically done via the insertion of pilots or reference signals into the modulation symbol stream. A *pilot* is nothing but a modulation symbol with known $\boldsymbol{c}_k$ and insertion position in the symbol stream. Since the radio channel is time-variant, hence making $\boldsymbol{\alpha}$ a function of time, i.e., $\boldsymbol{\alpha}(t)$, pilots need to be inserted periodically to cope with the wireless channel dynamics.

When the channel is a delay-spread or frequency-selective channel, $\boldsymbol{p}_k'(t)$ in (5.45) is no longer guaranteed a Nyquist pulse, resulting in inter-symbol interference at the matched filter's output. Estimation and "cancellation" of $\boldsymbol{\alpha}(t)$ from the received signal are a nontrivial process, known as *channel equalization*, and should be avoided in MTC through design. A common practice is the use of a multi-carrier waveform, in which a wide band radio channel is divided into multiple narrowband sub-channels or subcarriers such that each sub-channel is narrowband and hence frequency-flat. We will have more discussions on this topic later.

5.2.1.4 Error Vector Magnitude

In a perfect world, a signal sent by a transmitter would line up perfectly with the ideal signal in the corresponding signal space. Under the quadrature modulation, it would mean that all received constellation symbols land precisely at the corresponding constellation points. Unfortunately, in reality, various imperfections in the implementation, attributable to the nonlinearity of power amplifier (i.e., the in-band noise) and phase noise, as well as I/Q imbalance, carrier leakage (DC offset), low image rejection ratio, filter distortion, and carrier frequency offset, you name it, cause the actual modulation symbols to deviate from the original positions. As shown in Fig. 5.18, an error vector measures how far a modulated signal is deviated from its ideal position. The effect of RF frontend-related impairments is commonly measured by a metric known as *error vector magnitude* (EVM) for quantifying the RF quality of a digital radio transmitter (or receiver).

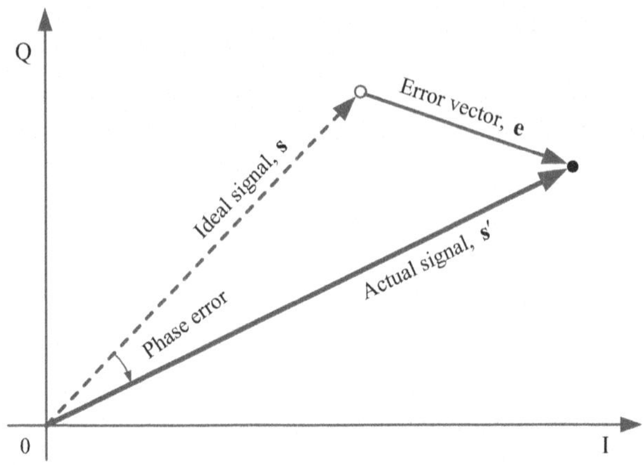

Fig. 5.18 Illustration of error vector magnitude (EVM) as a result of in-band noise in both phase and amplitude. Here, for ease of illustration, a two dimensional signal space is assumed

Specifically, an error vector is a distance vector in the signal space between the ideal signal **s** and the corrupted signal, $\mathbf{s}'$, i.e., $\boldsymbol{e} = \mathbf{s}' - \mathbf{s}$. EVM is thus defined as the root mean square ratio of the error vector to the ideal signal, i.e.,

$$\begin{aligned} \varepsilon_{\mathrm{EVM}} &\triangleq \left(E\left\{|\mathbf{e}|^2/|\mathbf{s}|^2\right\}\right)^{1/2} \\ &= \left(E\left\{|\mathbf{s}' - \mathbf{s}|^2/|\mathbf{s}|^2\right\}\right)^{1/2}. \end{aligned} \tag{5.52}$$

Therefore, the actual SNR of the radio channel is not determined only by the thermal noise and interference but is also affected by the imperfection of transceiver hardware, a part of the radio channel, represented by EVM. In effect, the EVM results in an equivalent noise power that can be factored into the *effective* SNR of the radio channel – the actual sample SNR seen at the output of the matched filter, i.e.,

$$\widetilde{\eta} = \left(\eta^{-1} + \varepsilon^2\right)^{-1}, \tag{5.53}$$

where η is the radio channel SNR with an ideal transceiver.

(5.53) reveals that a radio channel SNR is upper-bounded by the EVM,

$$\lim_{\eta \to \infty} \widetilde{\eta} = \varepsilon^{-2}, \tag{5.54}$$

a parameter closely related to the quality of the RF frontend of a transceiver. A high-end transceiver typically boasts an EVM of less than 4%, which delivers an SNR up to 28 dB, whereas a standard transceiver with an 8% EVM produces an SNR no greater than 22 dB. This fact determines the highest modulation order a transceiver can support, as we will see later.

5.2.1.5 PAPR

The PAPR defined in (5.14) shows how it is affected by the amplitude dynamics of an RF waveform. An unmodulated RF carrier has a 3-dB PAPR inherent to a sinusoidal waveform. Modulating an RF carrier introduces additional dynamics into the carrier and potentially increases the PAPR, depending on the modulation signal waveform.

Due to the random and discrete nature of the information bit sequence, $\{b_i\}$,under quadrature amplitude modulation, the resultant modulation symbol sequence $\{\boldsymbol{c}_k\}$ is inherently characterized by phase discontinuity (BPSK and QPSK) and even amplitude discontinuity (e.g., 16-QAM and 64-QAM) at a symbol boundary. The resultant modulation signal waveform, $\boldsymbol{s}(t)$, will be affected, depending on the modulation scheme and pulse shape, which will ultimately determine the amplitude dynamics of the RF carrier, $\widetilde{s}(t)$, and its PAPR.

It is obviously true that the QAM, like 16-QAM and 64-QAM, which exploits the amplitude to carry information, will change the sinusoidal amplitude of the RF carrier, thereby worsening/increasing the PAPR. It is seemingly true that the QAM, like QPSK and BPSK, which only involves phase changes and hence maintains a constant constellation amplitude, i.e., $|\boldsymbol{c}_k| = 1$, would not entail any additional carrier amplitude variation. However, due to pulse shaping, phase discontinuities between constellation symbols lead to unintentional amplitude variations in the resultant modulation symbols, even for modulation schemes with constant constellation amplitude. This pulse shaping effect can be seen from Figure 5.19, where the square-root raised cosine pulse-shaped BPSK symbols demonstrate large variations in amplitude, which changes the carrier waveform amplitude profile, ultimately degrading the PAPR of the modulated carrier waveform.

Figure 5.20 shows the effect of phase and amplitude modulation on the RF carrier PAPR. It is observed that phase modulation alone entails more than 3-dB additional PAPR (with respect to the unmodulated) and more if amplitude modulation is also involved. A multi-carrier QAM waveform (e.g., orthogonal frequency division

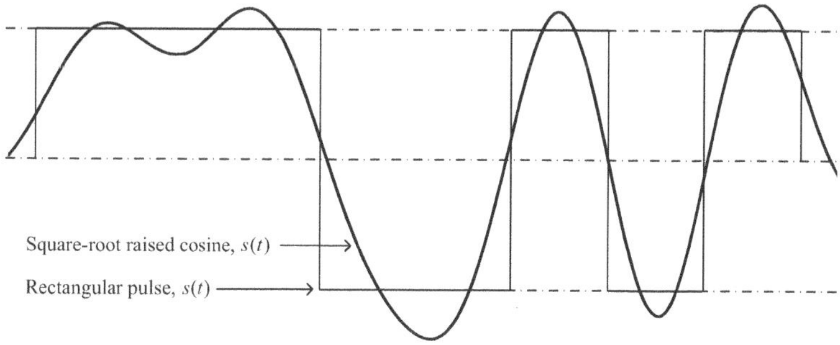

Fig. 5.19 The baseband modulation signal $s(t)$, composed of eight BPSK modulation symbols, $\{c_k\}=\{+1,+1,-1,-1,-1, +1,+1,-1\}$, based on the rectangular pulse and square-root raised cosine pulse with σ=0.35

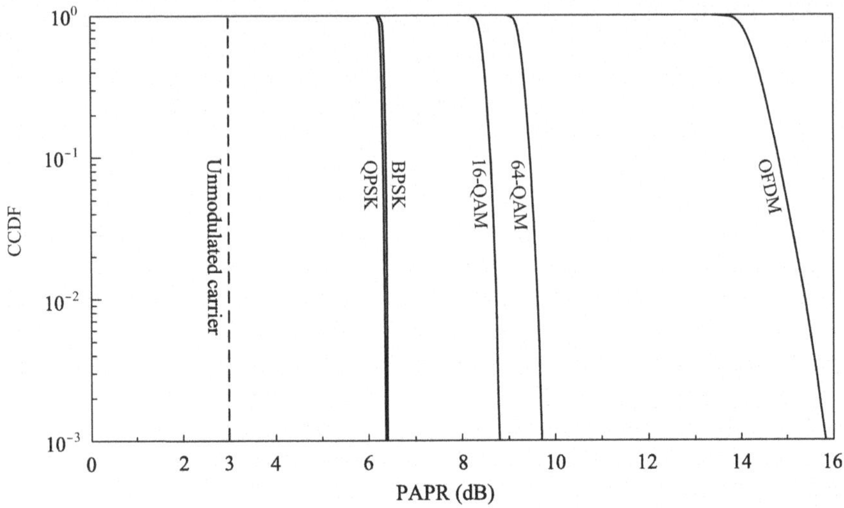

Fig. 5.20 Complementary cumulative distribution function (CCDF) of RF carrier PAPR with square-root raised cosine pulse shaping (σ=0.5) under different modulation schemes

modulation or OFDM, detailed later) effectively adds multiple QAM-modulated narrowband carriers together, not surprisingly producing even higher PAPR. Therefore, PAPR is probably the single most damaging factor that undermines QAM's dominance in wireless communication, leaving the door open a tiny crack for frequency modulation in certain applications like MTC, where energy-efficiency is often one of the primary concerns in addition to spectral efficiency.

5.2.2 *Frequency Modulation*

In frequency modulation, information bits are encoded into the frequency of an RF carrier. In this case, the baseband modulation signal in (5.20) takes the form of

$$s(t) = e^{j2\pi\theta(t)}, \tag{5.55}$$

where

$$\theta(t) = \int_{-\infty}^{t} \sum_{k=0}^{N_{\text{sym}}-1} s_k(\tau)d\tau, \tag{5.56}$$

and $s_k(t) \in \mathbb{R}$ is the modulation symbol.

In general, the modulation signal can no longer be represented as a linear combination of the modulation symbols, hence the name *nonlinear modulation*.

The bandpass signal is then

$$\begin{aligned}\tilde{s}(t) &= \mathcal{R}e\left\{s(t) \cdot e^{j2\pi f_{\text{carrier}} t}\right\} \\ &= \mathcal{R}e\left\{e^{j2\pi\left(f_{\text{carrier}} t + \theta(t)\right)}\right\}.\end{aligned} \tag{5.57}$$

The instantaneous frequency of the RF carrier is thus

$$\begin{aligned}f(t) &= f_{\text{carrier}} + \frac{d\theta(t)}{dt} \\ &= f_{\text{carrier}} + \sum_{k=0}^{N_{\text{sym}}-1} s_k(t) \\ &= f_{\text{carrier}} + \varsigma(t),\end{aligned} \tag{5.58}$$

where ς denotes the frequency offset or shift respective to f_{carrier}, and hence

$$\varsigma(t) \triangleq \sum_{k=0}^{N_{\text{sym}}-1} s_k(t) \tag{5.59}$$

is the frequency shift trajectory about the carrier frequency, which is a superposition of the modulation symbols – a function of the constellation symbols and the pulse shape. Substituting (5.26) into (5.56) and assuming $c_k \in \Omega^1$ result in

$$\begin{aligned}\theta(t) &= \int_{-\infty}^{t} \sum_{k=0}^{N_{\text{sym}}-1} c_k \cdot p(\tau - kT_{\text{sym}}) d\tau \\ &= \sum_{k=0}^{N_{\text{sym}}-1} c_k \cdot \int_{-\infty}^{t} p(\tau - kT_{\text{sym}}) d\tau,\end{aligned} \tag{5.60}$$

It is observed from (5.60) that the passband waveform in (5.57) has a constant envelope (i.e., a constant sinusoidal amplitude) and a guaranteed continuous phase. Both are particularly advantageous in power amplifier efficiency and beneficial for frequency localization.

We first consider a rectangular waveform as the baseband modulation pulse,

$$p(t) = AT_{\text{sym}}^{-1} \cdot \text{rect}\left(tT_{\text{sym}}^{-1} - \frac{1}{2}\right), \tag{5.61}$$

where rect(·) is the rectangular function defined in (5.33). Then substituting (5.61) into (5.60) gives

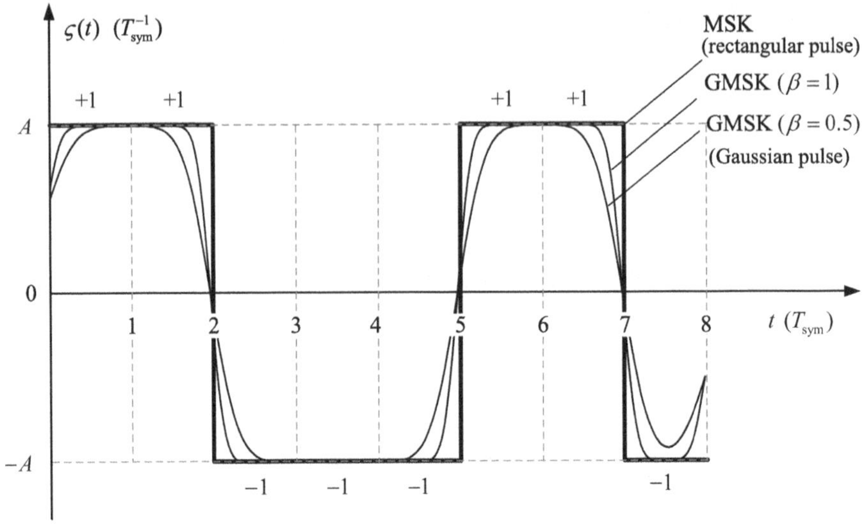

Fig. 5.21 The frequency trajectory composed of eight BPSK modulation symbols ($N_{\text{sym}} = 8$), based on the rectangular pulse with $A = 1/4$ (i.e., MSK) and Gaussian-filtered pulse with $\beta = 0.5$ and 1 (i.e., GMSK), where $\{c_k\}=\{+1,+1,-1,-1,-1,+1,+1,-1\}$

$$\begin{aligned}\theta(t) &= A\left(\sum_{n=0}^{k-1} c_n + \left(tT_{\text{sym}}^{-1} - k\right)\cdot c_k\right) \\ &= \varsigma_k \cdot t + \varphi_k, \quad kT_{\text{sym}} < t \le (k+1)T_{\text{sym}},\end{aligned} \tag{5.62}$$

where

$$\varsigma_k \triangleq c_k \cdot AT_{\text{sym}}^{-1}, \quad 0 \le k \le N_{\text{sym}} - 1, \tag{5.63}$$

is the frequency offset or shift away from the carrier frequency, f_{carrier}, during the kth symbol time of duration T_{sym}, and

$$\begin{aligned}\varphi_k &\triangleq A\left(\sum_{n=0}^{k} c_n - (k+1)c_k\right) \\ &= \varphi_{k-1} + Ak(c_{k-1} - c_k), \quad \varphi_0 \triangleq 0\ ,\end{aligned} \tag{5.64}$$

is the phase state at the beginning of the kth symbol time, a function of not only the current c_k, but the history, i.e., $\{c_{k-1}, \cdots, c_0\}$. Figure 5.21 plots an example of frequency shift, ς_k, from (5.63), with $c_k \in \Omega^1$ or $\{-1, +1\}$, i.e., binary *frequency shift keying* (FSK). Figure 5.22 is the corresponding $\theta(t)$ from (5.62).

Further, substituting (5.62) into (5.55) gives rise to

$$s(t) = e^{j2\pi\varsigma_k t} \cdot e^{j2\pi\varphi_k}, \quad kT_{\text{sym}} < t \le (k+1)T_{\text{sym}}, \tag{5.65}$$

i.e., the kth *effective* modulation symbol denoted as $\boldsymbol{s}(t)|_{c_k}$. The cross-correlation between $\boldsymbol{s}(t)|_{c_k=+1}$and $\boldsymbol{s}(t)|_{c_k=-1}$ are

$$\begin{aligned}\langle \boldsymbol{s}(t)|_{c_k=+1}, \boldsymbol{s}(t)|_{c_k=-1}\rangle &= \int_{kT_{\text{sym}}}^{(k+1)T_{\text{sym}}} \langle \boldsymbol{s}(\tau)|_{c_k=+1}, \boldsymbol{s}(\tau)|_{c_k=-1}\rangle d\tau \\ &= T_{\text{sym}} \text{sinc}(2\pi A) \cdot e^{j2\pi A}.\end{aligned} \tag{5.66}$$

To detect c_k with matched filters, we wish that

$$T_{\text{sym}} \text{sinc}(2\pi A) \cdot e^{j2\pi A} \cdot \boldsymbol{\alpha} = 0, \tag{5.67}$$

recalling that $\boldsymbol{\alpha}$ is the complex channel gain introduced by the radio channel during transmission.

With the knowledge of the radio channel gain, multiply (5.67) by $\boldsymbol{\alpha}^*$ and let

$$\begin{aligned}\mathcal{Re}\{T_{\text{sym}} \text{sinc}(2\pi \text{A}) \cdot e^{j2\pi A} \cdot \boldsymbol{\alpha} \cdot \boldsymbol{\alpha}^*\} &= T_{\text{sym}} |\boldsymbol{\alpha}|^2 \cdot \mathcal{Re}\{\text{sinc}(2\pi \text{A}) \cdot e^{j2\pi A}\} \\ &= T_{\text{sym}} |\boldsymbol{\alpha}|^2 \cdot \text{sinc}(4\pi A) \\ &= 0,\end{aligned} \tag{5.68}$$

to allow *coherent* orthogonal detection. The A that satisfies (5.66) is

$$A \in \left\{\frac{n}{4}, n = 1, 2, \cdots\right\}. \tag{5.69}$$

Obviously, from (5.63), $A = \frac{1}{4}$ corresponds to the minimum amount of frequency shift of

$$|\varsigma_k| = \frac{1}{4} T_{\text{sym}}^{-1}, \tag{5.70}$$

thereby occupying the minimum bandwidth, ergo the name *minimum shift keying* (MSK). (5.65) becomes

$$\boldsymbol{s}(t)|_{c_k} = e^{j\pi c_k T_{\text{sym}}^{-1} t/2} \cdot e^{j2\pi \varphi_k}, \tag{5.71}$$

where $\varphi_k = \varphi_{k-1} + \frac{1}{4}k(c_{k-1} - c_k) \in \{0, \pm\frac{1}{2}, \pm 1, \cdots\}$, i.e., $e^{j2\pi\varphi_k} = \pm 1$, which depends on previously transmitted symbols. In effect, it introduces a *differential phase modulation* into the modulation symbol, i.e., the information is carried in the change of the phase rather than the phase itself, giving rise to increased detection complexity or decreased detection performance. Pre-differential-coding the MSK signal at the transmitter is a common practice to remove the inherent differential encoding.

Alternatively, merely letting A be integer multiples of $\frac{1}{2}$, i.e.,

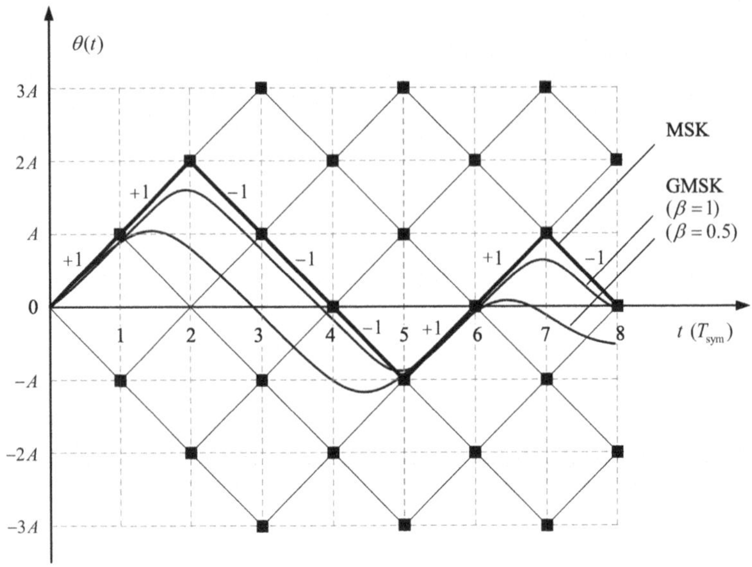

Fig. 5.22 The phase trajectory, $\theta(t)$ corresponding to the frequency trajectory, $\varsigma(t)$ in Fig. 5.21

$$A \in \left\{\frac{n}{2}, n = 1, 2, \cdots\right\}, \tag{5.72}$$

removes the phase dependency.

By selecting $A = \frac{1}{2}$,

$$\varphi_k = \varphi_{k-1} + \frac{1}{2}k(c_{k-1} - c_k) \in \{0, \pm 1, \pm 2, \cdots\}, \tag{5.73}$$

and hence, $e^{j2\pi\varphi_k} \equiv 1$. (5.65) then becomes

$$s(t)|_{c_k} = e^{j\pi c_k T_{\mathrm{sym}}^{-1} t}. \tag{5.74}$$

Furthermore, with $A = \frac{1}{2}$, (5.67) holds even without knowing $\boldsymbol{\alpha}$, allowing *non-coherent* orthogonal detection, nevertheless, at the expense of increased signaling bandwidth, i.e., as compared to (5.70), the frequency shift is doubled,

$$|\varsigma_k| = \frac{1}{2} T_{\mathrm{sym}}^{-1}. \tag{5.75}$$

The FSK devised this way is an orthogonal FSK, allowing either coherent or noncoherent detection – not true for MSK.

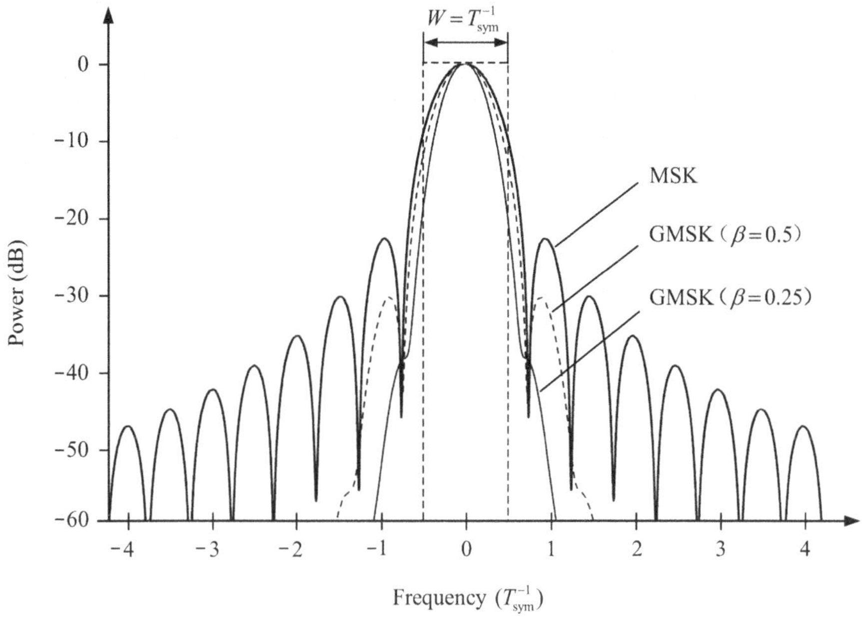

Fig. 5.23 Power spectral densities of MSK and GMSK (0.5, 0.25)

This binary FSK scheme can be readily extended to 2^m-FSK, where $c_k \in \{\pm1, \pm3, \pm5, \cdots, \pm(2^m - 1)\}$.

The continuous phase property gives FSK the same PAPR as the unmodulated sinusoidal waveform (referring back to Fig. 5.20), hence enjoying an edge over QAM in power amplifier efficiency. However, this advantage of FSK is overshadowed by its spectral inefficiency. As shown in Fig. 5.23, MSK occupies similar bandwidth as QPSK and QAM (see Fig. 5.12) while running at a lower bit rate (half the bit rate of QPSK and a quarter of 16-QAM). As the modulation order increases, the disadvantage becomes even more prominent, mainly due to the fact that information bits are represented by frequency shifts (one dimension) rather than both phase and amplitude (two dimensions) with a fixed frequency. As a result, a higher modulation order means wider bandwidth, as shown in Fig. 5.24, whereas, for QAM, the bandwidth remains the same regardless. Therefore, the spectral efficiency increases as the QAM order increases, well, at the cost of lower power amplifier efficiency as well as the need for channel knowledge. Here, spectral efficiency refers to the ratio of the information bit rate to the bandwidth of the modulation signal. Hence, the more information bits a modulation symbol carries on a given time-frequency resource, the higher the spectral efficiency.

However, just like the inter-symbol interference in linear modulation, nothing is complete without the receiver in the picture. When talking about the power efficiency of a modulation scheme, we also care about the minimum receive energy per information bit to successfully detect the bit at a specified bit error rate at a receiver

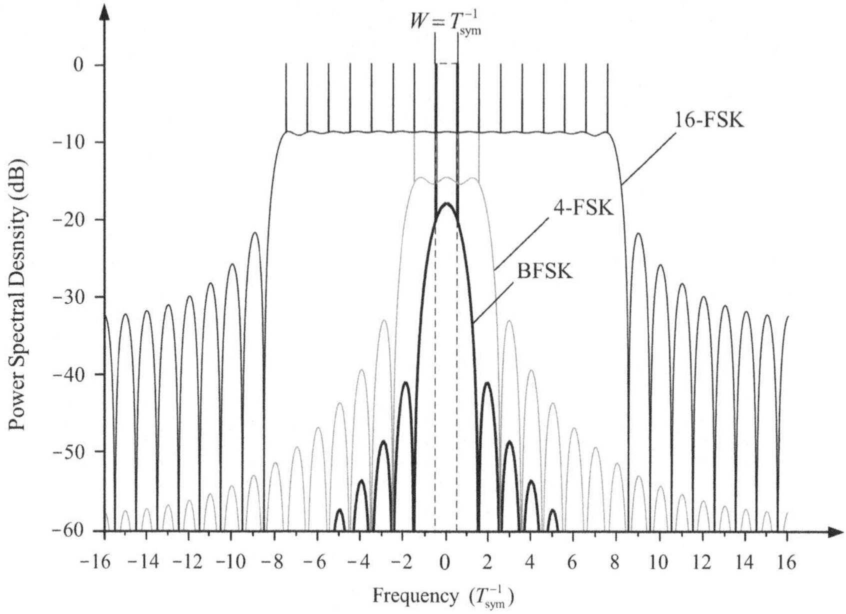

Fig. 5.24 The power spectral densities of BFSK, 4-FSK and 16-FSK

since different modulation schemes have different tolerances to radio channel impairments, i.e., noise. This issue can be better understood with the help of Fig. 5.25. It is observed that for the same bit error rate, the required energy per bit (E_b) under FSK is generally lower than QAM, i.e., more energy-efficient, especially for high-order modulations. Nevertheless, the price to pay is spectral efficiency, a typical tradeoff between energy and spectrum efficiencies in communications. Therefore, orthogonal frequency modulations like FSK are popular in applications where energy and cost are prioritized over spectral efficiency.

In practice, the rectangular pulse in (5.61) is shaped with a Gaussian filter to produce a smoother pulse,

$$p(t) = AT_{\mathrm{sym}}^{-1}\left[Q\left(\frac{\pi\beta}{\sqrt{\ln 2}}\left(tT_{\mathrm{sym}}^{-1}-\frac{1}{2}\right)\right)\text{-}Q\left(\frac{\pi\beta}{\sqrt{\ln 2}}\left(tT_{\mathrm{sym}}^{-1}+\frac{1}{2}\right)\right)\right]. \tag{5.76}$$

where $A=\frac{1}{4}$, $\beta \triangleq W_{3\mathrm{dB}}T_{\mathrm{sym}}$, $Q(\cdot)$ is the Q-function [3], and $W_{3\mathrm{dB}}=\beta/T_{\mathrm{sym}}$ the 3-dB bandwidth of the Gaussian filter. It produces a smoother phase transition at the symbol boundary, hence smoother frequency transition, as shown in Fig. 5.22, thereby better frequency localization (reduced bandwidth and sideband power), referring back to Fig. 5.23. However, the use of a Gaussian-filtered MSK pulse (commonly known as the GMSK pulse) causes inter-symbol interference – the price one pays for frequency localization. The smaller the value of β is, the better the spectrum becomes, but the severer the inter-symbol interference.

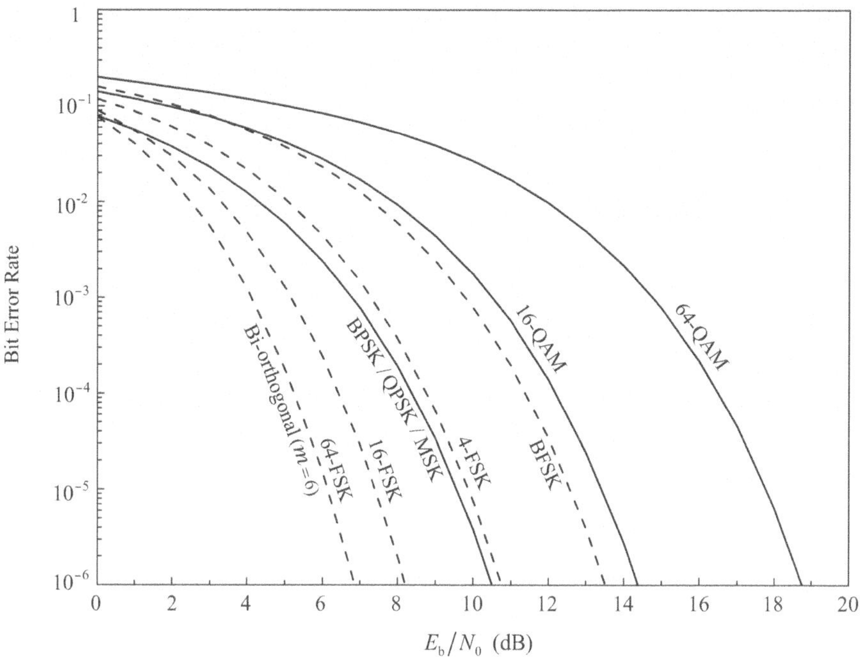

Fig. 5.25 Coherent demodulation bit error rate for various modulation schemes, where E_b/N_0 is the receive information bit SNR. Here, MSK is pre-differentially encoded

5.2.3 Orthogonal Code Modulation

In orthogonal FSK modulation, M information bits, $b_0, b_1, \cdots, b_{m-1}$, are represented by a set of 2^m orthogonal sinusoidal waveforms at 2^m discrete radio frequencies. *Orthogonal code modulation* is another form of orthogonal modulation, in which a set of 2^m orthonormal codes or vectors in $\mathbb{C}^{2^m}$ are used to modulate an RF sinusoidal waveform. Orthogonality means that they do not interfere with each other or are mathematically independent. This property not only maximizes the detection performance but simplifies the receiver design as well and hence is one of the most desirable properties for signals. We have already seen applications of the orthogonality property in the modulation signal design so far.

A well-known orthogonal code set is constructed via the Walsh-Hadamard matrix $\mathbf{H}_m$, which is a $2^m \times 2^m$ binary matrix that can be created recursively according to the following relation:

$$\mathbf{H}_n = \begin{bmatrix} \mathbf{H}_{n-1} & \mathbf{H}_{n-1} \\ \mathbf{H}_{n-1} & \overline{\mathbf{H}}_{n-1} \end{bmatrix}, \tag{5.77}$$

where $n = 2, 3, \cdots, m$, and $\mathbf{H}_1 = 1$. From (5.77), any pair of rows, $\mathbf{h}_i$ and $\mathbf{h}_j$, have an equal number of `1`s and `0`s. Hence, they are orthogonal after being mapped to BPSK constellation, i.e.,

$$\langle \mathbf{h}_i, \mathbf{h}_j \rangle = \delta(i-j), \forall i,j \in \{1,2,\cdots,2^m\}, \tag{5.78}$$

where $\delta(\cdot)$ is the Kronecker delta function. Here we retain the same notation for the BPSK-mapped Walsh-Hadamard row vectors. $\mathbf{H}_m$ thus provides 2^m orthogonal codes, each corresponding to a row of $\mathbf{H}_m$, i.e.,

$$\mathbf{h}_i = [h_{i,1}\ h_{i,2} \cdots h_{i,2^m}],\ \ h_{i,j} \in \Omega^1, i \in \{1,2,\cdots,2^m\}, \tag{5.79}$$

where Ω^1 denotes the BPSK constellation scaled by a normalization factor of $2^{-\frac{m}{2}}$. The corresponding baseband modulation waveform is thus

$$s_i(t) = \sum_{j=1}^{2^m} h_{i,j} \cdot p_j(t),\ \ h_{i,j} \in \Omega^1, i \in \{1,2,\cdots,2^m\}, \tag{5.80}$$

where $p_m(t)$ is the pulse waveform defined in (5.26).

At the matched filter output of a receiver with ideal sample timing in the absence of noise, we have the following samples, represented in a vector form as

$$\mathbf{h}_i' = \boldsymbol{\alpha} \cdot [h_{i,1}\ h_{i,2} \cdots h_{i,2^m}],\ \ i \in \{1,2,\cdots,2^m\}. \tag{5.81}$$

This idealized situation leads to

$$\langle \mathbf{h}_i', \mathbf{h}_j \rangle = \boldsymbol{\alpha} \cdot \delta(i-j), \forall i,j \in \{1,2,\cdots,2^m\}, \tag{5.82}$$

based on (5.78). Hence, either coherent or noncoherent demodulation is possible. Since both orthogonal vector modulation and orthogonal FSK are orthogonal modulations, they, in theory, have the same demodulation performance.

Like the orthogonal FSK, the required resource for orthogonal modulation increases exponentially with m. A close look at (5.82) infers that, with coherent demodulation, i.e., with channel knowledge of $\boldsymbol{\alpha}$ at the receiver, the dimension of $s_i(t)$ in (5.80) can be reduced to $m - n$, but convey the same m-bit information, only that $s_i(t)$ is multiplied by a constellation symbol, $\boldsymbol{c} \in \boldsymbol{\Omega}^n$, upfront, that contains n-bit information, i.e.,

$$s_i(t) = \boldsymbol{c} \cdot \sum_{j=1}^{2^{m-n}} h_{i,j} \cdot p_j(t),\ \ i \in \{1,2,\cdots,2^{m-n}\}. \tag{5.83}$$

Clearly, (5.82) still holds, with a multiplicative factor $\boldsymbol{c}_i$ in front, i.e.,

$$\langle \boldsymbol{h}'_i, \boldsymbol{h}_j \rangle = \boldsymbol{\alpha} \cdot \boldsymbol{c} \cdot \delta(i-j), \forall i,j \in \{1,2,\cdots,2^{m-n}\}. \tag{5.84}$$

Given that the complex channel gain $\boldsymbol{\alpha}$ is known, $\boldsymbol{c}$ can be recovered as well, thereby reducing the required resource (channel symbol time) by a factor of 2^n.

This scheme becomes natural if we think of $s_i(t)$ in (5.80) as a pulse,

$$p'_i(t) = \sum_{j=1}^{2^{m-n}} h_{i,j} \cdot p_j(t), \quad i \in \{1,2,\cdots,2^{m-n}\}, \tag{5.85}$$

which is orthogonal to each other, and $\boldsymbol{s}_i(t)$ in (5.83) is just a regular modulation symbol as in (5.26),

$$\boldsymbol{s}_i(t) = \boldsymbol{c} \cdot p'_i(t), \quad i \in \{1,2,\cdots,2^{m-n}\}, \tag{5.86}$$

except that $p'_i(t)$ now contains $m-n$ bits of information unknown to the receiver. Therefore, $m-n$ matched filters are needed at the receiver, which can be implemented efficiently with a fast Hadamard transform algorithm [4].

In particular, when $n = 1$, i.e., $c \in \Omega^1$ is a BPSK constellation symbol, this modulation scheme is commonly known as the *bi-orthogonal* modulation. Figure 5.25 plots the bit error performance for bi-orthogonal modulation with $m = 6$, i.e., $2^{m-1} = 32$ orthogonal codes. It has a similar performance as the orthogonal 64-FSK.

5.2.4 Single-Carrier and Multi-Carrier Modulations

So far, we have assumed single-carrier modulation, i.e., only one carrier frequency is used for modulation at a time. However, it does not stop a transceiver from exploiting more than one carrier frequency simultaneously. Therefore, multi-carrier modulation divides a given frequency band of width W into N_{subcar} subcarriers, and the modulation symbol stream is divided into N_{subcar} lower-rate modulation symbol streams transmitted in parallel on the N_{subcar} subcarriers. The benefit is obvious, i.e., a narrower band carrier is more likely frequency-flat, hence more multipath-resistant. A natural question is then, how large can N_{subcar} be for a given W without inter-carrier interference?

From the duality property of the Fourier transform, the answer is readily available from the previous Nyquist pulse criteria by interchanging the relationship between time and frequency. Now think of (5.26) as a modulation symbol in the frequency domain,

$$\begin{aligned} \boldsymbol{s}_\kappa(f) &= \boldsymbol{c}_\kappa \cdot p_\kappa(f) \\ &= \boldsymbol{c}_\kappa \cdot p(f - \kappa \cdot W_{\text{subcar}}), \end{aligned} \tag{5.87}$$

where $W_{\text{subcar}} \triangleq W/N_{\text{subcar}}$ is the subcarrier spacing, $\kappa \in \{0, 1, \cdots, N_{\text{subcar}} - 1\}$ the sub-carrier index, and $p_\kappa(f)$ a frequency-domain Nyquist pulse centered at the

κth sub-carrier. The Nyquist symbol rate principle asserts that the minimum subcarrier spacing is

$$W_{\text{subcar}} = T_{\text{sym}}^{-1}, \tag{5.88}$$

and the corresponding pulse is the sinc function. That is to say, the subcarrier spacing, or the number of subcarriers in a given bandwidth, depends on the symbol duration and vice versa.

To see that more clearly, we represent the multi-carrier modulation symbol in the frequency domain and then transform it to the time-domain,

$$\begin{aligned} s(t) &= \mathcal{F}^{-1}\left\{\sum_{\kappa=0}^{N_{\text{subcar}}-1} s_\kappa(f)\right\} \\ &= \mathcal{F}^{-1}\left\{\sum_{\kappa=0}^{N_{\text{subcar}}-1} c_\kappa \cdot p_\kappa(f)\right\} \\ &= \sum_{\kappa=0}^{N_{\text{subcar}}-1} c_\kappa \cdot \mathcal{F}^{-1}\left\{p\left(f - T_{\text{sym}}^{-1} \cdot \kappa\right)\right\} \\ &= \mathcal{F}^{-1}\{p(f)\} \cdot \sum_{\kappa=0}^{N_{\text{subcar}}-1} c_\kappa \cdot e^{j2\pi T_{\text{sym}}^{-1} \cdot \kappa \cdot t}. \end{aligned} \tag{5.89}$$

Since $p(f)$ is a sinc function, defined in (5.36), i.e.,

$$p(f) = \text{sinc}(\pi f T_{\text{sym}}), \tag{5.90}$$

(5.89) becomes

$$s(t) = \text{rect}\left(tT_{\text{sym}}^{-1}\right) \cdot \sum_{\kappa=0}^{N_{\text{subcar}}-1} c_\kappa \cdot e^{j2\pi T_{\text{sym}}^{-1} \cdot \kappa \cdot t}, \tag{5.91}$$

where rect(·) is the rectangular function or "window" defined in (5.32).

From (5.30), $\{p_\kappa(f), \kappa = 0, 1, \cdots, N_{\text{sym}} - 1\}$ in (5.87) are orthogonal, i.e.,

$$\left\langle p\left(f - iT_{\text{sym}}^{-1}\right), p\left(f - \kappa T_{\text{sym}}^{-1}\right)\right\rangle = \delta(i - \kappa), \quad i, \kappa = 0, 1, \cdots, N_{\text{sym}} - 1, \tag{5.92}$$

This particular multi-carrier modulation scheme is the well-known *orthogonal frequency division multiplexing* (OFDM) and (5.91) is referred to as the *OFDM symbol*. As illustrated in Fig. 5.26, under this framework, the kth OFDM symbol in the time domain can then be expressed as

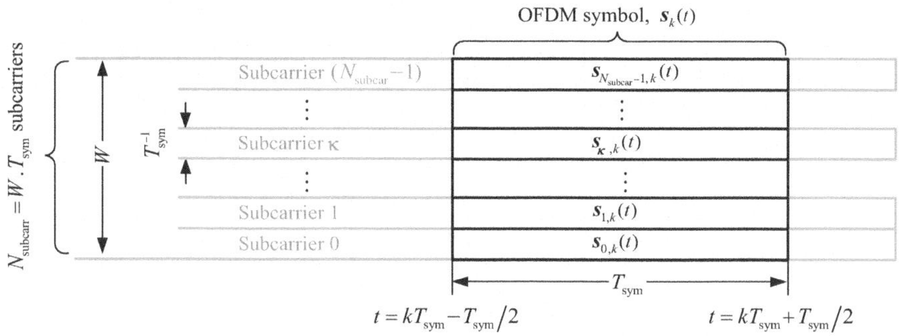

Fig. 5.26 Illustration of an OFDM symbol, (5.93), in the time and frequency domains

$$s_k(t) = \sum_{\kappa=0}^{N_{subcar}-1} s_{\kappa,k}(t) \tag{5.93}$$

where

$$s_{\kappa,k}(t) = c_{\kappa,k} \cdot p_\kappa\left(t - kT_{sym}\right), \tag{5.94}$$

is the kth modulation symbol on the κth subcarrier, and

$$p_\kappa\left(t - kT_{sym}\right) \triangleq e^{j2\pi\kappa T_{sym}^{-1}\cdot\left(t-kT_{sym}\right)} \cdot \mathrm{rect}\left(tT_{sym}^{-1} - k\right) \tag{5.95}$$

is a time-bounded sinusoidal pulse at a frequency of κT_{sym}^{-1} within the kth symbol interval of $[kT_{sym}-T_{sym}/2, kT_{sym}+T_{sym}/2)$, instead of a time-unbounded square-root raised cosine pulse as we see in the previous single-carrier modulation. Here, $p_\kappa(t)$ in (5.94) and $p_\kappa(f)$ in (5.87), a convenient abuse of notation, represent time and frequency domain pulses, respectively.

From one point of view, the kth OFDM modulation symbol as a whole is a superposition of N_{sucar} single-carrier modulation symbols, spaced $W/N_{subcar} = T_{sym}^{-1}$ apart in frequency. From another point of view, under OFDM, the radio resource is effectively divided into an orthogonal time-and-frequency grid. Each element in the grid is a resource with $T_{sym}^{-1} \cdot T_{sym} = 1$ degree of freedom, indexed by subcarrier and OFDM symbol, i.e., (κ, k), as depicted in Fig. 5.27, a useful and elegant feature for centralized resource management.

In recovering N_{subcar} constellation symbols from the kth OFDM symbol, a filter bank consisting of N_{subcar} matched filter,

$$h_\kappa(t) = \mathrm{rect}(tT_{sym}^{-1}) \cdot e^{j2\pi T_{sym}^{-1}\cdot\kappa\cdot t}, \quad \kappa \in \{0, 1, \cdots, N_{subcar} - 1\}, \tag{5.96}$$

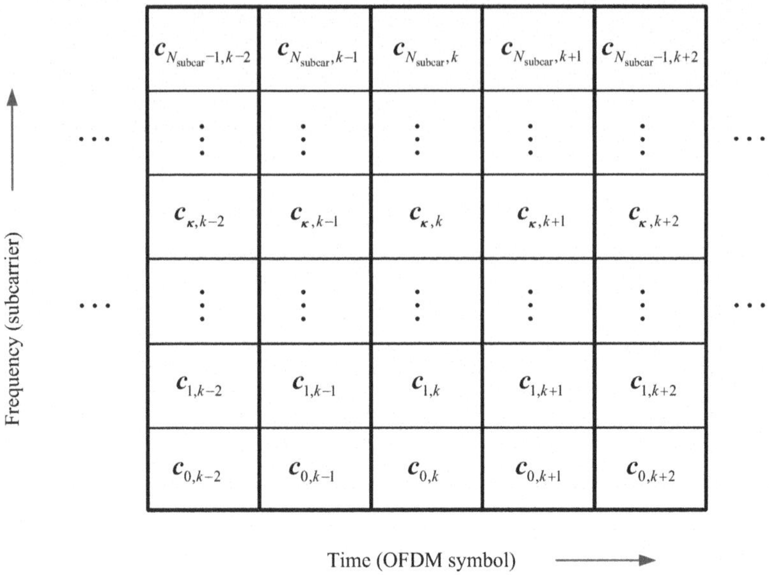

Fig. 5.27 Graphical illustration of the OFDM time and frequency resource grid, where each unit represents a resource element with $T_{\text{sym}}^{-1} \cdot T_{\text{sym}} = 1$ degree of freedom

is applied to the incoming OFDM symbol stream, and the output is sampled at $t = kT_{\text{sym}}$ to obtain

$$
\begin{aligned}
g_{\kappa}\left(kT_{\text{sym}}\right) &= \int_{-\infty}^{+\infty} \sum_{m} s_m(\tau) \cdot h_{\kappa}\left(kT_{\text{sym}} - \tau\right) d\tau \\
&= \int_{kT_{\text{sym}}-T_{\text{sym}}/2}^{kT_{\text{sym}}+T_{\text{sym}}/2} s_k(\tau) \cdot e^{-j2\pi T_{\text{sym}}^{-1} \cdot \kappa \cdot (\tau - kT_{\text{sym}})} d\tau, \quad \kappa \in \{0, 1, \cdots, N_{\text{subcar}} - 1\},
\end{aligned}
\tag{5.97}
$$

where the second equality is due to the rectangular function that isolates the OFDM symbols from each other.

Now substituting (5.93) into (5.97) and exchanging the order of summation and integration gives

$$
\begin{aligned}
g_{\kappa}\left(kT_{\text{sym}}\right) &= \sum_{i=0}^{N_{\text{subcar}}-1} c_{i,k} \cdot \int_{kT_{\text{sym}}-T_{\text{sym}}/2}^{kT_{\text{sym}}+T_{\text{sym}}/2} e^{j2\pi T_{\text{sym}}^{-1} \cdot (i-\kappa) \cdot (\tau - kT_{\text{sym}})} d\tau \\
&= \sum_{i=0}^{N_{\text{subcar}}-1} c_{\kappa,k} \cdot \delta(i - \kappa) \\
&= c_{\kappa,k}, \quad \kappa \in \{0, 1, \cdots, N_{\text{subcar}} - 1\}
\end{aligned}
\tag{5.98}
$$

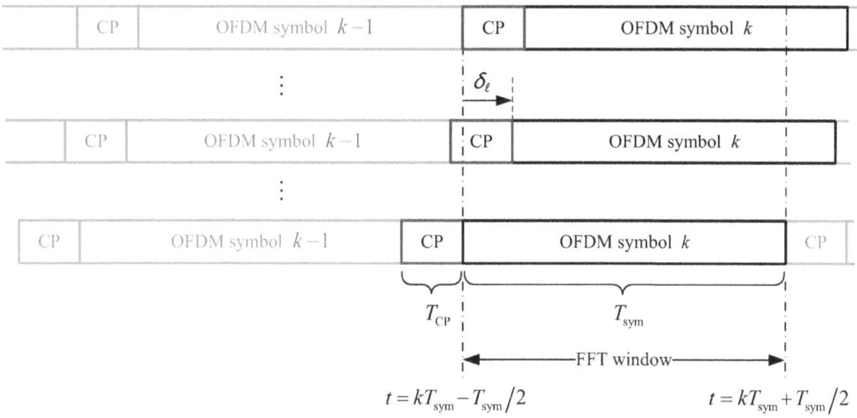

Fig. 5.28 OFDM symbols with cyclic prefix (CP) and the FFT window at a receiver, which includes multiple time-delayed replicas of an OFDM symbol from the transmitter through a multipath radio channel

where $\delta(\cdot)$ is the Kronecker delta function. Here perfect subcarrier frequency and symbol boundary synchronization are assumed.

A closer look at (5.97) indicates that the matched-filter bank operation corresponds to a Fourier transform evaluated at a set of N_{subcar} evenly spaced discrete frequencies, T_{sym}^{-1} apart, which can be computed efficiently with the N_{subcar}-point fast Fourier transform (FFT) algorithm.

The primary advantage of the multi-carrier modulation is that it divides the entire frequency band W into many subcarriers, each with a narrower bandwidth of $W_{\text{subcar}} = W/N_{\text{subcar}}$, centered at $T_{\text{sym}}^{-1} \cdot \kappa$, $\kappa \in \{0, 1, \cdots, N_{\text{subcar}} - 1\}$, hence more likely being frequency-flat, i.e., more multipath-resistant. However, since an OFDM symbol as a whole is still a wideband modulation symbol, a "guard period" as a time buffer to absorb the multipath-induced delay spread and prevent the temporal inter-symbol interference is needed.

First, although the rectangular-windowed time-domain pulse in (5.95) implies no inter-symbol interference between successive OFDM symbols, it does not protect an OFDM symbol from delay-spread-induced inter symbol interference, which destroys the orthogonality between OFDM symbols. For that, a time buffer or gap as large as the multipath delay spread is sufficient, as illustrated in Fig. 5.28.

Second, a close look at Fig. 5.28 shows that a chunk of the OFDM symbol from the ℓth path due to delay will be missing from the FFT window. A simple remedy is to *cyclically* extend the OFDM symbol waveform into the gap by preceding the OFDM symbol with a copy of the end part of that same symbol, hence the name, *cyclic prefix* (CP). Now the delayed path in the FFT window is a cyclically shifted version of the original OFDM symbol (and yet no inter-symbol interference). From (5.93), the following holds within the FFT window:

$$s_k(t - \delta_\ell) = \sum_{\kappa=0}^{N_{\text{subcar}}-1} c_{\kappa,k} \cdot e^{j2\pi T_{\text{sym}}^{-1} \cdot \kappa \cdot (t - kT_{\text{sym}} - \delta_\ell)}, \quad kT_{\text{sym}} - T_{\text{sym}}/2 \leq t < kT_{\text{sym}} + T_{\text{sym}}/2, \tag{5.99}$$

where $0 \leq \delta_\ell < T_{\text{CP}}$ is the ℓth path delay. Overall, what the receiver observes in the FFT window is

$$\begin{aligned} s_k^{\text{FFT}}(t) &= \sum_\ell \alpha^\ell \cdot s_k(t - \delta_\ell) \\ &= \sum_{\kappa=0}^{N_{\text{subcar}}-1} \alpha_\kappa \cdot c_{\kappa,k} \cdot e^{j2\pi T_{\text{sym}}^{-1} \cdot \kappa \cdot (t - kT_{\text{sym}})}, \; kT_{\text{sym}} - T_{\text{sym}}/2 \leq t < kT_{\text{sym}} + T_{\text{sym}}/2, \end{aligned} \tag{5.100}$$

where

$$\alpha_\kappa = \sum_\ell \alpha^\ell \cdot e^{-j2\pi T_{\text{sym}}^{-1} \cdot \kappa \cdot \delta_\ell}. \tag{5.101}$$

Therefore, multi-carrier modulation with CP insertion makes it possible to transform a frequency-selective wideband channel into multiple frequency-flat narrowband subchannels so that a single-tap channel estimation on each subchannel, as described earlier in this chapter, can be performed for channel equalization. The cost is the need for CP for *every* OFDM symbol.

5.3 Channel Coding

We now turn our attention to another important component of baseband processing: *channel coding* and automatic retransmission request or ARQ. Unlike the traditional ARQ, the ARQ protocol here is combined with channel coding, exploiting the full benefit of channel coding. This *hybrid* ARQ scheme provides a mechanism not just for error control but also for spectral efficiency and energy efficiency. Before delving into this topic, we justify the need for channel coding.

5.3.1 *The Need for Channel Coding*

Data integrity is the most critical aspect of the design and implementation of a communication system. Any unintended change to data as the result of transfer and processing is a failure of data integrity. The unintended change includes 1) missing data, e.g., dropped IP packets, and 2) altered data, i.e., bit errors (e.g., flipped bits by noise). Thus, a communication system must ensure that information is transferred intact from its transmitting peer to its receiving peer. A common method is to request a retransmission once a missing piece of data or a data error is detected. This technique is generally known as the automatic retrainsmission request or ARQ, but its implementation can differ. We have briefly discussed ARQ in Chap. 3; we revisit this subject in relation to channel coding.

Since bit errors are rare in *wired* communication systems, ARQ is normally not provided in lower layers, meaning that SDUs in error are detected but only canceled, not retransmitted by the lower layers. Retransmissions rely on higher layer protocols, i.e., the transport protocols of the network layer or the application (if there is such a need) in the application layer. This absence of ARQ in the lower layers is because the loss of data integrity in wired communication mainly results from the loss of data that happens in the network layer due to IP packet loss when a Layer-3 router's buffer overflows during network congestion or the packet's `time-to-live` expires during the delivery process. `Time-to-live` is a field in an IP packet header that indicates a datagram's time to exist inside a network before being discarded by a router – an Internet protocol to prevent a packet from lingering around longer than the sender intends.

Radio channels are arguably the most erroneous and unreliable communication media due to the factors like ambient interference, weather, shadowing, and fading, which impair the RF waveform, causing a high probability of data corruption or bit errors. Constant retransmissions invoked by the higher layer protocols entail excessive delays. As noted in Chap. 4, some transport protocols, such as the TCP of the Internet Protocol, treat frequent packet losses as an indication of network congestion and, in response, drastically slow down the transmission unnecessarily. Consequently, ARQ becomes necessary in the lower layers, i.e., the data link layer and physical layer, in a radio communication system.

The key to ARQ is the need for *error detection coding* (for determining if the SDU is received in error) and a signaling channel (e.g., DFCH) for feeding back acknowledgments (ACK) and negative acknowledgments (NACK) or timeouts (specified periods allowed to elapse before an ACK is received by the transmitter). Nevertheless, frequent requests for retransmission lead to inefficient use of radio resources and transmission latency, resulting from the frequent ACK/NACK signaling and retransmissions.

Thus, *forward error correction coding* is often employed in erroneous communication media for correcting errors (with best efforts) before resorting to retransmission, thereby reducing the error probability and the frequency of retransmissions. Even when correction fails, error correction coding makes retransmission more efficient as detailed in Sect. 5.5.2. Also, in some scenarios where retransmission is impossible, error correction coding is essential to building resilience to channel errors and robustness to channel impairments.

Both error detection coding and correction coding, collectively called *channel coding* (although error correction coding is often inferred if no confusion arises), exploit the added structured information in the data, commonly employed by the physical layer of a radio communication system to enable reliable transmissions over unreliable radio channels. A typical technique employed by the physical layer in a radio communication system is a combination of both.

At the transmitting peer, a physical layer SDU (i.e., the data link PDU) is protected by an error detection code to form a physical layer PDU. The PDU is then encoded by a forward error correction code. At the receiving peer, the physical layer PDU is decoded first by the forward error correction decoder in an attempt to correct any potential error introduced by the radio channel. The integrity of the recovered physical layer PDU is then checked by an error detection decoder at the receiver. If found, the corresponding data link layer PDU is flagged by the ARQ protocol of the data link layer and indicated to the transmitting peer, i.e., the ARQ protocol at the transmitter, which manages a retransmission. In some cases, the receiving data link layer may simply discard the corrupted data link PDU, depending on the type of service with which the PDU is associated. Such types include PDUs from broadcast services or MAC signaling, where ARQ is not possible.

5.3.2 *Error Detection Coding*

An error detection code is a channel code designed with the highest priority on error-detecting capability over error-correction so that the probability of undetected errors is minimal. It detects the presence of errors in the received bit sequence or code block but does not indicate where the errors are (otherwise, we could just correct them). The most commonly used error detection code is the *cyclic redundancy check* (CRC) with superior error detection capability but practically no error correction capability.

As the name implies, CRC codes belong to the class of cyclic codes with the cyclic shift property; when any code word is rotated left or right by any number of bits, the resultant bit sequence is still a code word in the code space. This property lends itself to easy encoding and decoding by using a simple shift register structure.

With the polynomial representation of a bit sequence, the coding process can be described succinctly. Specifically, assuming the message consisting of a bit sequence of u_{K-1}, u_{K-2}, $\cdots$, u_0, the corresponding polynomial representation in the binary Galois Field or GF(2) is

$$u_{K-1}(x) = u_{K-1}x^{K-1} + u_{K-2}x^{K-2} + \cdots + u_1x + u_0. \tag{5.102}$$

Similarly, the L-bit CRC parity bits can also be represented by the *parity polynomial*, $v_{L-1}(x)$,

$$v_{L-1}(x) = v_{L-1}x^{L-1} + v_{L-2}x^{L-2} + \cdots + v_1x + v_0. \tag{5.103}$$

Assuming $g_L(x)$ is the L-bit CRC *generator polynomial*:

$$g_L(x) = x^L + g_{L-1}x^{L-1} + \cdots g_2x^2 + g_1x + g_0, \tag{5.104}$$

the relationship between the original bit sequence in (5.102) and the parity bits in (5.103) can be expressed as

$$g_L(x) \cdot q_{K-1}(x) + v_{L-1}(x) = x^L \cdot u_{K-1}(x), \tag{5.105}$$

or equivalently,

$$g_L(x) \cdot q_{K-1}(x) = x^L \cdot u_{K-1}(x) + v_{L-1}(x), \tag{5.106}$$

which indicates that $v_{L-1}(x)$ is simply the remainder of the division of $x^L \cdot u_{K-1}(x)$ by $g_L(x)$.

The right side of (5.106) is the message bit sequence appended with CRC parity bits, i.e., the CRC codeword, denoted as

$$r_{K+L-1}(x) \triangleq x^L \cdot u_{K-1}(x) + v_{L-1}(x). \tag{5.107}$$

Per (5.106), the remainder of $r_{K+L-1}(x)/g_L(x)$ is zero, and any error in the above bit sequence represented by (5.107) results in a non-zero remainder, ergo a means for error detection.

More intriguingly, the polynomial division can be easily implemented by a feedback shift register structure, shown in Figs. 5.29 and 5.30. A shift register is a cascade of L *delay elements*, sharing the same clock (not shown in the figure), in which the output of each delay element is connected to the input of the next element in the chain. A delay element stores one element from GF(2), i.e., one bit. When the

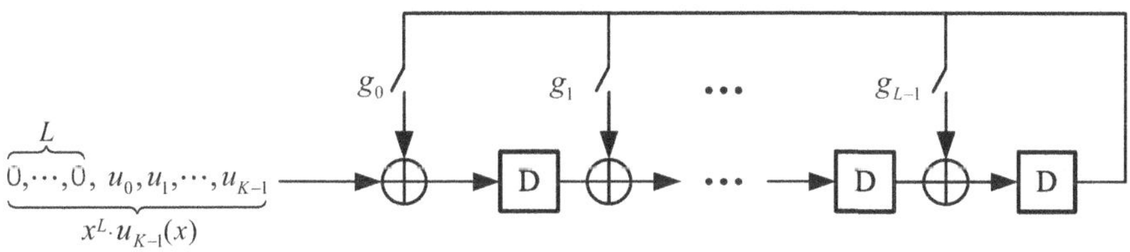

Fig. 5.29 A CRC encoder with a feedback shift register structure consisting of L delay elements, where $g_L = 1$ indicates a closed switch.

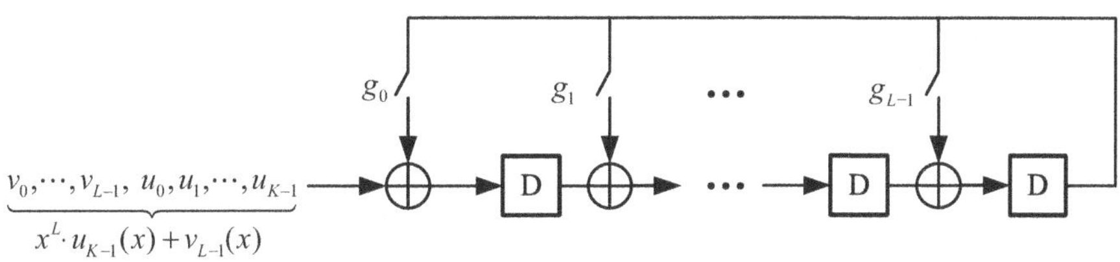

Fig. 5.30 A CRC decoder with a feedback shift register structure, where $v_0, \cdots, v_{L-1}$ are the CRC parity bits generated from Fig. 5.29

clock ticks, the bit value present at the output of a delay element is shifted into the next element on the right. The contents of the L delay elements are collectively called the *state* of the shift register, with the highest order corresponding to the rightmost element of the shift register.

The shift register is initialized with zeros. After $K + L$ total shifts into the register, the remainder resides in the register, i.e., the state of the shift register after $K + L$ shifts.

The selection of the generator polynomial is critical to the error detection capability of the CRC code. Commonly used generator polynomials for CRC are

$$g_{24}(x) = x^{24} + x^{23} + x^6 + x^5 + x + 1, \tag{5.108}$$

for a 24-bit CRC,

$$g_{12}(x) = x^{12} + x^{11} + x^3 + x^2 + x + 1, \tag{5.109}$$

for a 12-bit CRC. For rather short messages, a short CRC is useful for keeping the overhead low. For instance,

$$g_6(x) = x^6 + x + 1 \tag{5.110}$$

is for an 6-bit CRC. We will see its usage in the signaling channel design.

The miss-detection error probability of an L-bit CRC code is roughly, 2^{-L}, handy for selecting the right CRC lengths for the physical layer PDUs.

5.3.3 *Forward Error Correction Coding*

Wireless channels are error-prone. In this section, we will see that, for erroneous channels, forward error correction can bring appreciable gains. However, like error detection coding, forward error correction coding works by adding redundant bits to a bit sequence or an information block, structured to help the receiving peer correct potential transmission errors. The word "forward" stems from the fact that data flow is always in the forward direction (i.e., from transmitter to receiver) without feedback to the transmitter (like in ARQ) to correct the error. Therefore, it can be considered the "first attempt" to correct the error in the received information block to increase the resilience to radio channel errors, thereby less dependency on retransmission, especially in the scenarios where the retransmission is not an option, such as the broadcast channels and signaling channels.

Nevertheless, adding redundant bits means overhead. As we will see next, this overhead can be well justified if a "good" coding scheme is used.

5.3.3.1 Convolutional Code

A convolutional code is a type of forward error correction code. It generates parity bits via the sliding application of a *generator polynomial*, $g_L(x)$, to a bit sequence $u_k(x)$, adopting the same methodology used in error detection coding. The sliding application represents the convolution of $g_L(x)$ over $u_k(x)$, hence the term *convolutional coding*, generating a parity bit sequence, $v_k(x)$. The greater the L is, the more constraints are introduced into the parity bits, hence better error correction capability. Since each output parity bit depends on the current input bit and the L previous input bits, $L + 1$ is the *constraint length.* Consequently, the code is vulnerable to consecutive errors; evenly spread bit errors, ideally separated by more than L bits, are more likely to be corrected than otherwise. The purpose of *channel interleaving* (cf. Fig. 5.10) is to scramble the code bits of a code block, reducing the probability of consecutive errors after being descrambled at the receiver.

The shift register implementation with a non-recursive structure is depicted in Fig. 5.31. All delay elements are initialized with zero, i.e., the encoder starts from the all-zero state. Each shift of the input bit sequence into the shift register outputs a parity bit, v_k. A non-recursive convolutional code is typically non-systematic, and for non-systematic convolutional codes, only the parity bits are transmitted. For a rate $r_0 = 1/n$ non-systematic code, n generator polynomials are needed to generate n parity bits per shift in each sliding window:

$$g_L^i(x) = g_L^i x^L + \cdots + g_1^i x + 1, \quad i = 1, 2, \cdots, n. \tag{5.111}$$

However, the ending state of the encoder, i.e., the shift register state, is determined by the input information bit sequence, which can be any of the 2^L states. An open ending state causes a loss in decoding performance and an increase in

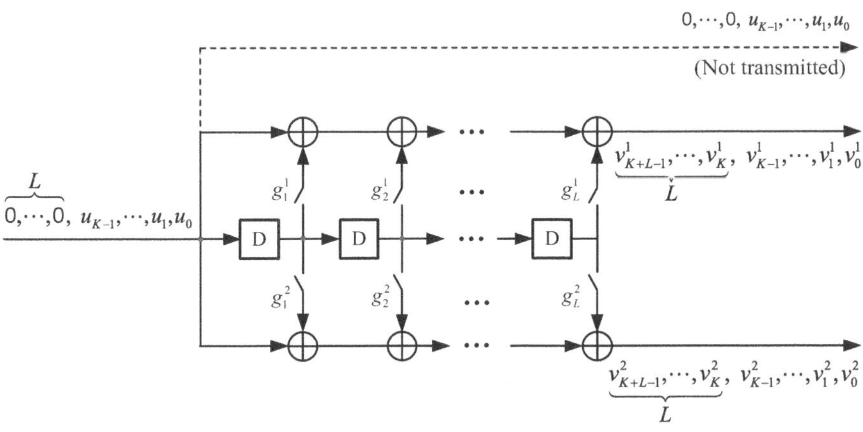

Fig. 5.31 Rate 1/2 non-recursive, non-systematic convolutional encoder structure with a constraint length of $L + 1$. The initial state is set to zero. L zero tail bits are appended to the end of the bit sequence to drive the ending state to zero

complexity since, for a maximum likelihood decoder, it means that every ending state has to be tested, resulting in 2^L times decoding complexity. Therefore, the encoder state is forced to the all-zero state by feeding the encoder additional L all-zero "tail bits" (see Fig. 5.31). Consequently, instead of $r_0 = 1/n$, the *effective code rate* becomes

$$r' = \frac{K}{(K+L)\cdot n} = \frac{r_0}{1+\frac{L}{K}}. \tag{5.112}$$

For large length of bit sequence, i.e., $K \gg L$, (5.112) approaches the *base code rate* $r_0 = 1/n$ and the rate loss due to the terminating tail bits is negligible. Nevertheless, for short information blocks, this overhead can be significant.

Since the transmission of additional redundant bits, non-information-bearing traffic in terms of overall system throughput is a system overhead or cost, the question remains: Will coding actually bring a gain after taking into account the cost?

Figure 5.32 shows the benefit of convolutional coding. For a fair comparison, the energy spent on the additional bits from coding is evenly divided among the information bits to reflect the additional costs paid for coding. We observe that the *coding gain* (over no forward error correction coding) is about 6 dB for large block size (496 bits) and 4 dB for short block size (12 bits) at a 1% block error rate. As expected, the coding gain is less for a small block size. Thus, the coding gain represents a code's ability to lower the required E_b/N_0 to correct errors.

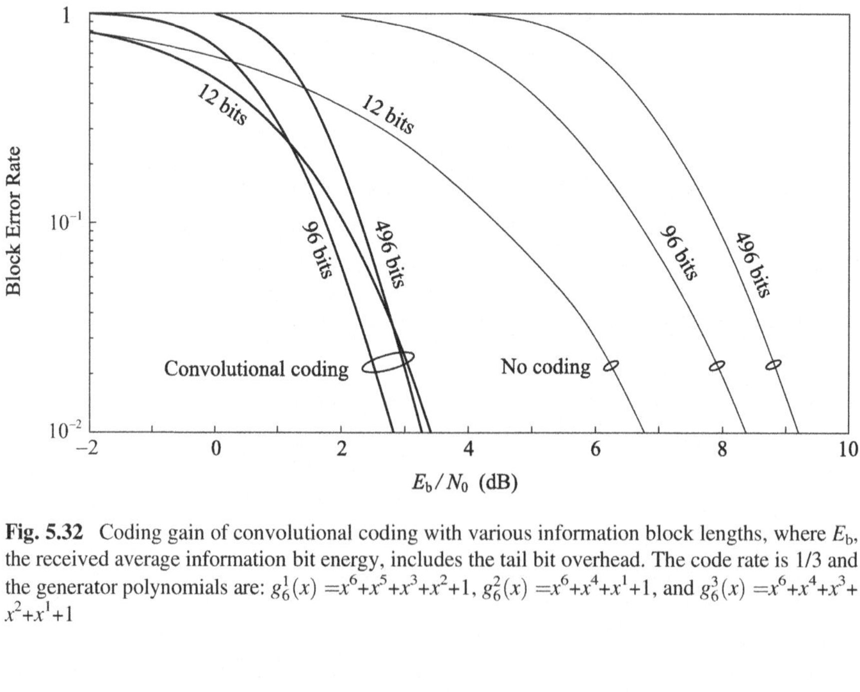

Fig. 5.32 Coding gain of convolutional coding with various information block lengths, where E_b, the received average information bit energy, includes the tail bit overhead. The code rate is 1/3 and the generator polynomials are: $g_6^1(x) = x^6+x^5+x^3+x^2+1$, $g_6^2(x) = x^6+x^4+x^1+1$, and $g_6^3(x) = x^6+x^4+x^3+x^2+x^1+1$

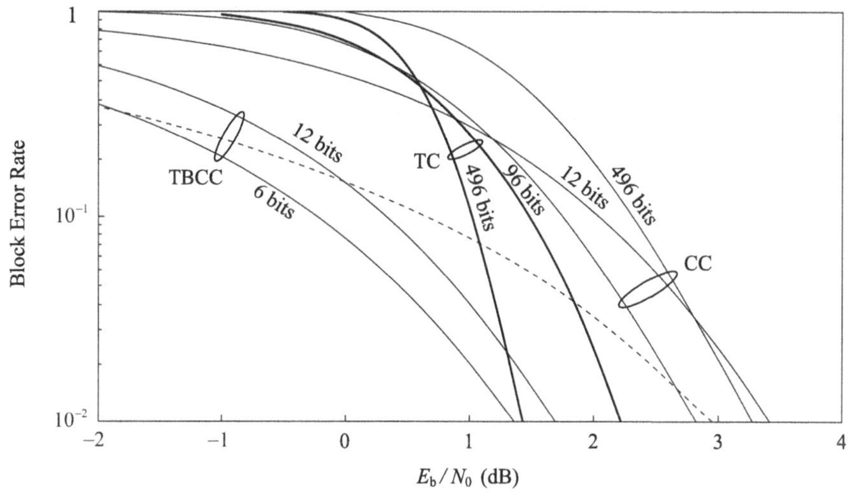

Fig. 5.33 Performance comparison for various information block lengths, among 1/3 convolutional code (CC), 1/3 tail-biting convolutional code (TBCC), 1/3 turbo code (TC), and bi-orthogonal modulation with Walsh-Hadamard codes (dashed line), where E_b, the average energy per information bit, includes the tail bit overhead (if any)

To ameliorate the rate deficiency for short information blocks caused by the zero-tail termination, the tail-biting convolutional encoder ensures that the starting state of the encoder is the same as its ending state, hence the name "tail biting." It is achieved by setting the initial state with the last L information bits of the bit sequence, i.e., $u_{K-L}, \cdots, u_{K-2}, u_{K-1}$, which naturally is the ending state when all K bits are shifted in, thereby saving L all-zero tail bit overhead. However, even with the added constraint of the same initial and final states, the ending state is still undetermined, resulting in an intermediate decoding performance and complexity between the zero state and open state.

Figure 5.33 plots the decoding performance of the tail-biting convolutional code (TBCC) for short information blocks against the original convolutional code (CC). It is observed that the tail-biting convolutional code is superior to the convolutional code, indicating the significance of tail bit overhead. It also outperforms the bi-orthogonal modulation/code introduced earlier (Sect. 5.2.3). In later chapters, we will see the tail-biting convolutional code applications in protecting some signaling channels, e.g., PUSCH in centralized medium access control (Chaps. 8 and 9) and PSCH in distributed medium access control (Chap. 7), for transporting mobile-generated MAC Signaling PDUs. These PDUs are typically small-sized, rendering TBCC a more efficient code than CC. From Fig. 5.33, we also observe that, both tail-biting and regular convolutional codes are inferior to the turbo code (TC) for large-size information blocks, typical for Traffic PDUs, which brings us to our next topic.

5.3.3.2 Turbo Code

Turbo codes are a class of the most powerful forward error correction codes. Figure 5.34 shows a typical turbo code encoder. It is a systematic parallel concatenated convolutional code with two 8-state constituent encoders and one turbo code internal *interleaver*. The constituent encoder is a recursive systematic convolutional encoder obtained from the non-recursive convolutional encoder by feeding back one of the parity bit outputs to its input.

An information bit sequence is encoded by a constituent encoder, and an interleaved version of the information sequence is encoded by another constituent encoder. For an input block size of K bits, the output of a turbo encoder consists of three length-K sequences,

$$\begin{array}{lllll} \text{Systematic bits}: & u_0 & u_1 & \cdots & u_{K-1}, \\ \text{Parity bits}: & v_0^1 & v_1^1 & \cdots & v_{K-1}^1, \\ \text{Parity bits}: & v_0^2 & v_1^2 & \cdots & v_{K-1}^2, \end{array} \tag{5.113}$$

corresponding to the systematic bit sequence (the information bit sequence) and two parity bit sequences, respectively.

The initial states of both encoders are set to all zeros. The states are forced to the all-zero state by feeding the encoders additional bits, three zero bits for each encoder.

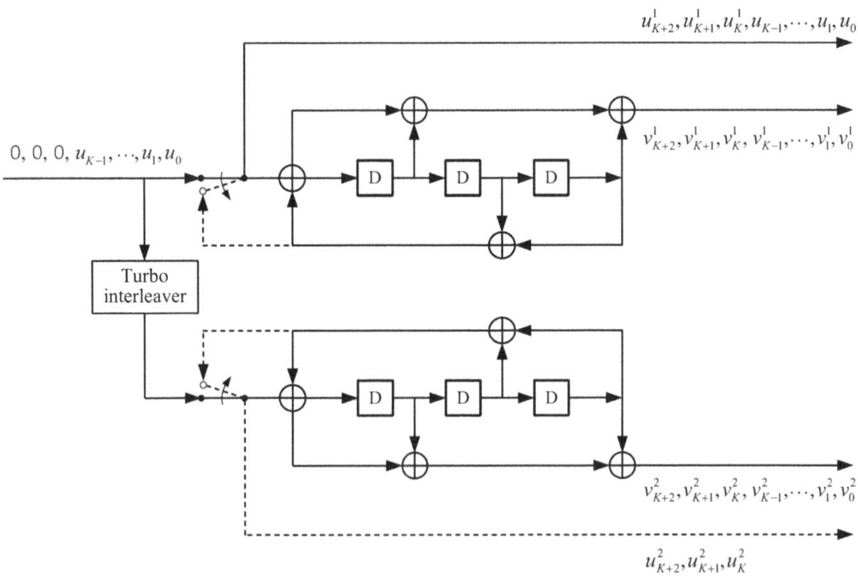

Fig. 5.34 Rate 1/3 turbo code structure with two constituent encoders, both are recursive systematic convolutional encoders

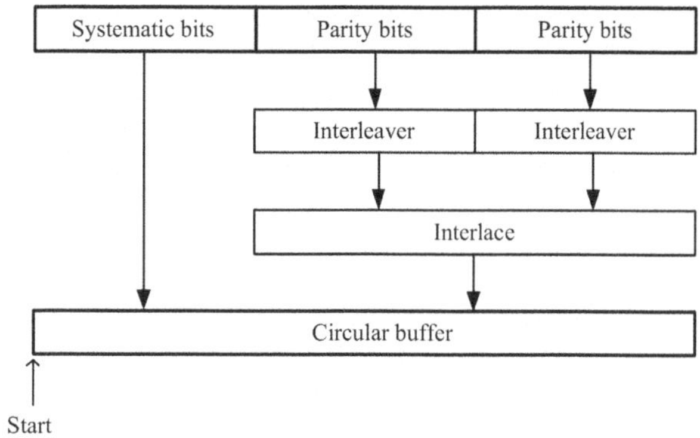

Fig. 5.35 Circular buffer and rate matching for base code rate 1/3 turbo codes

The first three zero bits terminate the first encoder by switching the input to the dashed line, giving rise to the following outputs:

$$\begin{aligned} &\text{Systematic bits}: \quad u_K^1 \quad u_{K+1}^1 \quad u_{K+2}^1, \\ &\text{Parity bits}: \quad v_K^1 \quad v_{K+1}^1 \quad v_{K+2}^1, \end{aligned} \tag{5.114}$$

while the second three zero bits terminate the second encoder, producing:

$$\begin{aligned} &\text{Systematic bits}: \quad u_K^2 \quad u_{K+1}^2 \quad u_{K+2}^2, \\ &\text{Parity bits}: \quad v_K^2 \quad v_{K+1}^2 \quad v_{K+2}^2, \end{aligned} \tag{5.115}$$

therefore, a total of 12 tail bits from termination.

The tail bits are multiplexed to the end of the three streams, each of length $K + 4$ bits:

$$\begin{array}{llllllllll} \text{Systematic bits}: & u_0 & u_1 & \cdots & u_{K-1} & u_K^1 & v_K^1 & u_{K+1}^1 & v_{K+1}^1, \\ \text{Parity bits}: & v_0^1 & v_1^1 & \cdots & v_{K-1}^1 & u_{K+2}^1 & v_{K+2}^1 & u_K^2 & v_K^2, \\ \text{Parity bits}: & v_0^2 & v_1^2 & \cdots & v_{K-1}^2 & u_{K+1}^2 & v_{K+1}^2 & u_{K+2}^2 & v_{K+2}^2. \end{array} \tag{5.116}$$

The systematic bits are stored in a *circular buffer* as shown in Fig. 5.35; the two streams of parity bits are first interleaved individually and then interlaced bit by bit into the buffer. Depending on the physical channel payload (configured by the MAC protocol of the data link layer), not all code bits (parity bits in particular) in the buffer may be transmitted, leading to an *actual code rate*, r, higher than the base code rate, $r_0 = 1/3$. Clearly, $r \geq r_0$. Interleaving allows equal level protection of the parity bits, and interlacing of the parity sequences allows equal levels of protection for each constituent code to minimize the effect of puncturing parity bits on the turbo code performance.

The number of bits given by the payload is read out sequentially from the start of the buffer. If the end of the buffer is reached, simply wrap around to the start until the required number of bits is acquired. This process, henceforth referred to as *rate matching*, provides the flexibility for achieving an *arbitrary* channel coding rate to match the desired effective code rate, r', defined as the ratio of the total number of transmitted bits with respect to the number of information bits.

Two different types of interleavers are used in turbo encoder and rate matching. Interleaving is a process of rearranging the ordering of a bit sequence. The interleavers used in the rate matching in Fig. 5.35 are based on a simple row-column permutation: bits are sequentially written into an, e.g., 32-row matrix by rows and read out by columns. The interleaver of the turbo encoder in Fig. 5.34 is much more complex. In fact, it is the most critical component of the turbo encoder. It is used to permute the input bits such that the constituent encoders are operating on the same sequence of input bits but in different bit order. The interleaver provides "scrambled" information bits to the second encoder, which allows the two constituent decoders of a receiver to work in tandem through uncorrelated information exchange in an iterative fashion, hence the name "turbo code." As such, a structured interleaver, like the one used in rate matching, typically fails miserably (manifested by an error floor). A random interleaver is statistically best in destroying the coherence present in a sequence; however, it is practically much more complex than a structured interleaver. Obviously, the shorter a sequence is, the more difficult it is to design an interleaver to achieve such an effect. As a result, turbo codes are suited for large information blocks, typically more than 100 bits. The overhead introduced by the tail bits is thus negligible. The design of such an interleaver is rather involved and hence is out of the scope of this book. There are more details on this subject in the literature. For interested readers, see [5].

Figure 5.33 shows a comparison of the decoding performance of Walsh-Hadamard bi-orthogonal (see Section 5.2.3), convolutional, and turbo codes. It also shows the impact of information block sizes on the performance, where the transmitter is assumed to communicate with the receiver through an AWGN channel. It is observed that the tail-biting convolutional code is most efficient for small-sized information blocks, whereas the turbo code has a clear advantage over convolutional codes for large-sized information blocks due to the more effective turbo interleavers for larger information blocks. The advantage fades as the block size decreases.

Here, soft-decision decoding is assumed for convolutional and turbo codes. In contrast to a hard-decision decoder that operates on data that takes on a fixed set of possible values, i.e., `0` or `1`, the inputs to a soft-decision decoder take on a whole range of values in-between. This extra information indicates the likelihood of each received sample being `0` or `1` and is used to form a better estimate of the original code bit for decoding. Here *code bits* refer to the output bits from the encoder, which include both parity bits and information bits for systematic codes, whereas parity bits only for non-systematic codes.

A commonly used soft-decision metric of a code bit is the so-called *log likelihood ratio*, defined as

$$\lambda_b \triangleq \log \frac{\Pr(b = 1|y)}{\Pr(b = 0|y)}, \tag{5.117}$$

where λ_b is a log likelihood ratio of a code bit, $b \in \{0, 1\}$, being 1 over 0, given a received sample, $\boldsymbol{y}$, from the output of a matched filter of a demodulator [6]. Therefore, it is understood that, under the transceiver structure of Fig. 5.10, b is the code bit after interleaving, i.e., the input to the constellation mapper. Without loss of generality, we use the representation in (5.50) as an example. The output sample of the kth modulation symbol, $\boldsymbol{s}'_k(t)$, from the matched filter under ideal sample timing is

$$\boldsymbol{y}_k \triangleq \boldsymbol{\alpha} \cdot \boldsymbol{c}_k + \boldsymbol{n}_k, \tag{5.118}$$

where $\boldsymbol{n}_k$ is zero-mean complex Gaussian with variance N_0. From (5.27), $\boldsymbol{c}_k \in \Omega^M$ is a constellation point that maps to an M-bit sequence, $b_0^k, b_1^k, \cdots, b_{M-1}^k$. The modulation symbol index, k, is omitted for notation simplification, and the mth bit is denoted as b_m, whose decision metric can be represented as

$$\lambda_m^k \triangleq \log \frac{\Pr(b_m = 1|\boldsymbol{y}_k)}{\Pr(b_m = 0|\boldsymbol{y}_k)}, \tag{5.119}$$

according to (5.117), where

$$\Pr(b_m = \ell|\boldsymbol{y}_k) = \sum_{\boldsymbol{c} \in \Omega^M_{b_m=\ell}} \Pr(\boldsymbol{c}|\boldsymbol{y}_k), \quad \ell \in \{0, 1\}, \tag{5.120}$$

is the probability of $b_m = \ell$ given that $\boldsymbol{y}_k$ is observed, and $\Omega^M_{b_m=\ell} \triangleq \{\Omega^M : b_m = \ell\}$ the subset of Ω^M that contains all the constellation points whose mth bit is ℓ. Substitute (5.120) into (5.119) to obtain

$$\begin{aligned} \lambda_m^k &= \log \sum_{\boldsymbol{c} \in \Omega^M_{b_m=1}} \Pr(\boldsymbol{c}|\boldsymbol{y}_k) - \log \sum_{\boldsymbol{c} \in \Omega^M_{b_m=0}} \Pr(\boldsymbol{c}|\boldsymbol{y}_k) \\ &= \log \sum_{\boldsymbol{c} \in \Omega^M_{b_m=1}} \Pr(\boldsymbol{y}_k|\boldsymbol{c}) - \log \sum_{\boldsymbol{c} \in \Omega^M_{b_m=0}} \Pr(\boldsymbol{y}_k|\boldsymbol{c}), \end{aligned} \tag{5.121}$$

where the second equality assumes all constellation points in Ω^M are equally likely to be transmitted. From (5.118),

$$\Pr(\boldsymbol{y}_k|\boldsymbol{c}) = \frac{1}{2\pi N_0} \exp\left(-\frac{1}{2N_0} |\boldsymbol{y}_k - \boldsymbol{\alpha} \cdot \boldsymbol{c}|^2\right). \tag{5.122}$$

Here, for the purpose of demonstration, perfect channel estimation is assumed, i.e., both $\boldsymbol{\alpha}$ and N_0 are known. In practice, both need to be estimated via pilot signals on the fly.

5.4 Adaptive Modulation and Coding

Radio link adaptation denotes matching the modulation, coding, resources, and other signal and protocol parameters to the radio channel state (e.g., the path loss, the interference, the receiver sensitivity, the available transmitter power margin and so forth). In its simplest form, link adaptation is performed by the medium access control protocol of the data link layer through an adaptive modulation and coding function. As the name implies, it exploits the channel state information by adapting the modulation and coding scheme or MCS, discussed in the previous sections, according to the radio link quality, e.g., the matched filter output SNR, η, in (5.51), for transmission robustness and spectral efficiency maximization.

5.4.1 Spectral Efficiency and MCS

The radio link spectral efficiency of a combined modulation and coding scheme is equal to the spectral efficiency of the modulation (i.e., bits per modulation symbol) multiplied by the effective code rate. Adaptive modulation and coding aim to cope with constant changes in radio link conditions by dynamically selecting the optimum MCS for each transmission. A low link quality channel needs to be sustained by a robust MCS with a low modulation order and a low code rate. Therefore, as link conditions worsen, a robust MCS is employed, whereas a spectrally efficient MCS is used when link conditions become benign. Achievable user throughput largely depends upon proper MCS selection. If the MCS is too conservative, i.e., lower modulation order and code rate, the instantaneous channel capacity is not fully exploited. In contrast, if the MCS is too aggressive, the decoding error rate increases, and the link spectral efficiency is degraded.

The above concept is better justified by Fig. 5.36, where the spectral efficiency, achieved through various practical modulation and coding schemes, is plotted against the ideal Shannon limit. Each curve in the plot represents a specific modulation order coupled with turbo coding at a specific code rate, i.e., a specific MCS, over a typical SNR range. Different MCS's provide a finer spectral efficiency granularity to achieve various spectral efficiencies under various channel conditions. Among these, for a given channel quality or SNR, there exists an MCS that produces the highest spectral efficiency. For example, for SNRs less than 5 dB, QPSK is better than any higher-order QAM. In the SNR range from 5 to 13 dB, 16-QAM is the best modulation scheme, whereas from 13 dB to 20 dB, 64-QAM is a better choice. Above 20 dB, 256-QAM must be used to attain the spectral efficiency offered by the

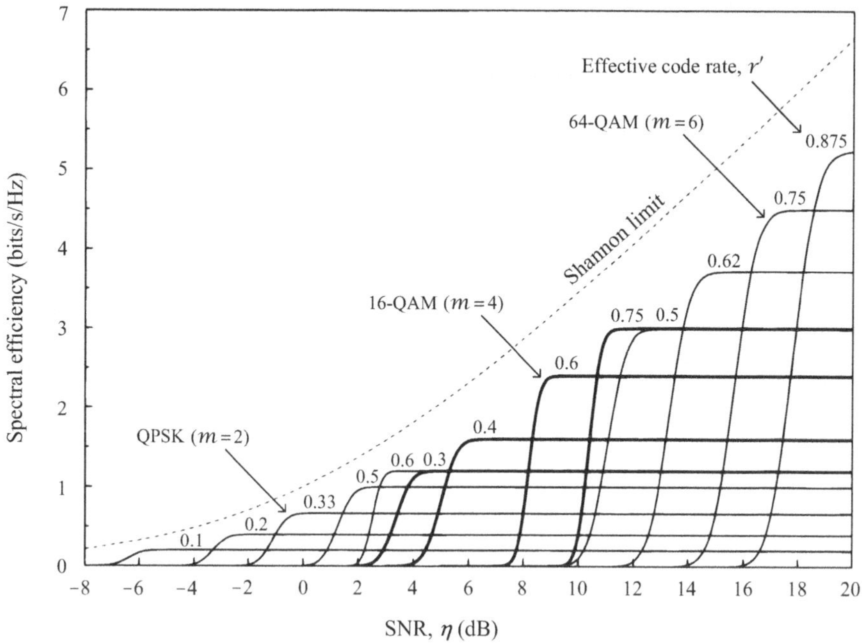

Fig. 5.36 Spectral efficiencies for various modulation and coding schemes (combinations of modulation orders and effective code rates), assuming base code rate 1/3 turbo coding and rate matching, under an AWGN channel and ideal channel estimation

channel. Nonetheless, higher-order QAM requires better (i.e., lower EVM hence more costly) transceivers and may not be supported by all devices – a factor that needs to be considered in practical.

5.4.2 MCS Selection

Given the channel condition, SNR or η, and a target decoding error rate, e.g., 1%, MCS is thus the selection of the right modulation order and effective code rate pair, i.e., (m, r'), such that the target decoding error probability is met. For each (m, r'), there exists an E_b/N_0 under which the target decoding error probability can be achieved. We refer to this E_b/N_0 as the target E_b/N_0 denoted as E_b^+/N_0, a function of (m, r'). Therefore, for a given η, we look for the (m, r') such that the corresponding E_b/N_0, ie.,

$$E_b/N_0 = \eta - m - r' \text{ (dB)}, \tag{5.123}$$

is greater than E_b^+/N_0 under the same (m, r'). That is,

$$(m, r') = \text{argmin}\{E_b/N_0 | E_b/N_0 = \eta - m - r' \geq E_b^+/N_0\}, \tag{5.124}$$

There can be many such choices satisfying this condition, but we want one that maximizes the spectral efficiency. Referring to Figure 5.36, given a channel condition, $\eta = 3$ dB, the MCS that maximizes the spectral efficiency is (2, 0.5).

Furthermore, (5.124) indicates that the larger the deficit between η and E_b^+/N_0 is, the lower the r' needs to be, hence the more the required resources. Therefore, we wish to minimize E_b^+/N_0 for every (m, r') so that the selected (m, r') from (5.124) can be more spectrally efficient.

The effective code rate of a signal represents the ability of the signal to reject channel noise, achieved by the added structure in the transmitted signal, like coding and repetition, so that a receiver can leverage the embedded structure to reject the noise – assumed to be structure-less and uncorrelated with the desired signal, like the thermal noise (AWGN). In this sense, the effective code rate is better known as the *processing gain*, defined as the negation of the effective code rate, i.e., $\mathcal{g} \triangleq -r' \geq 0$ (dB), i.e., a non-negative gain. (5.123) becomes

$$E_b/N_0 = \eta - m + \mathcal{g} \text{ (dB)}. \tag{5.125}$$

Clearly, both coding and repetition lead to a less-than-one effective code rate ($r' < 1$), i.e., a positive processing gain ($\mathcal{g}$>0 dB). However, as observed earlier, a forward error correction code provides a coding gain that essentially lowers the required E_b/N_0, which effectively narrows the deficit, leading to a less processing gain needed to close the deficit from (5.125). The coding gain reaches its maximum when the code rate equals the base code rate (the lowest code rate associated with an encoder), i.e., $r = r_0$, or $r' \leq r_0$. On the other hand, the gain through repetition at the transmitter and combining the replicas at a receiver, termed *combining gain*, does not change the required E_b/N_0. Therefore, repetition can be considered the simplest coding that brings in a combination gain but no coding gain. For the same η, it requires more processing gain (more resources) to close the link. Hence the processing gain through coding that reaps both coding and combining gains is more resource- and energy-efficient (less energy per information bit).

In this regard, the transmitter should exhaust all code bits (parity bits in particular) before repeating to profit from the full intrinsic coding gain provided by the encoder. The circular buffer is just devised for this purpose by ensuring that all code bits are consumed before repeating.

5.4.3 Radio Link Condition and Feedback

Wireless radio links are inevitably subject to random fluctuations in receive signal strength due to both large-scale fading (i.e., propagation loss) and small-scale or multipath fading. Adaptation of the MCS to the radio link condition makes the communication link more robust against changing radio link conditions and benefits the communication system in overall spectral efficiency and the mobile station in energy efficiency. However, these benefits do come with a condition: the knowledge of the current radio link condition, which is typically not available *at the transmitter* and needs to be acquired through feedback. This is the case for FDD, in which the uplink and downlink are far separated in frequency, hence likely uncorrelated. In the centralized medium access control system, the downlink quality between the control station and a mobile station is measured at the mobile station from the downlink physical signaling channels, such as PBBCH and PDSCH, and communicated to the control station through the uplink signaling channel, i.e., RACH at the beginning of communication or USCH during communication. The uplink quality is measured at the control station from the uplink physical channels, such as PRACH that carries the RACH from the mobile station at the beginning of the communication and PUSCH and PUTCH during communication, and signaled to the mobile station through ACH on PDSCH.

Intuitively, the SNR, η, measured at the receiver, is a natural channel quality indicator (CQI) to the transmitter. However, an SNR itself does not indicate the capability of a receiver. For instance, a higher-end receiver is capable of supporting a higher spectral-efficiency MCS than a lower-end receiver under the same SNR. A CQI represented by the spectral efficiency that the receiver can provide under this SNR reflects a combination of both.

Obviously, the finer the granularity of MCS is, the closer the spectral efficiency gets to the Shannon limit. However, this also means higher CQI measurement accuracy, not to mention more resources or overhead for signaling between the transmitter and receiver.

Under TDD, the uplink and downlink share the same frequency, and the CQI feedback is less of an issue in light of the radio link reciprocity property. However, since the CQI represents the quality of the radio channel that involves different RF chains and receiver capabilities of the transceivers under different communication directions, CQI feedback is still helpful in resolving the disparities.

5.5 Automatic Retransmission Request

During wireless transmissions, it is common that some information blocks are received with bit errors, as indicated by the error detection code (e.g., CRC). The error-correction decoder (e.g., the turbo code) may fail to correct them. Failed decoding of a code block results from either collisions or insufficient received bit SNR, i.e., E_b/N_0, resulting in two retransmission schemes. As noted earlier, in maritime MTC, automatic retransmission request or ARQ is an error control protocol that resides in the data link layer (with the help from the physical layer) of the air interface for recovering the information block corrupted during transmission.

5.5.1 Retransmission with Simple-Repetition

As illustrated in Fig. 5.37, a turbo coder described in Fig. 5.34 is used for forward error correction coding. The encoder generates a code block of code rate 1/3 for a given information block and stores it in a circular buffer (Fig. 5.35). Based on the modulation and coding scheme or MCS, a specific combination of a modulation order with an effective code rate, prescribed by the data link layer's medium access control protocol, the rate-matched code block, including the information bits and a portion of the parity bits, is modulated and transmitted. Depending on whether the decoding is successful, the receiver sends ACK or NACK on a feedback channel (not shown in Fig. 5.37).

In distributed medium access control communication (e.g., proximity communication), the error is mainly from a head-on collision due to the contention-based random access nature (see Chap. 2). Retransmission (Fig. 5.37) and decoding from a fresh start is indeed the proper approach, meaning that the data stored in the decoder buffer is not used in the subsequent recovery attempt and is thus cleared since the received data is likely garbled beyond repair. Thus, the receiver overrides the decoder buffer with the newly received copies and attempts to decode again. This process repeats until the information block is successfully decoded or the maximum number of retransmissions is reached.

Under centralized medium access control (e.g., satellite and terrestrial communications), collisions are prevented. A decoding failure is mostly due to insufficient processing gain in the signal to close the link.

It is typically a result of a mismatch between the MCS used in the transmission and the channel condition (i.e., η) caused by inaccuracy in channel prediction and MCS selection by the adaptive modulation and coding protocol. In this case, ARQ can be done more efficiently by directly targeting the "processing gain deficiency" issue, which can be solved more efficiently, both spectrally and energy-wise, by exploiting the processing gains embedded in *both* transmissions.

Therefore, the data stored in the decoder buffer, i.e., the soft metric of code bits from the demodulator in (5.121), is reusable with the retransmission in resolving the processing gain deficiency. Instead of discarding the data stored in the decoder and starting from scratch, a soft-decision decoder exploits both transmissions by combining the previous transmission with the retransmission (a simple repetition of the previous transmission as shown in Fig. 5.37). The processing gain of the received signal as a whole is doubled, i.e., increased by 3 dB. This processing gain is from

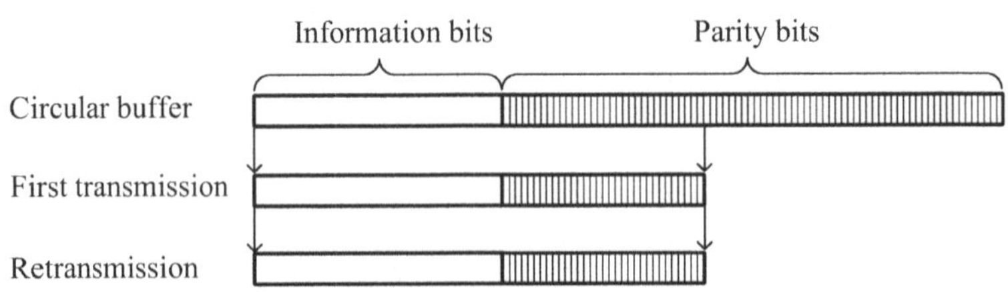

Fig. 5.37 Illustration of ARQ retransmission scheme: Simple repetition for exploitation of combining gain

combining the repetitions, hence the combining gain. The chance of successful decoding is appreciably increased, thereby saving us from the extra resources that otherwise would be committed in the retransmission.

The medium access control protocol typically targets a decoding error rate between 1% and 10% and relies on retransmissions to clean up the residual errors, although the actual target error rate depends on the application profile that the data is associated with. A lower target error rate imposes a higher constraint on channel prediction accuracy, which is practically difficult by virtue of a wireless channel's random nature. On the other hand, a higher error rate not only costs extra resources on retransmissions and signaling but incurs excessive delays.

5.5.2 *Circular Retransmission*

Both repetition and coding increase processing gain. However, the processing gain from coding is more powerful (due to the coding gain) against noise through a more complex encoding and decoding process. In this regard, the transmitter should exhaust all code bits before repetition to fully exploit the intrinsic coding gain provided by the turbo encoder. The circular buffer is just devised for this purpose by ensuring that all code bits are consumed before starting repetition.

A transmitted code block after rate-matching typically contains only a portion of the full-rate code block stored in the circular buffer (Fig. 5.37), constrained by the available resources assigned according to the predicted channel condition as a part of the MCS selection; thus, not all the parity bits are transmitted. The actual code rate, r, is then higher than the base code rate, r_0, offered by the turbo encoder. Therefore, rather than repeating the previous transmission, the retransmission picks up where the previous transmission is left off in the circular buffer, as depicted in Fig. 5.38, and wraps around to the beginning when the end of the buffer is encountered. Simply put, the transmitter exhausts the buffer, i.e., the leftover parity bits from the previous transmission, before repeating.

Transmitting the leftover (not yet transmitted) parity bits further lowers the actual code rate, producing more coding gains, but repeating does not. This approach is most effective when the first transmission has a high code rate. As demonstrated in Fig. 5.39, the circular repetition scheme harvests more than 1 dB extra gain (coding

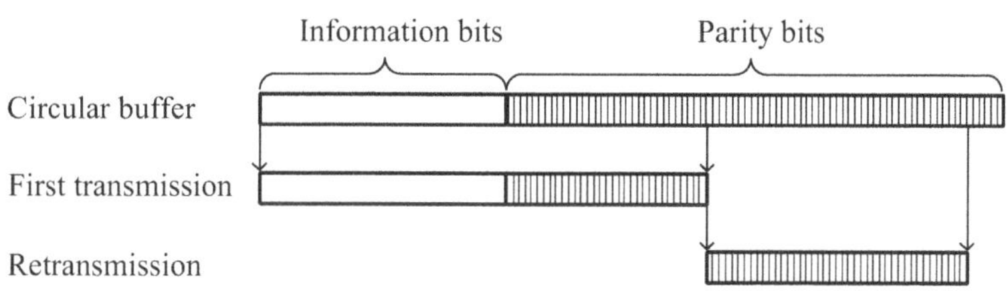

Fig. 5.38 Illustration of ARQ retransmission scheme: Circular repetition for code rate reduction and exploitation of coding gain

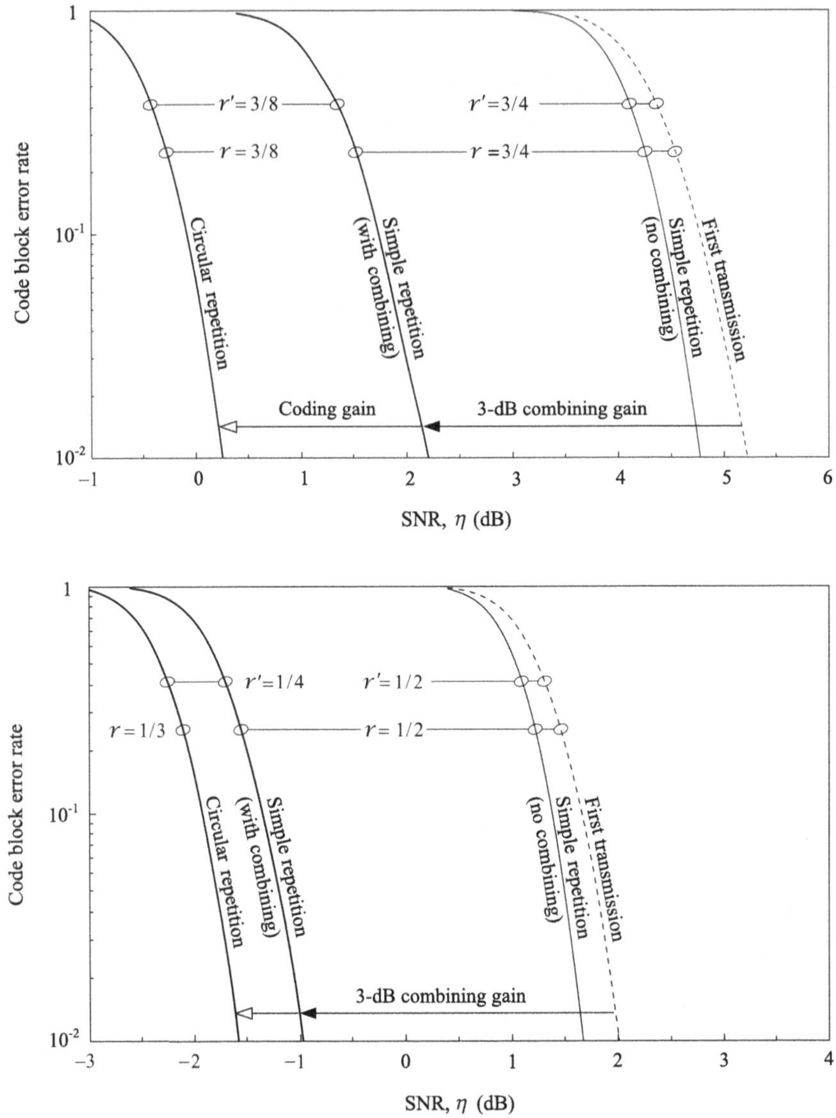

Fig. 5.39 Lower layer (data link layer and physical layer) ARQ performance under different retransmission schemes (QPSK modulation under AWGN channel): code block decoding error rate plotted against the sample SNR, i.e., the matched filter output SNR, η in (5.51), where the turbo encoder base code rate is r_0 =1/3, the information block size is 496 bits, and the same amount of resources are used in the retransmission (solid line) as in the first transmission (dashed line). Top: first transmission code rate r =3/4, and bottom: first transmission code rate r =1/2. Among these three retransmission approaches, circular repetition attains the lowest code rate, r =3/8 (top) and 1/3 (bottom), ergo, the highest gain in lowering the decoding SNR, $\eta \approx$ 0.25dB and -1.6 dB, respectively

gain) on top of the combining gain at an initial code rate of $r = 1/2$ and 4 dB extra gain under an initial code rate of 3/4. In essence, both have the same effective code rate but different actual code rates.

By now, we have seen that both "combined repetition" and circular repetition require the physical layer assistance and hence are collectively referred to as *data link layer controlled physical layer ARQ*. We also see how adding ARQ in the lower layers (close to where the error happens) can reap the full benefits of error correction efficiency and latency, a privilege not available for the high layer ARQ. It helps compensate for the shortcomings of a wireless system, allowing seamless integration into the IP network driven by the wired communication protocols that are inherently incompatible with the wireless world.

5.6 Timing and Carrier Synchronization

For stations to communicate with each other, they must rendezvous not just in the space domain (i.e., in range) but in time and frequency domains. The unison or coincidence in time and frequency is termed synchronization. Specifically, *clock synchronization* is the process of achieving and maintaining coordination among independent local clocks of mobile stations to provide a common notion of time across the network. The representations and derivations that we have touched upon in Sect. 5.2 have been made under the assumption that radio channels introduce no delay in conveying the emitted signal to the receiver and that synchronous clocks are available at the transmitter and receiver for determining the symbol time boundary. Hence, practical considerations have been suppressed so that we can concentrate on the theoretical issues. However, *timing synchronization* and *carrier synchronization* must be performed at a receiver in practical wireless communication systems in which the above assumptions are seldom warranted.

5.6.1 Timing Synchronization

Timing synchronization is the process by which the receiver determines the correct instants of time at which to sample the incoming signal. So far, we have assumed that the matched filter is synchronized to the incoming $s_k(t)$, i.e., to the ideal sample time, $t_k = kT_{\text{sym}}$, under perfect clock synchronization with the transmitter. The significance of the ideal sample time is that it guarantees not only maximum sample SNR against AWGN noise but also free of inter-symbol interference, ergo, maximum signal to interference and noise ratio or SINR, assuming $s_k(t)$ is a Nyquist pulse. However, propagation through a radio channel inevitably introduces a delay, δ, into the received signal even if the signal is transmitted following the local clock of the transmitter that is perfectly synchronized with the receiver.

At the input of the matched filter, we have

$$\begin{aligned} \boldsymbol{s}'(t) &= \boldsymbol{s}(t-\delta) + \boldsymbol{n}(t) \\ &= \sum_{k=0}^{N_{\text{sym}}-1} \boldsymbol{s}_k(t-\delta) + \boldsymbol{n}(t), \end{aligned} \tag{5.125}$$

where $\boldsymbol{n}(t)$ is AWGN with zero-mean and power spectral density of N_0. Using the local clock synchronized with the transmitter, the sample time at the matched filter output is $t_k = kT_{\text{sym}}$, which is off by δ. The output sample is thus no longer SNR-optimal. Therefore, timing synchronization to the incoming signal (not the transmitter) must be established before communication can start. An established timing means that a receiver acquires δ and uses it to establish a new sample time, $t'_k = kT_{\text{sym}} + \delta$.

To facilitate the estimation of δ, a synchronization signal, $\boldsymbol{s}_{\text{sync}}(t)$, known to the receiver, is transmitted at the time t=0 by a transmitter. A receiver employs a correlator or a matched filter with an impulse response that matches the synchronization waveform, i.e., $\boldsymbol{h}(t) = \boldsymbol{s}^*_{\text{sync}}(-t)$, similar to the matched-filter used for demodulation in Section 5.2.1. What is different here is that, in the current case, the optimal sample time is not yet known, given the received signal of

$$\boldsymbol{s}'_{\text{sync}}(t) = \boldsymbol{s}_{\text{sync}}(t-\delta) + \boldsymbol{n}(t). \tag{5.126}$$

A blind search for the optimal timing involves the search for the maximum output of the matched filter within a search window of width $\delta_{\max}$,

$$\widehat{\delta} = \max_{\delta \in (0,\, \delta_{\max})} \left| \int_0^{T_{\text{sync}}+\delta_{\max}} \boldsymbol{s}'_{\text{sync}}(\tau) \cdot \boldsymbol{s}^*_{\text{sync}}(\tau-\delta) \cdot d\tau \right|^2, \tag{5.127}$$

where T_{sync} is the length of $\boldsymbol{s}_{\text{sync}}(t)$, and $\delta_{\max}$ the maximum propagation delay. For the AWGN channel, i.e., the received signal is corrupted by AWGN, (5.127) is a maximum-likelihood estimate, and $T_{\text{sync}} \gg T_{\text{sym}}$ for large processing gain to boost the output SNR for estimation accuracy. Apparently, a synchronization waveform, $\boldsymbol{s}_{\text{sync}}(t)$, with a good autocorrelation function is desirable.

5.6.2 *Carrier Synchronization*

Carrier synchronization is the process by which a receiver locks on both the frequency and phase of the incoming RF signal. Although both communicating ends may maintain synchronized clocks via the method described in Section 5.1.6, the radio channel may introduce a random phase in addition to a propagation loss, called channel gain, i.e., $\boldsymbol{\alpha} = |\alpha| e^{j\varphi}$, where the channel phase, φ, is needed for coherent demodulation, and both channel phase and amplitude are needed for QAM, as discussed in detail in Sect. 5.2.1.

Moreover, in a mobile environment, there is a frequency offset, $\varsigma_{\text{Doppler}}$, between the *received* RF signal and the local clock, called *Doppler shift*, due to the well-known Doppler effect. This frequency offset introduces a phase ramp,

$$\varphi(t) = 2\pi\varsigma_{\text{Doppler}}t, \tag{5.128}$$

to the baseband signal according to (5.24). Thus,

$$\boldsymbol{s}'(t) = \boldsymbol{s}(t-\delta) \cdot e^{j\varphi(t-\delta)} + \boldsymbol{n}(t). \tag{5.129}$$

In the presence of both the Doppler shift and clock synchronization error,

$$\varphi(t) = 2\pi\left(\zeta_{\text{mobile}} f_{\text{c}} + \varsigma_{\text{Doppler}}\right)t. \tag{5.130}$$

Clearly, this time-varying phase offset introduces a phase distortion into the received synchronization signal,

$$\boldsymbol{s}'_{\text{sync}}(t) = \boldsymbol{s}_{\text{sync}}(t-\delta) \cdot e^{j\varphi(t-\delta)} + \boldsymbol{n}(t), \tag{5.131}$$

causing a mismatch between the incoming signal and the local matched filter and compromising the matched filter performance. Therefore, (5.127) is replaced by a joint maximum-likelihood estimator to take both frequency and time offsets into account,

$$\left(\widehat{\delta}, \widehat{\varsigma}\right) = \max_{\substack{\delta \in (0,\, \delta_{\max}) \\ \varsigma \in (-\varsigma_{\max},\, \varsigma_{\max})}} \left| \int_0^{T_{\text{sync}}} \boldsymbol{s}'_{\text{sync}}(\tau) \cdot \boldsymbol{s}^*_{\text{sync}}(\tau-\delta) \cdot e^{j2\pi\varsigma\tau} d\tau \right|^2. \tag{5.132}$$

A synchronization signal is typically transmitted at the beginning of a transmission burst for receivers to perform timing and carrier synchronization when the timing and frequency offsets are assumed to be large. Once the offsets are estimated and compensated, communication can then start.

However, this one-time estimation may not be sufficient since both time and phase may drift as time elapses due to the residual estimation errors and channel dynamics. Estimation refinement and channel tracking are thus a continuous effort by the receiver during communication to maintain synchronous with the incoming signal [7, 8]. A frequency-locked loop and a timing-locked loop are often employed for this purpose. For the sake of overhead, lightweight synchronization signals or pilot symbols are often intermittently inserted into the modulation symbol stream for updating the frequency and timing-locked loops as well as for channel estimation (i.e., estimation of $\boldsymbol{\alpha}$) on the fly. We will see design examples of such synchronization signals in the second half of the book.

5.7 Mobile Station Categorization

As noted in Chap. 1, depending on maritime IoT application and deployment scenarios, IoT devices may differ a great deal with respect to their communication capabilities, such as cost, maximum transmit power, energy supply, synchronization source, and form factor, and must be taken into consideration at every level of the protocol design. To address these various maritime application requirements more efficiently, the MTC devices or mobile stations that serve the communication needs of maritime IoT applications and services are classified into different categories in terms of their transceiver characteristics that may affect the communication strategy and dominate mobile station behavior.

5.7.1 Category 1 Mobile Stations

This type of mobile station is constrained by its small form factor, low cost, and low power. Small form factor means low antenna efficiency and low battery capacity. Low cost and power imply low-quality RF frontend (i.e., high EVM), unstable synchronization source (no GPS or the like), and low processing capability. As battery replacement is difficult and costly in most deployment scenarios, e.g., mass deployment scenarios (e.g., some sensor networks), the device is disposable, and the battery life essentially determines the overall lifespan of the network. Therefore, the battery's energy must be used at its optimum to ensure a long time operation, from months to years, in most settings. The lim-ited available energy resources, either batteries or energy harvesting from random energy arrivals, dictates the mobile station's overall operation behavior.

Specifically, this type of mobile station operates only in a half-duplex mode for reduced complexity and cost. It allows for bi-directional communications but cannot be paged by the network either through PCH or ACH and wakes up only when local application data arrive. Since the radio transceiver is the most energy-consuming component, and because idle listening over an extended period can be just as expensive as transmitting, the transceiver must be switched off as long as possible to conserve energy and hence extend battery life. When the transeiver is switched off, however, it is not able to receive. Therefore, for this type of mobile station, the only opportunity for the network to transmit downlink data to the device is a narrow time window right after the uplink transmission is completed, within which the mobile station is expected to listen in on ACH. Consequently, this type of station has the most relaxed downlink latency requirement but is also the most energy-efficient, ideal for battery-operated maritime IoT devices.

This type of mobile station has low transmit power, typically 100 mW for battery-powered devices but can be in the range of a few milliwatts, e.g., 5 mW, for self-sustaining energy harvesting applications. This limitation may significantly limit the air interface link budget, restrict its communication range, and impact the overall system spectral efficiency and capacity.

Low-cost transceivers are the default for these devices. It has the following implications: The synchronization source is expected to have low precision, e.g., 20 ppm, most likely not equipped with TCXO or GPS, and hence timing and frequency acquisitions become more challenging. EVM is anticipated to be greater than 15%, and hence only the lowest order modulation (e.g., BPSK and QPSK) can be supported. The average antenna gain is assumed to be a lump sum of -3 dB, including other RF coupling losses.

Collectively, these limitations will surely become a bottleneck of the MTC system design, therefore one of the primary focal points of the design.

5.7.2 Category 2 Mobile Stations

This type of mobile station is similar to Category 1 mobile stations except that it can be paged by the network for pending downlink data. One method to reduce energy consumption due to paging is to duty-cycle the station's transceiver. In this method, the mobile station stays awake for short periods and stays asleep most of the time. So the active energy consumption is considerably reduced. During awake periods, the mobile station listens in on the paging channel, i.e., PCH, in the assigned time slots for pending downlink messages.

Duty-cycling is thus employed for this category of mobile station for energy consumption reduction while maintaining reasonable downlink delay. Depending upon the duty cycle, the transceiver operates only for a small percentage of the time. Therefore, the duty-cycled mobile station is more flexible in balancing the energy consumption and downlink latency and hence suitable for battery-operated and energy-harvesting maritime IoT devices with certain downlink latency requirements.

5.7.3 Category 3 Mobile Stations

This mobile station category is the same as Category 2, except that it has higher transmit power (>500 mW) and the capability of higher-order modulations up to 64-QAM, which requires a less than 8% EVM of a transceiver. It is also capable of full-duplexing, i.e., bi-directional communication at the same time.

5.7.4 Category 4 Mobile Stations

This mobile station category is similar to Category 3, except that its pending downlink message is not notified through PCH. Instead, it continuously checks the downlink assignment channel, ACH, for downlink data destined to it on the traffic channel, DTCH, eliminating the paging delay at the cost of power efficiency. It has the capability of spectrally efficient higher-order modulation up to 256-QAM, which requires a less than 4% EVM transceiver. Mobile stations of this type have the highest transmit power (e.g., greater than 1 W or 30 dBm) and are typically provided with continuous power supply from, e.g., a mains supply aboard a large ship.

References

1. Masami Kihara, *Digital Clocks for Synchronization and Communications* (Artech, 2002)
2. J. Proakis, M. Salehi, *Digital Communications* (McGraw-Hill Education, 2007)
3. B. Sklar, *Digital Communications: Fundamentals and Applications* (Pearson, 2013)
4. M. Cohn, A. Lempel, On fast M-sequence transforms. IEEE Trans. Inf. Theory **23**, 135–137 (1977)
5. A. Nimbalker, Y Blankenship, B. Classon, T. Fuja, T. K. Blankenship, ARP and QPP Interleavers for LTE Turbo Coding, in Proc. IEEE Wireless Communications and Networking Conference, 2008
6. M.M. Wang, W. Xiao, T. Brown, Soft decision metric generation for QAM with channel estimation error. IEEE Trans. Commun. **50**, 1058–1061 (2002)
7. M.M. Wang, L. Xiao, T. Brown, M. Dong, Optimal symbol timing for OFDM wireless commun. IEEE Trans. Wirel. Commun. **8**, 5328–5337 (2009)
8. M.M. Wang, A. Agrawal, A. Khandekar, S. Aedudodla, Preamble design, system acquisition, and determination in modern OFDMA cellular communications: an overview. IEEE Commun. Mag. **49**(7), 164–175 (2011)

Chapter 6
Radio Spectrum

Abstract In the first part of the book, we have gone to great lengths to develop the maritime MTC concept, architecture, and framework. In the second part, we materialize this concept through a detailed practical design example based on the radio spectrum assigned by ITU for *e*-Navigation. The radio spectrum is undoubtedly the most critical component for any practical wireless communication system and is more so for maritime MTC, whose worldwide existence is contingent on the "internationality" of the radio spectrum. Therefore, the spectrum must be assigned and managed on a global scale through an international regulatory agency, like ITU. Since the spectrum characteristics (such as frequency, bandwidth, and in-band and out-of-band interference) have a direct tie with the characteristics of the radio channel, and ergo the air interface, it is imperative to determine the radio spectrum and understand its characteristics before plunging into the design. Otherwise, it would be like designing a house without knowing the land it will sit on. (Well, you can, but there is a good chance it will not be a good fit.) This chapter elaborates on the international assignment of radio spectrum for maritime MTC and explores the solutions focusing on interference, regulations, and sharing techniques.

Keywords Maritime IoT · Radio spectrum allocation · Allotment · Assignment · Regulation · Spectrum sharing · Maritime mobile communication band · VHF frequency band · VDES frequency band · AIS · ASM · VDE-SAT · VDE-TER · FDD band · TDD band · Power flux density mask

6.1 Spectrum and Interference

If maritime MTC is the backbone of maritime IoT, radio frequency spectrum is, by all means, the lifeblood of maritime MTC. As noted in Chap. 5, the radio frequency spectrum (or simply radio spectrum for short) is the continuum of frequencies of electromagnetic radiation, roughly from 3 kHz to 300 GHz.

M. M. Wang, J. Zhang, *Machine-Type Communication for Maritime Internet-of-Things*, Wireless Networks, https://doi.org/10.1007/978-3-030-77908-5_6

The radio spectrum is a limited natural resource, like land, water, forests, minerals, and energy resources. Even though the electromagnetic spectrum is upper-unbounded, its applicability to wireless communication is limited mainly due to the fact that higher frequencies result in increased path losses according to Friis' equation and that the effects of blockage and shadowing at higher frequencies become more significant. However, it has been and will continue to be in increasing demand from a large and growing number of services, mainly: (1) government and research use, which include defense, public safety (for the safety of life on land, in the skies, and at sea), transport, and radio astronomy; (2) commercial services to the public (e.g., mobile telephony, mobile Internet); and (3) industrial, scientific, and medical services.

6.1.1 Spectrum Management

The scarcity of radio spectrum requires it to be carefully managed for efficiency. Spectrum management is the oversight of electromagnetic radiation characteristics, which vary with frequency, and even space and time. The objective is to enable the optimum number and types of service to share and coexist in the same spectrum while preventing harmful interference. Like the land use that involves zoning regulations and restrictions, the radio spectrum regulator leverages several mechanisms to make the spectrum available for communication services through allocation, allotment, and assignment.

Through zoning, the land is divided into residential, commercial, and industrial districts (or zones) with specific requirements as to the type of buildings allowed and restrictions on accessory buildings, building setbacks from the streets, and the size and height of buildings. Likewise, through spectrum allocation, the entire radio spectrum is divided into frequency bands identified by an upper and lower frequency limit, each of which is earmarked for specific uses under specified conditions such as emission power and spectral masks. These uses are referred to as radio services, e.g., fixed, mobile, broadcasting, amateur, satellite, radio positioning, radio navigation, radio astronomy, meteorology, emergency, environmental monitoring, and much more.

Each frequency band possesses unique physical characteristics. For example, line-of-sight is the most pronounced propagation mode at the VHF band (30 MHz to 300 MHz) and above. Unlike the lower frequency waves, these waves can penetrate the earth's atmosphere and reach the orbiting satellite and hence have earned the nickname of "space wave." Yet, compared to the higher frequency waves of this kind, VHF has the best propagation range and is thus optimal for communications at sea and in space, of which the spectrum that covers 156 MHz to 174 MHz is internationally allocated for maritime mobile communications at sea.

Unlike other natural resources (such as land and minerals), the radio spectrum is spatially reusable. However, radio propagation does not stop at national boundaries as there are no borders to the radio waves, thereby giving rise to interference between neighboring countries, much like the natural resource air. Just as transboundary flows of polluted air occur between neighboring countries, the radio spectrum is degradable due to various forms of "pollution" or mismanagement. Effective spectrum management is thus necessary, which requires regulation not only at the national level but also at a global level. Thus, another goal of spectrum management is to mitigate radio spectrum pollution and maximize the benefit of the usable electromagnetic spectrum, done via spectrum allotment and coordination at regional and global levels. Through allotment, portions of an allocated frequency band are designated to individual countries or geographical areas for a particular radio communication service, managed by different administrations, so that interference is prevented along borders. For satellite communication services, specific orbital positions may also be included in the allotment. Like the radio spectrum, the earth orbits are limited natural resources.

An assignment is the final subdivision of a spectrum: a license of a frequency band to an assignee or licensee (a service or a group service). The issue of the associated license to the licensee gives the licensee the authority for transmitting at the assigned frequency and the assigned bandwidth of emission for stated purposes using stated emission parameters. Some assignments are exclusive, meaning a licensed band allows the exclusive use of particular frequencies in particular geographic areas. When a service is granted the right to communicate at a specific band and in certain locations, any other service is prohibited from using that band in these locations. Other assignments are issued on a nonexclusive basis, requiring the users to cooperate on an informal basis; for instance, a *secondary user* should, in principle, avoid "harmful" interference to the *primary user*. Some assignments are available for use without a license. Users on the unlicensed bands do not have exclusive use of the spectrum and are subject to interference.

Most countries deem the radio spectrum a national resource, hence an exclusive property of the state, and can make rules about spectrum usage over their jurisdiction. For instance, in the United States of America, the spectrum is jointly managed by the Federal Communications Commission (FCC) for non-governmental applications and by the National Telecommunications and Information Administration (NTIA) for governmental applications. They are independent US government agencies overseen by Congress, responsible for implementing and enforcing America's communications law and regulations. The FCC administers spectrum for non-Federal use, i.e., state, local government, commercial, private internal business, and personal use, including radio frequency use, media responsibility, and public safety across the United States, whereas the NTIA administers spectrum for Federal use, e.g., use by the Army, the FAA, and the FBI [1].

The European Electronic Communications Committee (ECC) is an organization of the European Conference of Postal and Telecommunications Administrations, coordinating commercial, operational, regulatory, and technical standardization issues among its 48 member countries. The ECC develops standard European policies and regulations in electronic communications and related applications in the European context, taking account of European and international legislation and regulations. Its primary objective is to harmonize the efficient use of the radio spectrum and satellite orbits across Europe [2].

On top of the national and regional regulatory agencies, like FCC and ECC, is the International Telecommunication Union or ITU – the United Nations specialized agency for information and communication technologies, currently having 193 member states. ITU coordinates the shared global use of the radio frequency spectrum, including satellite orbits, and oversights the efficient use of these natural resources by all radiocommunication services, all through the World Radiocommunication Conferences (WRC) held every 3–4 years. Furthermore, ITU produces standards covering all fields of telecommunication and information communication technology, establishing "recommendations" to assure the necessary performance and quality in operating radio communication systems, which become mandatory only when adopted as part of a national/regional law.

6.1.2 Interference

Due to the radio spectrum scarcity, sharing is fundamental to communications, wired or wireless. However, the broadcast nature of wireless transmissions makes interference ubiquitous and inevitable, which deteriorates communication reliability and quality of service provisioning. Interference management is thus crucial and challenging in wireless communications.

Interference is present at every level of wireless communication on the radio spectrum.

At the spectrum level, for communications on an unlicensed spectrum, the coexistence of different systems and technologies makes interference unpredictable (random) and essentially unavoidable. For communications on a licensed spectrum designated to specific services, interference is mitigated through air interface standardization so that the transmission behavior becomes more predictable and the interference more controllable to some manageable extent. However, when the spectrum is allocated geographically to different kinds of services, for instance, terrestrial services and maritime services, interference between these services and systems could still exist.

Interference exists even within the same system. For distributed or autonomous medium access control, the transmission is controlled autonomously by individual stations. Without knowing the other station's intentions beforehand, transmissions are conducted in a random contention-based fashion, and severe interference caused by frequent collisions is inevitable, as noted in Chap. 2. For centralized medium access control, the transmissions are controlled by the control station so that intra-cell interference is maximally avoided or controlled systematically by a control station through the MAC resource assignment or scheduling function at the data link layer. Transmissions on both downlink and uplink are assigned resources for communications between the control station and mobile stations. The assigned resources for uplink and downlink traffic are typically orthogonal to avoid interference. Nevertheless, inter-cell interference is present, especially at the edge of a cell, because cells share the same spectrum for spatial spectrum reuse. Coordination between neighboring cells through static or dynamic resource partitioning or coordinated transmission and reception is possible for inter-cell interference mitigation at the network layer.

The interference mentioned above is caused by more than one transmission from different stations on the same spectrum and hence is called co-channel or *co-frequency interference*. Since the radio spectrum is divided into frequency bands and channelized into *frequency channels*, interference is present between frequency channels, caused by unwanted or unintentional out-of-band emissions, i.e., the adjacent channel interference touched upon in Chaps. 2 and 5. It is unavoidable by virtue of the practically unbounded emission spectrum and imperfect isolation between a transmitter and a receiver operating simultaneously on two adjacent channels. The severity of the adjacent channel interference is accentuated by the near-far effect—the effect of a strong signal from a near source in making it hard for a receiver to hear a weaker signal from a distant source. This is a direct result of the inverse-square law of Frii's equation, a fact that even a small difference in the distance creates a huge difference in receive signal strength, capable of making a nearby adjacent channel leakage bigger than the distant intended signal.

This chapter examines the assignment of such a spectrum dedicated to the *e*-Navigation radio services within the VHF maritime mobile communication band, most recently by ITU, in alignment with the maritime MTC concept developed in the previous chapters. This radio spectrum is depicted in Fig. 6.1. The realization of such a concept on this particular assignment is referred to as the VHF Data Exchange (VDE) System or simply VDES.

This assignment seems to be ordinary and straightforward; however, there are many practical issues due to (1) interference emanated from the coexistence of multiple air interfaces (including the legacy AIS) and (2) potential interference to incumbent *land* systems, which warrant special attention and effective solutions.

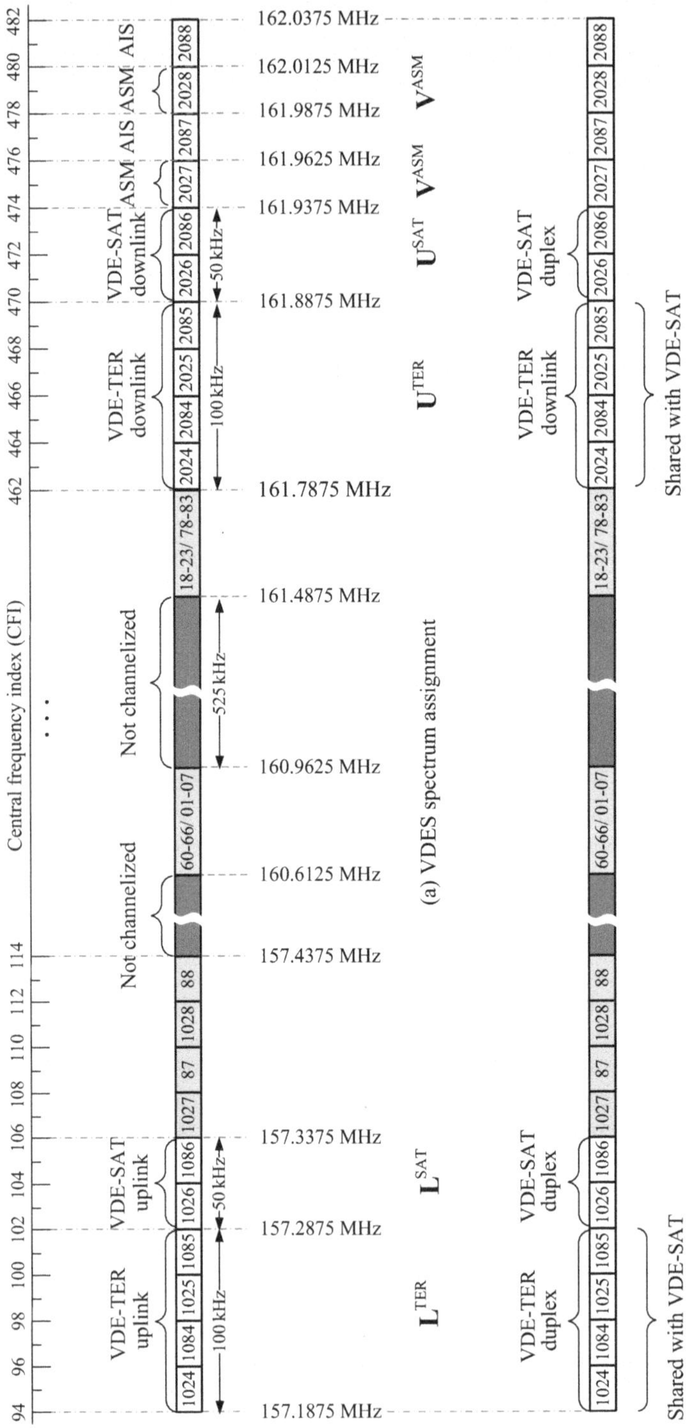

Fig. 6.1 Radio spectrum allocation for VDES, where the maritime mobile spectrum is channelized into 25-kHz individual frequency channels. Recall from Chap. 2, a frequency channel is defined as duplex if both uplink (from mobile stations to a control station) and downlink (from a control station to mobile stations) transmissions are allowed, whereas a simplex channel only allows one-way transmissions, either uplink or downlink

6.2 The Proximity Communication Spectrum

As shown in Fig. 6.1, the VHF maritime mobile communication band is channelized into 25-kHz individual subbands or frequency channels, of which Channels 2027 and 2028, denoted as $\mathbf{V}_{475}^{\mathrm{ASM}}$ and $\mathbf{V}_{479}^{\mathrm{ASM}}$, are dedicated to the maritime proximity communication, coined Application Specific Messages or ASM under the VDES terminology.

Next to these two ASM channels, Channels 2087 and 2088, are the legacy Automatic Identification System (AIS) channels, formerly introduced by ITU and IMO in the mid-1990s to provide vessel identification, position reporting, and tracking (see Chap. 1).

AIS can be deemed an air interface for proximity communication, although its original intention was more than proximity-based services. However, its outdated design hinders its use in maritime IoT applications (details in Chap. 7). A new proximity air interface with the enhanced communication capability for improved spectral efficiency and capacity like ASM, structured in alignment with the maritime MTC architecture, is imperative. Nonetheless, as a legacy air interface, AIS will continue to exist and will be protected.

The ASM spectrum assignment side-by-side with AIS presents a challenge in the air interface design. The out-of-band emission from an ASM transmitter may block the co-located or nearby AIS receiver due to the huge potential difference in signal strength between the ASM and AIS signals at the AIS receiver, i.e., the near-far effect in wireless communication. The out-of-band emissions from ASM may enter the AIS band and affect the whole receive path, degrading its sensitivity. The adjacent channel interference issue is one of the main topics of this chapter and remains so in Chap. 7.

6.3 The Terrestrial Communication Spectrum

In the same maritime mobile communication band, paired frequency channels 24, 84, 25, and 85 are assigned to the maritime MTC terrestrial air interface, referred to as VDE-TER under VDES, all as simplex channels as shown in Fig. 6.1a. A simplex channel allows for "one-way" transmission only, either uplink or downlink, recalling that the terrestrial component of VDES employs the infrastructure-based communication mechanism for a reason explained in Chap. 2. This traditional paired assignment facilitates frequency-division duplexing or FDD (cf. Chap. 2), which requires two simplex frequency channels at each communicating end, one for transmit and one for receive with the remote end configured as the opposite.

Following the FDD spectrum allocation tradition, the lower leg, i.e., channels 1024, 1084, 1025, and 1085, is assigned for uplink transmissions, denoted as

$$\mathbf{L}^{\mathrm{TER}} \triangleq \{1024, 1084, 1025, 1085\},$$

and the upper leg, i.e., channels 2024, 2084, 2025, and 2085, is for downlink,

$$\mathbf{U}^{\mathrm{TER}} \triangleq \{2024, 2084, 2025, 2085\}.$$

There is a reason for assigning the lower leg to the uplink in typical FDD spectrum allocations. The lower frequency is advantageous in the propagation range, which benefits the mobile station that is typically power- and energy-limited.

6.4 The Satellite Communication Spectrum

Spectrum allocation for the maritime MTC satellite component (i.e., VDE-SAT) is more complex than that for VDE-TER, mainly due to the satellite *downlink* interference to the incumbent land systems on the same frequency band.

As shown in Fig. 6.1a, paired frequency channels 26 and 86 are assigned to VDE-SAT, with the lower leg including channels 1026 and 1086 denoted as,

$$\mathbf{L}^{\mathrm{SAT}} \triangleq \{1026, 1086\},$$

and the upper leg, channels 2026 and 2086,

$$\mathbf{U}^{\mathrm{SAT}} \triangleq \{2026, 2086\}.$$

Both are *simplex* channels, facilitating the traditional FDD transmissions with $\mathbf{L}^{\mathrm{SAT}}$ for the uplink and $\mathbf{U}^{\mathrm{SAT}}$ the downlink.

As noted before, the radio spectrum as a natural resource is geographically owned by nations, just like other natural resources. Although the VDE-SAT downlink channel is on the international VHF maritime mobile band (156–174 MHz), *on land*, this same band in most countries is allocated to conventional and trunked land mobile systems by safety agencies, and utilities and transportation companies, e.g., police, fire, ambulance services, dispatched services, and railroad services. Many businesses and industries throughout the world use land mobile services as their primary means of communication, especially from a fixed location to mobile users (i.e., from a base station to a fleet of mobile stations). Therefore, coordination between the VDE-SAT systems and the *victim* systems (i.e., the co-frequency incumbent land systems) is of utmost importance.

The challenge lies in the fact that there is no existing regulatory rule directly established for the protection of the land system against the satellite system or the like, and hence the assessment of the potential impact on the incumbent land systems (primary users) becomes difficult if not impossible. A method adopted in the current analysis is to place general restrictions on the emissions from space stations (secondary users) inferred from the existing regulatory rules for interference protection between legacy land systems specified by ECC and ITU. The restrictions are expressed in terms of values of the maximum allowed electromagnetic *power flux density* (PFD) emitted by any space station to the earth's surface at all possible incident angles in a reference bandwidth. In a nutshell, it serves as a "protection mask" for the land system such that the *actual* interference that the land system experiences is no worse than that from a *land* mobile system permitted by these already existing regulations.

This concept is graphically illustrated in Fig. 6.2, henceforth referred to as the *electromagnetic PFD mask* or simply *PFD mask.* The PFD mask is derived from the original land system regulatory restrictions and used as the constraint on the emission energy from the space stations for protection against harmful interference to the incumbent co-frequency land communication systems. The derivation follows.

6.4.1 PFD Mask Derivation

Specifically, ECC recommends an *indicative coordination threshold* for interference avoidance among land mobile systems from neighboring countries. For the VDE-SAT downlink frequency band, $\mathbf{U}^{\mathrm{SAT}}$, the corresponding coordination field strength is 12 dBμV/m per 25 kHz.

ITU recommends an *energy per bit to noise plus interference* power spectral density ratio of 10 dB for digital land mobile systems and a typical *total received power to the unwanted power* (SINAD) value of 12 dB for analog land mobile systems for establishing degradation protection in the frequency band between 138 and 174 MHz.

The corresponding protection masks or PFD masks inferred from the corresponding ECC and ITU regulations, i.e., $\widehat{\Psi}^{\mathrm{FSC}}(\theta)$ and $\widehat{\Psi}^{\mathrm{C/I}}(\theta)$, are derived next.

6.4.1.1 Field Strength Constraint (FSC)

The ECC recommendation T/R 25-08 defines a field strength threshold for interference coordination between land mobile systems from neighboring countries in the frequency band from 29.7 to 470 MHz. Five indicative thresholds with a reference bandwidth of 25 kHz are specified, each of which indicates the allowed average interference field strength, E, impinging on a land station antenna from an interfering land station. It translates to an electromagnetic PFD value, ψ,

$$\psi = \sqrt{\sigma\mu^{-1}} \cdot E^2, \tag{6.1}$$

via Maxwell's equations, where σ and μ are the electric and magnetic constants, respectively.

For the VDES frequency range from 156.7625 to 162.0375 MHz, the corresponding field strength E is 12 dBμV/m per 25 kHz, which translates to a PFD value of $\psi = -104$ dBm/m^2 per 25 kHz from (6.1).

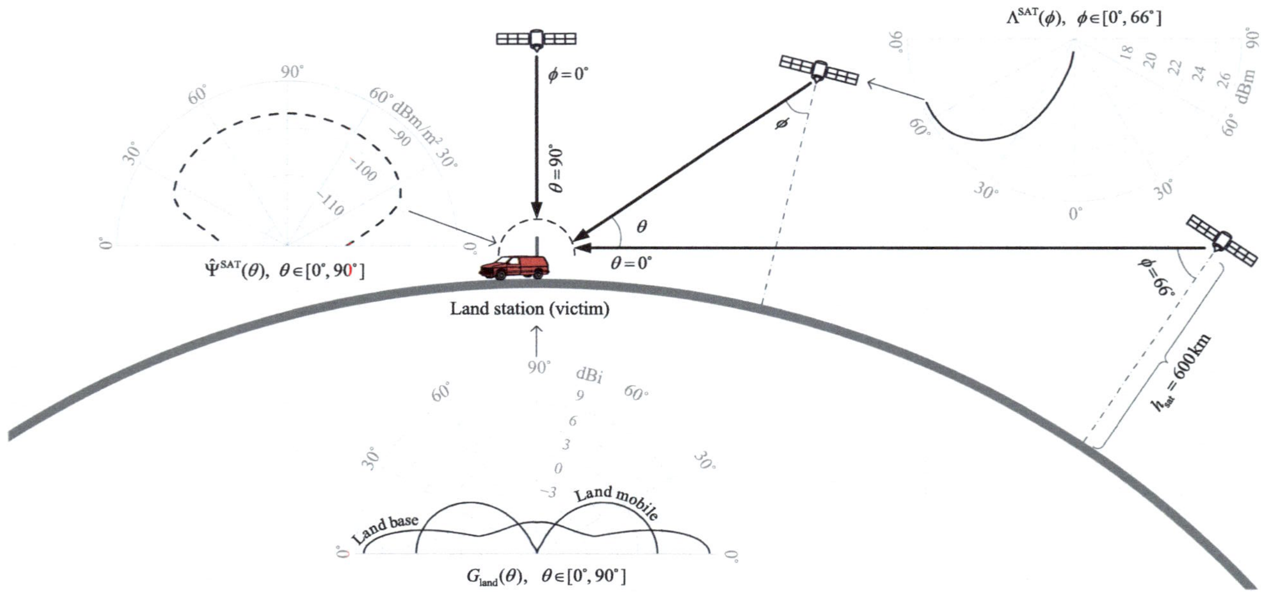

Fig. 6.2 Illustration of the concept of a "PFD mask" for protection of the *land* stations: the interference into a co-frequency land station receiver from a VDE-SAT space station (the aggressor) at the elevation angle from $\theta = 0^\circ$ to 90°, corresponding to the off-nadir angle from $\phi = 66^\circ$ to 0°, where the EIRP (per 25 kHz) of a typical antenna (Yagi antenna) of the space station is constrained by the PFD mask (per 25 kHz)

The corresponding interference from an aggressor (e.g., the VDE-SAT space station) into the victim land station receiver depends on the antenna gain of the station, which can be shown to be

$$I_{\text{land}}(\theta) = \psi \frac{\lambda^2 \iota^{-1}}{4\pi} G_{\text{land}}(\theta), \tag{6.2}$$

where λ is the wavelength of the interference signal, ι the receiver feeder loss, and $G_{\text{land}}(\theta)$ the antenna gain of a receiving land station at a given elevation angle, θ. As noted in Chap. 5 (Fig. 5.4), an elevation angle is the angle between the horizontal plane and the line of sight, measured in the vertical plane. An average antenna gain pattern (relative to an isotropic antenna) for both land mobile and base stations is provided in [3] and is plotted in Fig. 6.2.

The resultant interference that the victim actually sees depends on the incident angle of the interfering electromagnetic wave. (In this book, for the sake of simplification, the incident angle is defined the same as the elevation angle.) Consequently, the largest interference from an aggressor into the victim's receiver permitted by FSC is given by

$$\begin{aligned} \widehat{I}_{\text{land}} &\triangleq \max_{\theta \in [0^\circ, 90^\circ]} I_{\text{land}}(\theta) \\ &= \psi \cdot \frac{\lambda^2 \iota^{-1}}{4\pi} \cdot \max_{\theta \in [0^\circ, 90^\circ]} G_{\text{land}}(\theta). \end{aligned} \tag{6.3}$$

Now, applying the same criterion to a VDE-SAT station (the aggressor) indicates that the PFD from a VDE-SAT space station must be less than

$$\begin{aligned} \Psi^{\text{FSC}}(\theta) &\triangleq 4\pi\lambda^{-2}\iota\vartheta_{\text{land}}^{\text{sat}} \cdot \widehat{I}_{\text{land}} \cdot G_{\text{land}}^{-1}(\theta) \\ &= \psi \cdot \vartheta_{\text{land}}^{\text{sat}} \cdot \overline{G}_{\text{land}}^{-1}(\theta), \end{aligned} \tag{6.4}$$

where $\vartheta_{\text{land}}^{\text{sat}}$ is the antenna polarization loss between the VDE-SAT space station transmitter (circular) and the victim land station receiver (vertical), typically 3 dB, and

$$\overline{G}_{\text{land}}(\theta) \triangleq G_{\text{land}}(\theta) / \max_{\theta \in [0^\circ, 90^\circ]} G_{\text{land}}(\theta) \tag{6.5}$$

is the normalized land mobile antenna gain. However, differently from the land system case (in which $\theta \approx 0^\circ$ since land service transceivers typically direct to the horizon), the elevation angle in the current case can be anywhere from 0 to 90°, i.e., $\theta \in [0^\circ, 90^\circ]$. Equation (6.4) indicates that, for VDE-SAT to comply with the FSC, the PFD that a VDE-SAT space station impinges upon the land systems must be less than $\Psi^{\text{FSC}}(\theta)$.

Substituting $G_{\text{land}}(\theta)$ in (6.4) with the respective antenna gains of the land mobile station and base station in Fig. 6.2, i.e.,$G_{\text{mobile}}(\theta)$ and $G_{\text{base}}(\theta)$, obtains cases for land mobile stations, i.e., $\Psi^{\text{FSC}}_{\text{mobile}}(\theta)$ and base stations, $\Psi^{\text{FSC}}_{\text{base}}(\theta)$, respectively. The maximum allowed PFD irradiating the earth's surface by a VDE-SAT satellite inferred from the FSC of ECC is thus whichever is more restrictive:

$$\widehat{\Psi}^{\text{FSC}}(\theta) \triangleq \min\left\{\Psi^{\text{FSC}}_{\text{mobile}}(\theta), \Psi^{\text{FSC}}_{\text{base}}(\theta)\right\}, \forall\theta \in [0^\circ, 90^\circ]. \tag{6.6}$$

6.4.1.2 Carrier-to-Interference Constraint (C/I)

The criterion used by ITU for protection between land mobile systems is from a very different perspective from that of ECC. It is from the standpoint of the ultimate system performance requirement.

For digital land mobile systems in the frequency band between 138 and 174 MHz, a bit error rate of 2–5% with 4-level frequency modulation (C4FM) or 4-FSK (see Chap. 5) is targeted, as specified by [4]. The corresponding required (or minimum) per bit signal-to-interference plus noise ratio (SINR) can be found in [5] as

$$\frac{E_{\text{b}}}{N_0 + I} \approx 10 \text{ dB}, \tag{6.7}$$

where E_{b} is the received energy per information bit and $N_0 + I$ is the interference (I) plus noise (N_0) power spectral density. Since C4FM modulation carries two bits per symbol, it translates to a minimum carrier-to-interference plus noise ratio of

$$\begin{aligned}\eta_{\text{d}} &\triangleq \frac{E_{\text{b}}}{N_0 + I} \cdot R_{\text{b}} \\ &= \frac{E_{\text{b}}}{N_0 + I} \cdot 2R_{\text{s}} \\ &\approx 13 \text{ dB},\end{aligned} \tag{6.8}$$

where R_{b} is the bit rate per Hz (bits/s/Hz) and R_{s} is the per Hz modulation symbol rate.

Apparently, the condition of (6.7) is met as long as

$$\frac{C}{N_0 + I} \geq \eta_{\text{d}} \tag{6.9}$$

where C is the received per channel carrier power spectral density. Therefore,

$$\frac{C}{I} \geq \left(\eta_{\text{d}}^{-1} - \left(\frac{C}{N_0}\right)^{-1}\right)^{-1}. \tag{6.10}$$

Table 6.1 Typical parameters of land mobile systems

Land station type	Base station		Mobile station	
	Digital	Analog	Digital	Analog
Modulation type	C4FM	FM	C4FM	FM
Output power (typical/min)	60W/20W	30W/5W	30W/1W	30W/1W
Antenna gain	8 dBi	8 dBi	2 dBi	2 dBi
Feeder loss	2 dB	2 dB	1 dB	1 dB
Antenna height	65 m		2 m	
Distance to horizon	29 km		5 km	
Free space loss	107 dB			
Additional path loss	34 dB			
Noise figure	7 dB			

The performance of an analog system is commonly measured by the ratio of the total received power to the unwanted power (SINAD). ITU recommends a typical SINAD value of 12 dB for establishing degradation protection for analog land mobile systems, i.e.,

$$\frac{C + N_0 + I + D}{N_0 + I + D} \geq \eta_a \tag{6.11}$$

or

$$\frac{C}{I} \geq \left((\eta_a - 1)^{-1} - \left(\left(\frac{C}{N_0} \right)^{-1} + \left(\frac{C}{D} \right)^{-1} \right) \right)^{-1}, \tag{6.12}$$

where C/D is the signal to distortion ratio, roughly 20 dB, and $\eta_a = 12$ dB.

According to the representative parameters of technical and operational characteristics of conventional and trunked land mobile systems operating in the frequency band 138–174 MHz, listed in Table 6.1, the typical antenna height of a base station and a mobile station is 65 m and 2 m, respectively, which determines the maximum line-of-sight distance (limited by the radio horizon) between the land base station and land mobile station, i.e., 29 km + 5 km = 34 km. This distance leads to a maximum free space loss of $\widehat{\ell} \approx 107$ dB in the frequency band from 156.7625 to 162.0375 MHz (cf. Fig. 6.1) according to Friis' equation in Chap. 5, plus an *additional* path loss (land path in the VHF band), estimated as $\Delta\ell \approx 34$ dB [6], giving rise to a maximum total propagation loss of

$$\begin{aligned} \widehat{L}_{\text{land}} &= \widehat{\ell} + \Delta\ell \\ &\approx 141 \text{ dB}. \end{aligned} \tag{6.13}$$

Further taking into account the receiver feeder loss ι, the minimum carrier power received by the land station receiver, i.e., the receiver sensitivity, is

$$C_{\min} = P \cdot \iota^{-1} G_{\text{base}}(\phi) \cdot \widehat{L}_{\text{land}}^{-1} \cdot \iota^{-1} G_{\text{mobile}}(\theta), \tag{6.14}$$

where P is the minimum transmit power of a land transmitter, $G_{\text{base}}(\phi)$ and $G_{\text{mobile}}(\theta)$ are the respective antenna gains (excluding the feeder loss) of a land base station and a land mobile station. Here, the land base station off-nadir angle is $\phi \approx 90^\circ$ and the mobile station elevation angle $\theta \approx 0^\circ$, recalling that land transceivers direct to the horizon at near 0°elevation angle (mobile station) or 90°off-nadir (base station).

Given the 15 kHz channel bandwidth and the noise figure ξ of 7 dB of the legacy land stations (cf. Table 6.1), the system noise level is

$$\begin{aligned} N_0 &= \kappa T \xi \\ &= -125 \text{ dBm}/15 \text{ kHz}, \end{aligned} \tag{6.15}$$

where $T = 290$ K and κ is the Boltzmann constant.

For the digital land systems, the corresponding minimum carrier to interference ratio is

$$\eta^{\text{ITU-d}} \triangleq \left(\eta_{\text{d}}^{-1} - \left(\frac{C_{\min}}{N_0} \right)^{-1} \right)^{-1}, \tag{6.16}$$

and the maximum allowed interference level follows as

$$\widehat{I}^{\text{ITU-d}} = \frac{C_{\min}}{\eta^{\text{ITU-d}}}, \tag{6.17}$$

which is -104 dBm per 15 kHz and -118 dBm per 15 kHz for land mobile stations and land base stations, respectively.

Similarly, for the analog land systems, we define

$$\eta^{\text{ITU-a}} \triangleq \left((\eta_{\text{a}} - 1)^{-1} - \left(\left(\frac{C_{\min}}{N_0} \right)^{-1} + \left(\frac{C_{\min}}{D} \right)^{-1} \right) \right)^{-1}, \tag{6.18}$$

and the maximum allowed interference per channel is therefore

$$\widehat{I}^{\text{ITU-a}} = \frac{C_{\min}}{\eta^{\text{ITU-a}}}, \tag{6.19}$$

which is -109 dBm per 15 kHz and -117 dBm per 15 kHz for land mobile and base stations, respectively.

Combining the digital and analog scenarios, we have the maximum permitted interference caused by an interfering land station,

$$\widehat{I}^{\mathrm{C/I}} = \min\left\{\widehat{I}^{\mathrm{ITU\text{-}d}}, \widehat{I}^{\mathrm{ITU\text{-}a}}\right\}, \tag{6.20}$$

i.e., −109 dBm per 15 kHz for a victim mobile station and −118 dBm per 15 kHz for a base station.

If we take into consideration the polarization loss between the space station antenna and the land station antenna, for the case where the interfere is a satellite space station, the PFD constraint inferred from (6.7) and (6.11) is therefore

$$\Psi^{\mathrm{C/I}}(\theta) = 4\pi\lambda^{-2}\imath\vartheta_{\mathrm{land}}^{\mathrm{sat}} \cdot \widehat{I}^{\mathrm{C/I}} \cdot G_{\mathrm{land}}^{-1}(\theta), \forall\theta \in [0^{\circ}, 90^{\circ}]. \tag{6.21}$$

Substituting different antenna gains into (6.21) produces the VDE-SAT PFD masks corresponding to both land mobiles and base stations, i.e.,

$$\Psi_{\mathrm{mobile}}^{\mathrm{C/I}}(\theta) = 4\pi\lambda^{-2}\imath\vartheta_{\mathrm{mobile}}^{\mathrm{SAT}} \cdot \widehat{I}^{\mathrm{C/I}} \cdot G_{\mathrm{mobile}}^{-1}(\theta), \tag{6.22}$$

and

$$\Psi_{\mathrm{base}}^{\mathrm{C/I}}(\theta) = 4\pi\lambda^{-2}\imath\vartheta_{\mathrm{base}}^{\mathrm{SAT}} \cdot \widehat{I}^{\mathrm{C/I}} \cdot G_{\mathrm{base}}^{-1}(\theta), \forall\theta \in [0^{\circ}, 90^{\circ}]. \tag{6.23}$$

The PFD governed by (6.7) and (6.11) is then determined by

$$\widehat{\Psi}^{\mathrm{C/I}}(\theta) \triangleq \min\left\{\Psi_{\mathrm{mobile}}^{\mathrm{C/I}}(\theta), \Psi_{\mathrm{base}}^{\mathrm{C/I}}(\theta)\right\}, \theta \in [0^{\circ}, 90^{\circ}]. \tag{6.24}$$

To ensure that the design of VDE-SAT meets both ECC and ITU requirements, the electromagnetic PFD irradiating the surface of the earth from the VDE-SAT satellite must satisfy the ultimate mask deduced from these two requirements, i.e.,

$$\widehat{\Psi}^{\mathrm{SAT}}(\theta) \triangleq \min\left\{\widehat{\Psi}^{\mathrm{FSC}}(\theta), \widehat{\Psi}^{\mathrm{C/I}}(\theta)\right\}, \forall\theta \in [0^{\circ}, 90^{\circ}], \tag{6.25}$$

where θ is the elevation angle of a land station, as plotted in Fig. 6.2. It serves as the protection mask against the interference from the space station from the perspective of a victim land station. It guarantees the satisfaction of the ECC and ITU regulatory constraints in the sense that a VDES space station produces no more impairment on the performance of a victim land system than any other co-frequency land system would.

To conform to the ultimate PFD mask in (6.25), the effective isotropic radiated power (EIRP) of the satellite space station as a function of the off-nadir angle of the space station, ϕ, must satisfy

$$\Lambda^{\mathrm{SAT}}(\phi) \leq \frac{1}{4\pi}\lambda^{2}\ell_{\mathrm{earth}}^{\mathrm{sat}}(\phi) \cdot \widehat{\Psi}^{\mathrm{SAT}}(\theta), \quad \forall\theta \in [0^{\circ}, 90^{\circ}], \tag{6.26}$$

where λ is the wavelength, and $\ell_{\mathrm{earth}}^{\mathrm{sat}}(\phi)$ is the propagation loss between the VDE-SAT space station and a land station receiver on the surface of the earth at an off-nadir angle of ϕ corresponding to the elevation angle θ of the land station. Denoting R_{earth} as the radius of the earth, h_{sat} the orbital altitude of the space station, the relationship between ϕ and θ can be expressed by

$$\phi = \arcsin\left(\left(1+\frac{h_{\text{sat}}}{R_{\text{earth}}}\right)^{-1}\cos(\theta)\right),\ \ \forall\theta \in [0^\circ, 90^\circ]. \qquad (6.27)$$

Clearly, $\phi \in [0^\circ, 66^\circ]$ for a LEO satellite orbiting at h_{sat}= 600 km, as depicted in Fig. 6.2.

6.5 Spectrum Sharing

As aforementioned in Chap. 2, VDE-TER is mainly for limited coastal areas, i.e., for nearshore deployments; its spectrum, $\{\mathbf{L}^{\text{TER}}, \mathbf{U}^{\text{TER}}\}$, is thus largely underutilized, leading to an inefficient and wasteful use of scarce spectral resources. To maximize the utilization of the extremely precious VHF spectrum, the VDES spectrum assignment allows VDE-SAT to utilize the VDE-TER spectrum (both upper and lower legs) as long as the transmission does not, in principle, cause harmful interference to VDE-TER. In practice, however, it is non-trivial for real-time coordination of a satellite space station with the shore stations in its field of view, alone its traveling path over the world. In the current design, we take advantage of the PFD mask concept to negate the need for such coordination.

Recall that the PFD mask concept ensures that the interference from the VDE-SAT *downlink* can be absorbed by the incumbent land systems. Naturally, this assurance should also remain true for VDE-TER by design. In fact, it can be factored into the VDE-TER link budget (Chap. 8). A VDE-SAT space station is hence free to transmit on the VDE-TER downlink channels, $\mathbf{U}^{\text{TER}}$, any time anywhere as long as the PFD mask given by (6.25) or the EIRP constraint specified in (6.26) is satisfied. As illustrated in Fig. 6.3, a space station is free to use both VDE-SAT and VDE-TER spectrum, i.e., $\mathbf{U}^{\text{SAT}}$ and $\mathbf{U}^{\text{TER}}$, *anywhere* for *downlink* transmissions with the constrained EIRP in (6.26).

As for the VDE-SAT *uplink*, the mobile station may *opportunistically* use the VDE-TER uplink frequency channel, $\mathbf{L}^{\text{TER}}$, for VDE-SAT uplink transmissions wherever the absence of the VDE-TER service can be assured (e.g., offshore). For that, an effective mechanism to facilitate such assurance must be embedded in the air interface design. An example of such a design is depicted in Fig. 6.3, where the presence of a VDE-TER system is indicated by a special "beacon signal" transmitted on the VDE-TER downlink channel, $\mathbf{U}^{\text{TER}}$. Only outside the beacon coverage is the VDE-TER uplink channels, $\mathbf{L}^{\text{TER}}$, available for VDE-SAT uplink transmissions. The range of the beacon is extended beyond the VDE-TER cell coverage to create a "buffer zone" that protects the VDE-TER uplink, especially for those of mobile stations at the edge of the cell. The buffer zone should be sufficiently large such that

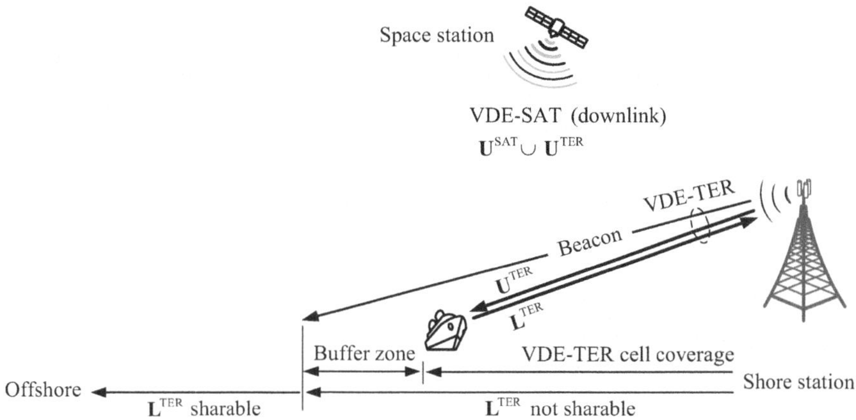

Fig. 6.3 VDE-TER spectrum sharing scheme with VDE-SAT: (1) $\mathbf{U}^{TER}$is sharable with VDE-SAT downlink anywhere; and (2) $\mathbf{L}^{TER}$cannot be shared in the no-share zone

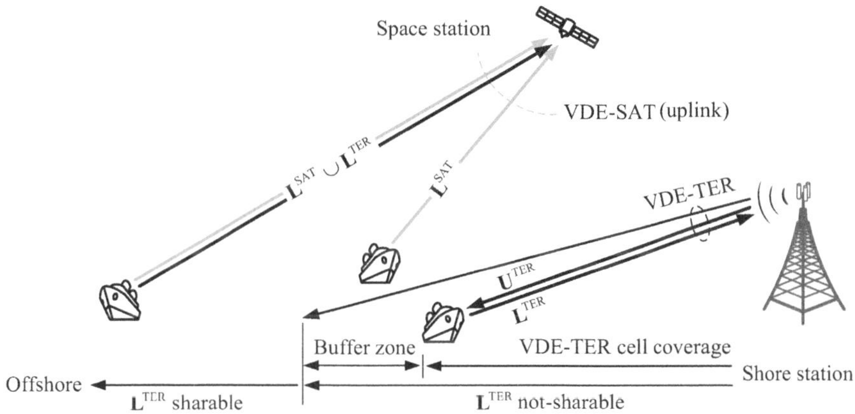

Fig. 6.4 Illustration of VDE-TER spectrum sharing with VDE-SAT *uplink*. Outside the "no-share zone" a mobile station is free to use both VDE-SAT spectrum and VDE-TER spectrum, i.e.,{$\mathbf{L}^{SAT}$, $\mathbf{L}^{TER}$}, for uplink transmissions; whereas within the no-share zone, $\mathbf{L}^{TER}$ is restricted from use in VDE-SAT uplink

the VDE-SAT signals on $\mathbf{L}^{TER}$ seen by the shore station are below the noise level. For that, the VDE-SAT downlink transmission should avoid overlapping with the beacon signals to assure that the beacon is protected for maximum detectability. An ideal candidate for such a beacon signal is the downlink system broadcast signal, i.e., the Bulletin Board broadcast signal, PBBCH. Chapter 8 provides great details on the design.

As illustrated in Fig. 6.4, a mobile station thus refrains itself from transmitting VDE-SAT signals on $\mathbf{L}^{TER}$ whenever a VDE-TER beacon signal is detected;

instead, the dedicated VDE-SAT channels, $\mathbf{L}^{\mathrm{SAT}}$, is preferred for VDE-SAT uplink transmissions if satellite communication is needed within the no-share zone. Apparently, the buffer zone diminishes as the VDE-TER cell radius becomes limited by the "radio horizon," and the no-share zone coincides with the cell coverage. In that case, the VDE-TER uplink is naturally protected by the horizon.

Indeed, the restriction of VDE-SAT uplink transmissions on VDE-TER spectrum in the "no-share zone" protects the VDE-TER system; however, it does not mean that the VDE-SAT uplink transmission on VDE-TER uplink spectrum, $\mathbf{L}^{\mathrm{TER}}$, in the "share zone" is free of "interference" from the VDE-TER uplink transmissions since a space station has a much larger field of view than that of a mobile station on earth. A space station is thus responsible for monitoring the interference on the VDE-TER uplink channels, $\mathbf{L}^{\mathrm{TER}}$, and scheduling the appropriate resources, and modulation and coding scheme or MCS for the VDE-SAT uplink transmissions, exploiting the centralized resource control mechanism that VDE-SAT employs.

This simple scheme provides a seamless spectrum sharing mechanism that minimizes the interference between these two systems and maximizes the spectrum usage and overall spectral efficiency and, most importantly, negates the need for coordination between space and shore stations. A detailed VDE-TER and SAT air interface design to enable this sharing mechanism is deferred to Chaps. 8 and 9.

Direct "mobile-to-mobile" (e.g., ship-to-ship) communication is allowed on the VDE-TER channels, $\{\mathbf{L}^{\mathrm{TER}}, \mathbf{U}^{\mathrm{TER}}\}$ in the absence of the VDES network (VDE-TER and VDE-SAT) coverage. The lower leg $\mathbf{L}^{\mathrm{TER}}$ is preferred to reduce the interference with the AIS/ASM channels and, more importantly, to protect the control station (shore or space station) transmissions, especially the system broadcast signals like PBBCH. To that end, mobile-to-mobile communication on the VDE-TER or SAT spectrum can only be engaged when there is no evidence of control station transmissions, but the resources for broadcasting system control information are excluded from transmission *at all times*. Mobile stations monitor these resources and refrain from transmission whenever the broadcast signal is detected.

6.6 FDD vs. TDD

As touched upon in Chap. 2, an important issue for FDD communication is protecting the receiver from the (co-located) transmitter at each communicating end. The *out-of-band* emissions may enter the receiver band and affect the whole receive path, degrading its sensitivity. This type of *cross-link interference* is caused by the leakage of the out-of-band emissions of a transmitter into the co-located receiver. A full-duplex FDD transceiver relies on the emission power spectrum mask, the isolation between the transmit and receive RF chains (including *spatial* separation) to bring the interference down to a manageable level. Since spatial separation is costly or physically difficult to achieve, most full-duplex FDD systems depend primarily on *frequency* separation between the two paired frequency bands to ease the transceiver design to the extent that the same antenna can be used for both transmit and receive with the help of an RF duplexer. This protection is especially crucial for systems operating in the lower end and middle section of the radio spectrum, including the VHF maritime mobile band that requires the use of sizable antennas.

6.6.1 VDE-TER

The frequency separation of 4.5 MHz between the uplink and downlink, i.e., the lower leg, $\mathbf{L}^{\mathrm{TER}}$, and the upper leg, $\mathbf{U}^{\mathrm{TER}}$, is barely sufficient in some deployments; nevertheless, it is still manageable for a shore station and large ships. For certain low-cost and dimension-limited maritime devices, it is up to the shore station to ensure the device operating in half-duplex via careful scheduling under the centralized resource allocation framework.

However, for VDES with its peculiar spectrum allocation, the protection of the VDE-TER receiver is not the only issue. The protection of the legacy AIS (on channels 2087 and 2088), as well as ASM (channels 2027 and 2028), is also of importance. Both are in close proximity to $\mathbf{U}^{\mathrm{TER}}$, 75 kHz apart from AIS and 50 kHz from ASM. For shore stations with *co-located* AIS and ASM receivers, the transmission of VDE-TER on $\mathbf{U}^{\mathrm{TER}}$ will severely interfere with its reception of AIS and ASM signals under such tight frequency separation, which is more difficult to deal with than the cross-link interference issue in FDD.

Specifically, assuming the receiver noise temperature is 30 dBK, hence the noise floor of -125 dBm per 25 kHz. The required isolation between the VDE-TER transmitter and the AIS receiver is thus at least 96 dB for a 5W per 25 kHz VDE-TER transmitter in order to knock the interference down to 6 dB below the noise floor, assuming 75-dBc attenuation within a 75-kHz stop band of the transmit spectrum.

Although such stringent RF isolation can be achieved using high-end RF filter systems combined with spatial separation of VDE-TER and AIS antennas (e.g., 10-m vertical separation may provide an extra 50 dB isolation), there still could be scenarios where such a degree of isolation between VDE-TER and AIS transceivers *co-located* at a shore station is too costly. Due to this concern, the simplex channels of the lower leg, $\mathbf{L}^{\mathrm{TER}}$, for VDE-TER uplink have been revised as *duplex* channels in the WRC 2019 (see Fig. 6.1b) to leave the time-division duplexing or TDD option available for this scenario. Referring back to Chap. 2, under TDD, a frequency channel is temporally divided into multiple time slots, and both downlink and uplink transmissions are duplexed onto the same frequency channel but in different time slots with the help of an RF switch. A shore station now at least has an option to use the lower leg, $\mathbf{L}^{\mathrm{TER}}$, for both uplink and downlink transmissions in a TDD fashion to avoid downlink transmissions on $\mathbf{U}^{\mathrm{TER}}$. The larger frequency separation between $\mathbf{L}^{\mathrm{TER}}$ and AIS greatly eases the isolation necessary to bring the interference to the co-located AIS receiver down to a more manageable level. However, the spectrum usage for VDE-TER is halved.

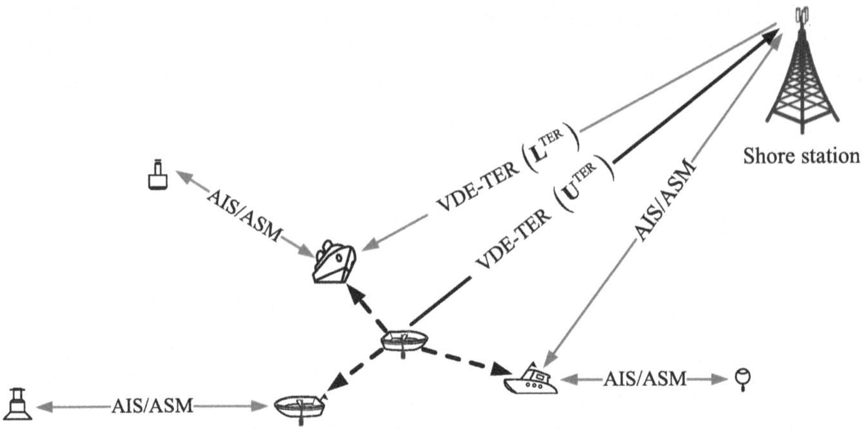

Fig. 6.5 Illustration of the potential adjacent-channel interference from the neighboring VDE-TER uplink, where the shore station is equipped with both VDE-TER and AIS transceivers. The dashed line denotes the interference of the VDE-TER uplink to the AIS/ASM reception, had the upper leg, $\mathbf{U}^{TER}$, been used for VDE-TER uplink transmissions

A natural question is then, why not simply use the lower leg, $\mathbf{L}^{TER}$, for downlink, and the upper leg, $\mathbf{U}^{TER}$, for uplink in the first place? The reason is that it would cause the same problem at the mobile stations and would be even more difficult to manage, considering the diverse types of mobile stations and the fact that a low-cost transceiver for mobile stations is a default. As illustrated in Fig. 6.5, the reception of the AIS/ASM signals could be interfered with by the out-of-band emissions from the VDE-TER uplink transmissions in the environment of dense traffic near shore, had $\mathbf{U}^{TER}$ been allowed for VDE-TER uplink transmissions. Simply leaving the "burden" to shore stations is a more manageable choice.

6.6.2 VDE-SAT

Like VDE-TER, VDE-SAT also has the out-of-band emission issue. However, the downlink interference to the AIS and ASM reception (as seen in VDE-TER) is no longer a problem since ASM is not meant for satellite space station reception, remembering that ASM is for proximity communication by design. The reason that the legacy AIS is not suitable for space station reception is that the `guard period` of the AIS transmission burst (0.6 ms) is far from sufficient for absorbing the propagation time differences between mobile stations at the space station receiver (e.g., 8 ms). We will see more details on the transmission burst when reviewing the AIS interface design in Chap. 7. Nevertheless, the FDD cross-link interference presents a more serious challenge for VDE-SAT than that for VDE-TER. As aforementioned, one important and difficult issue for FDD communication is the protection of the receiver from the co-located transmitter at each communicating end. The 4.6-MHz frequency separation between the uplink and downlink, i.e., the lower leg, $\mathbf{L}^{SAT}$, and the upper leg, $\mathbf{U}^{SAT}$, in Fig. 6.1a.y be fine for VDE-TER shore stations but could be problematic for VDE-SAT space stations.

From the EIRP constraint in Fig. 6.2, the space station transmit power can be up to 26 dBm/25 kHz−8 dBi = 8 dBm/25 kHz; more than 78 dB of isolation is thus needed in order to bring the out-of-band emissions from the output of the transmitter down to 6 dB below the noise floor, assuming the noise temperature is 26 dBK/25 kHz. This magnitude of isolation could be difficult for both payload-limited and dimension-limited LEO space stations (e.g., 20 kg in weight and 20 × 30 × 40 cm in dimension), especially in the 160-MHz frequency band. So, realistically, only half-duplex FDD can be supported for VDE-SAT, under which the space station and mobile stations take turns to transmit on the downlink and uplink frequencies, i.e., $\mathbf{U}^{SAT}$ and $\mathbf{L}^{SAT}$, respectively, thereby negating the need for a transceiver to transmit and receive simultaneously. Although the leakage from the transmitter to the receiver is avoided, the duty cycle per channel is halved, and the spectral efficiency is, in effect, cut down to half, which is detrimental to VDE-SAT, whose bandwidth is already thin.

A viable solution to circumvent this cross-link self-interference without sacrificing spectral efficiency is to employ the TDD transmission technique. With the *time* separation provided by TDD, the transceiver of a station never needs to transmit and receive at the same time (like half-duplex FDD), but still maintains full channel utilization, and yet allows to reuse or "time-share" the RF resources, as noted Chap. 2. This means a significant reduction in complexity and cost, although in this particular case, two sets of RF frontend filters/amplifiers may be needed for the paired (FDD) bands, which is not natural for TDD.

Nonetheless, this concern is captured in [7] and has convinced ITU to revise the simplex assignment plan. Both lower and upper legs of VDE-SAT, $\mathbf{L}^{SAT}$ and $\mathbf{U}^{SAT}$, are now assigned as duplex channels as a remedy to this potential issue, as depicted in Fig. 6.1b, allowing both uplink and downlink transmissions in these channels for optional TDD transmissions.

In addition to the advantage of RF isolation, a further advantage of TDD is the extra degrees of freedom for the network to allocate communication resources in proportion to the traffic demand in both directions by varying the time partition of the uplink and downlink transmissions to better align with the service-centric networking concept. Nevertheless, the downside is that TDD complicates the interference management, which requires the synchronization of uplink and downlink between stations (cf. Chap. 2), as well as between VDE-SAT and VDE-TER. Another caveat is that the uplink transmission (from mobile stations) on the upper leg $\mathbf{U}^{SAT}$ will interfere with AIS and ASM receptions in the vicinity, although it is less of a concern since VDE-SAT is meant for offshore communication in open oceans where maritime traffic is sparse. For nearshore or mobile stations with co-located AIS/ASM transceivers, the lower leg of VDE-SAT $\mathbf{L}^{SAT}$ is preferred for VDE-SAT uplink transmissions.

Overall, despite these imperfections in spectrum allocation, including very limited bandwidth, the VHF frequencies are ideal for maritime MTC in terms of transmission range, which also translates to potential power saving for power-limited VDES devices. The RF frontend, such as the VHF amplifiers, mixers, bandpass filters, and local oscillators, is evolving rapidly. Smaller, lighter, and more efficient VHF antenna systems are gaining traction as the RF technology continues to evolve, which will further benefit maritime MTC, especially for Category 1 mobile stations. The limited bandwidth means that VDES systems have limited capacity and are only suitable for short burst data services but covering primary maritime IoT use cases.

In the remainder of the book, we use a unified design without deviating from its core concept that is governed by the maritime requirements and constrained by the physical radio spectrum. The following chapters are thus devoted to such a design, specifically tailored to the requirements outlined in Chap. 1 and the radio spectrum specified in this chapter.

References

1. Federal Communications Commission (FCC). https://www.fcc.gov
2. European Electronic Communications Committee (ECC). https://ec.europa.eu/jrc/en/event/workshop/ecc-meetings
3. Reference radiation patterns of omnidirectional, sectoral and other antennas for the fixed and mobile devices for use in sharing studies in the frequency range from 400 MHz to about 70 GHz, document Rec. ITU-R F.1336-4, ITU, Geneva, Switzerland (2014)
4. Technical and operational characteristics of conventional and trunked land mobile systems operating in the mobile service allocations below 869 MHz to be used in sharing studies, document Rec. ITU-R M.1808, ITU, Geneva, Switzerland (2007)
5. Equipment characteristics for digital transmission in the land mobile services, Report 1021, ITU, Geneva, Switzerland (1986)
6. Method for point-to-area predictions for terrestrial services in the frequency range 30 MHZ to 3000 MHz, document Rec. ITU-R P.1546-5, ITU, Geneva, Switzerland (2013)
7. T. Xia, M.M. Wang, X. You, Satellite machine-type communication for maritime Internet of Things: an interference perspective. IEEE Access **7**, 76404–76415 (2019)

Chapter 7
Proximity Air Interface Design

Abstract The proximity-based maritime IoT service is a type of service that becomes available once a mobile station comes near other mobile stations, such as route exchange and most aids-to-navigation services. This type of service typically involves proximity communication that allows direct communication between neighboring mobile stations. We refer to this type of maritime MTC as *proximity MTC* under the maritime MTC concept presented in Chap. 2. When realized on the International Maritime Mobile Communication VHF spectrum specified in Chap. 6, there are two such air interfaces using distributed medium access control under an ad hoc topology to provide the proximity communication service, including the ASM and the legacy AIS although the original intention of AIS was not just for proximity services. Channels 2087 and 2088 are assigned to AIS, and Channels 2027 and 2028 are dedicated to ASM, as replotted in Fig. 7.1. The legacy AIS is characterized by simplicity but suffers from spectral and energy inefficiency and a lack of adaptability to emerging maritime applications and services; with hindsight, ASM is poised to broaden the service capability and capacity via the well-defined protocol structure outlined in Chap. 3 and improved air interface design. Starting from this chapter, we present design examples of the three types of air interfaces introduced in Chap. 2, based on the VHF spectrum in Fig. 7.1. They are hence collectively referred to as VHF Data Exchange (VDE) System or VDES, constituted of ASM for proximity MTC, and VDE-TER and VDE-SAT for wide-area terrestrial MTC and satellite MTC. This chapter is devoted to ASM as a proximity MTC air interface under VDES.

Keywords Maritime IoT · Proximity-based service · Proximity machine-type communication air interface · Distributed medium access control · Ad hoc networking · Random medium access · Sensing-based medium access · AIS · ASM · VDES

M. M. Wang, J. Zhang, *Machine-Type Communication for Maritime Internet-of-Things*, Wireless Networks, https://doi.org/10.1007/978-3-030-77908-5_7

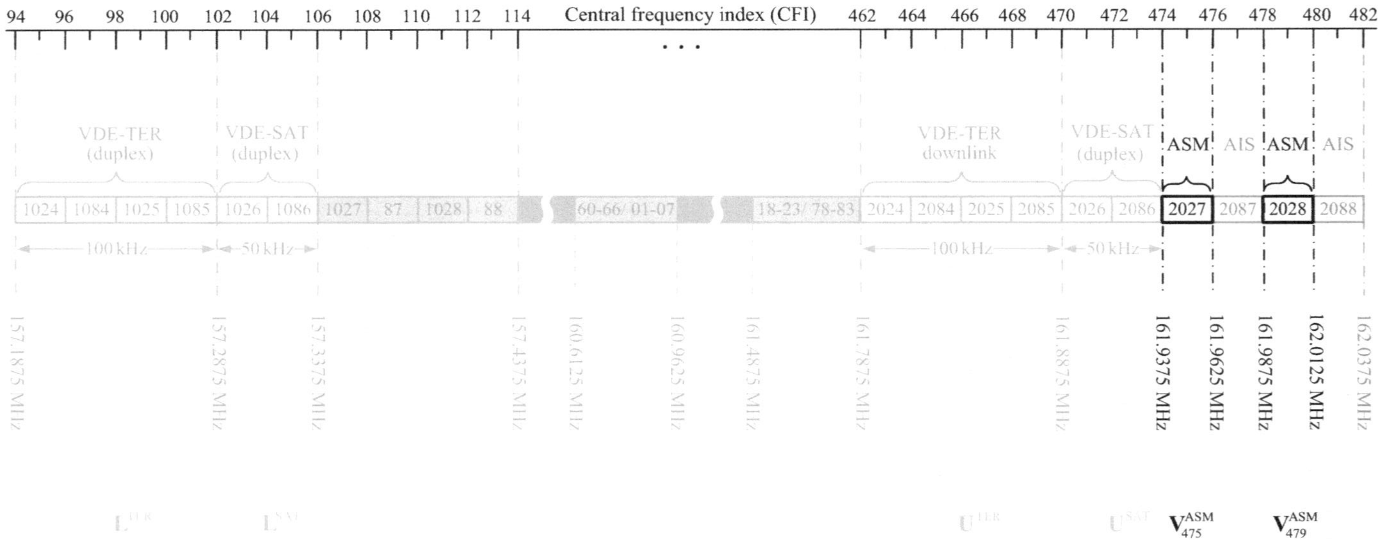

Fig. 7.1 ASM radio spectrum assignments, Channels 2027 and 2028 (CFIs 475 and 479) on the VHF maritime mobile band, denoted as $\mathbf{V}_{475}^{ASM}$and $\mathbf{V}_{479}^{ASM}$. The legacy AIS occupies Channels 2087 and 2088 (CFIs 477 and 481). (Replotted from Fig. 6.1 in Chap. 6)

7.1 Medium Access for Proximity Communication

In a practical wireless communication system, users or stations share the communication medium or radio resources based on *multiplexing*. As touched upon in Chap. 2, a medium access method (or a multiple access method) allows more than two mobile stations connected to the same transmission medium to transmit over it and share its capacity. As discussed in Chap. 3, this multiple access method is a part of the data link layer's medium access protocol, in charge of assigning multiplex resources to different mobile stations and avoiding or minimizing collisions. It is the physical layer's job to provide the physical means for transporting information over the physical multiplex resources.

For the given radio spectrum (Channels 2087 and 2088 of the VHF Maritime Mobile Band), the physical multiplex resource of the legacy AIS is organized into non-overlapping (hence orthogonal) *time slots*, which divide a UTC minute into exactly 2250 slots, as illustrated in Fig. 7.2. Here, UTC stands for the Coordinated Universal Time, also known as Greenwich Mean Time, an international time standard by which the world regulates the clock or time. A physical realization of UTC is provided by GPS (cf. Chap. 5), for example. For unification, the same slot structure is adopted for VDES, i.e., ASM, VDE-TER, and VDE-SAT; a slot is the smallest medium access resource element in the current design. A station typically obtains UTC from the GPS signal, allowing synchronizing the time slot over the entire network to ease medium access control and interference management. A GPS satellite has an atomic clock synchronized to UTC and broadcasts this information all over the world.

As discussed in Chap. 2, the ad hoc communication topology is favorable to proximity communication. Transmission resources, i.e., the orthogonal time slots of a radio channel, are shared among stations in a time division multiple access (TDMA) fashion. Slot assignment among mobile stations is thus the core component of the TDMA-based medium access control protocol. At the heart of the protocol is the distributed medium access mechanism that allows stations to share the same radio channel and operate autonomously through contention-based transmissions without the benefit of a centralized controller of an infrastructure. Before plunging into the detailed air interface design, we review the distributed medium access mechanisms for a mobile station to autonomously allocate its communication resources, i.e., the slots, under the slotted TDMA transmission framework.

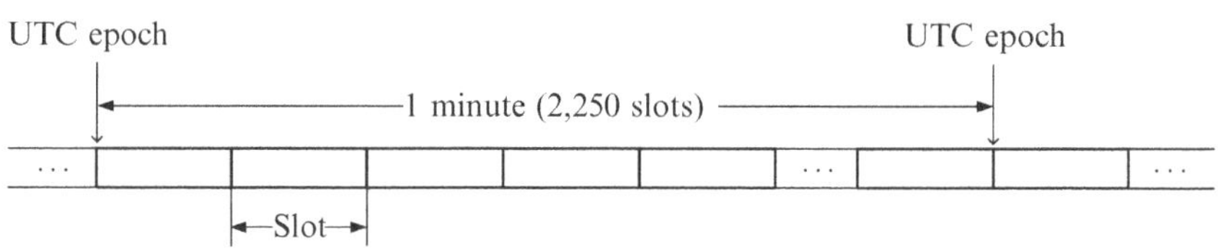

Fig. 7.2 Time slots as temporal multiplex resources, which divide a UTC minute into exactly 2250 non-overlapping slots

As touched upon in Chap. 2, distributed medium access mechanisms can be classified into two main categories: *random access* and *sensing-based access*. The classic random medium access strategy is the well-known slotted Aloha, and the most commonly used sensing-based medium access strategy is the carrier-sensing medium access. These two methods are particularly suited for data services that possess a short bursty traffic pattern. Random TDMA (RA-TDMA) and carrier-sensing TDMA (CS-TDMA) are the two variants of these two methods employed by AIS and adopted by ASM in this chapter for short burst data services. For periodic data services, like aid-to-navigation and ship position reporting, the sensing-based method is modified to produce improved performance by considering the periodicity characteristic of the service. It is commonly known as the *self-organized TDMA* (SO-TDMA), a variant of virtual channel sensing.

7.1.1 Random TDMA

Random TDMA or slotted Aloha is the simplest medium access strategy by which each transmitter, sharing the time slot medium with others, randomly selects one or more slots and transmits a burst on the slot(s). All stations share a common time reference (derived from UTC) to ensure that transmissions are time-aligned with the TDMA slot boundary. When the burst collides with other bursts at the intended receiver, the receiver may not be able to recover the message contingent on the signal strength relative to the interferers. Retransmissions may be needed depending on the nature of the application and service. The collision probability is low when the radio channel loading is light, offering lower latency than other schemes. As the traffic density increases, so do the collision probability and, ergo, the latency. From Chap. 2, the overall throughput reaches its maximum at a 37% load.

7.1.2 Carrier-Sensing TDMA

Carrier-sensing TDMA or CS-TDMA uses the "listen before talk" rule to reduce the collision probability under RA-TDMA by performing an additional check. Simply put, a mobile station verifies the absence of other traffic before transmitting on the shared TDMA slots after the slot is selected through RA-TDMA.

Stations using CS-TDMA continuously monitor the designated radio channel's background noise level. This background level is used as a reference for a received signal strength measurement at the start of each TDMA slot. When a transmission is triggered, a slot is randomly selected following RA-TDMA, and clear channel assessment or CCA (detailed in Chap. 2) is conducted at the beginning of the slot. For energy-based CCA, the energy is measured within the CCA window. If the energy level is significantly above the background noise level, the slot is assumed to be in use or busy, and the transmission is deferred. Otherwise, the slot is deemed idle, and the transmission is made, as illustrated in Fig. 7.3. As mentioned in Chap. 2,

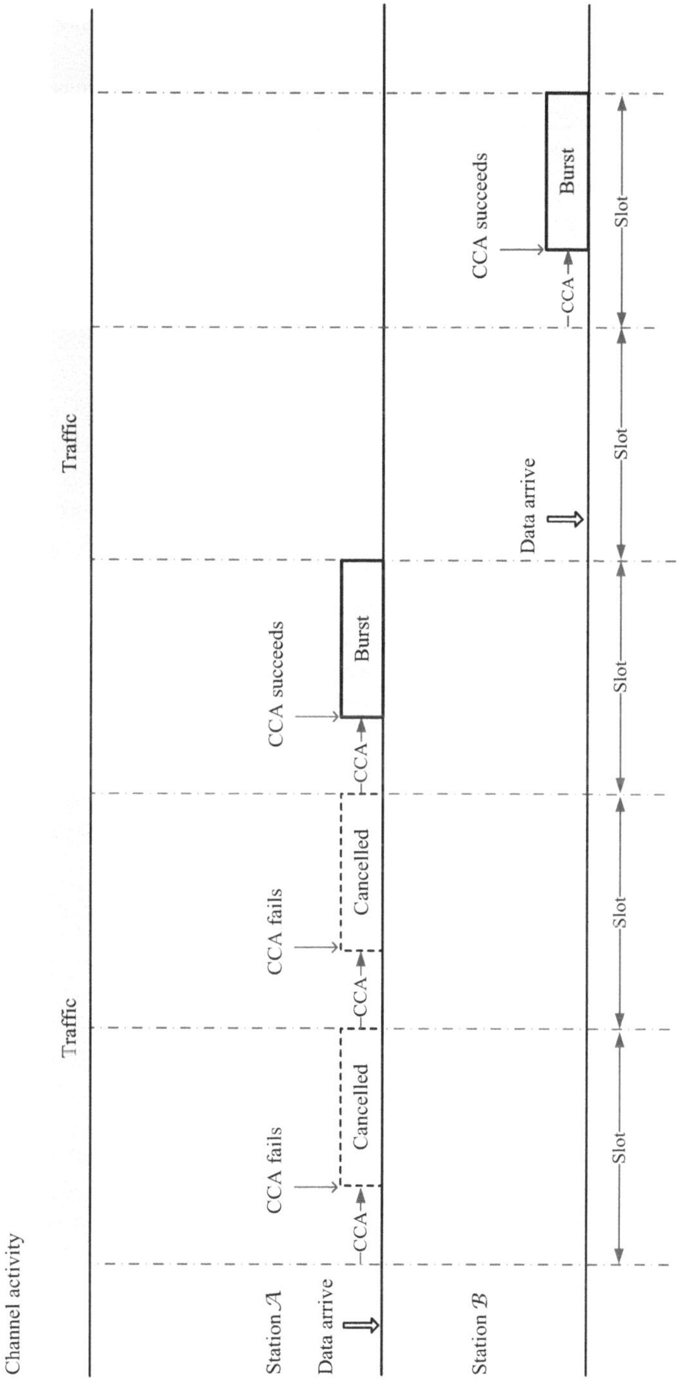

Fig. 7.3 Illustration of carrier-sensing TDMA, where transmission obeys a fixed timeline within a slot. CCA is performed before each transmission burst. A successful CCA warrants the transmission, whereas a failed CCA cancels the transmission

CCA can also be based on waveform detection, i.e., the `preamble` for AIS (Sect. 7.2.1) and the `sync word` in ASM (Sect. 7.3.3). However, it does not work for detecting multi-slot bursts.

CCA leaves fewer resources (i.e., time) available for transmission since a part of the slot is used for CCA. Moreover, the delayed transmission will fail to be detected by other CS-TDMA stations, causing more than one station to succeed CCA. Therefore, collision happens when multiple CS-TDMA stations perform CCA on the same slot.

7.1.3 Self-organized TDMA

Self-organized TDMA or SO-TDMA is a variant of virtual channel sensing, a sensing-based access method that uses information embedded in the detected transmissions (e.g., a MAC Signaling PDU) to predict future channel status. It is thus particularly suitable for services with *periodic* transmission patterns. Since it needs a station to "decipher" the transmissions from other stations, SO-TDMA only works among stations that operate under the same air interface, which is the case for AIS and ASM, each operating on the assigned channels.

In a nutshell, each transmission includes an indication of the TDMA slots to be used by the transmitting station for subsequent transmissions. This embedded information allows stations to build up a "channel map" in terms of which slots are to be occupied by other stations. Each station avoids busy slots per the map and *randomly* selects a slot among the slots predicted to be available, subject to the periodicity and latency requirements of the service. This created randomness prevents two or more stations from consistently colliding with each other.

7.1.3.1 Selection Window

For applications and services with periodic message transmissions, a station needs to reserve multiple resources or slots separated by an interval of I slots that equals its message transmission period. For ship reporting, it is determined by the speed and heading of a ship. For instance, a message period of 30 s means the desired transmission rate is two transmissions per minute, corresponding to I=1125 slots. This seemingly trivial matter becomes not so trivial when there are potentially many contending stations doing the same thing *autonomously*.

When the resource acquisition/selection process is triggered by, e.g., a request from higher layers to transmit a message from an application at a time instant (indicated by the hollow arrow in Fig. 7.4) corresponding to slot ℓ, denoted as θ_ℓ, on a given radio channel, a mobile station autonomously selects the transmission resource, i.e., slot(s), using a distributed channel access mechanism.

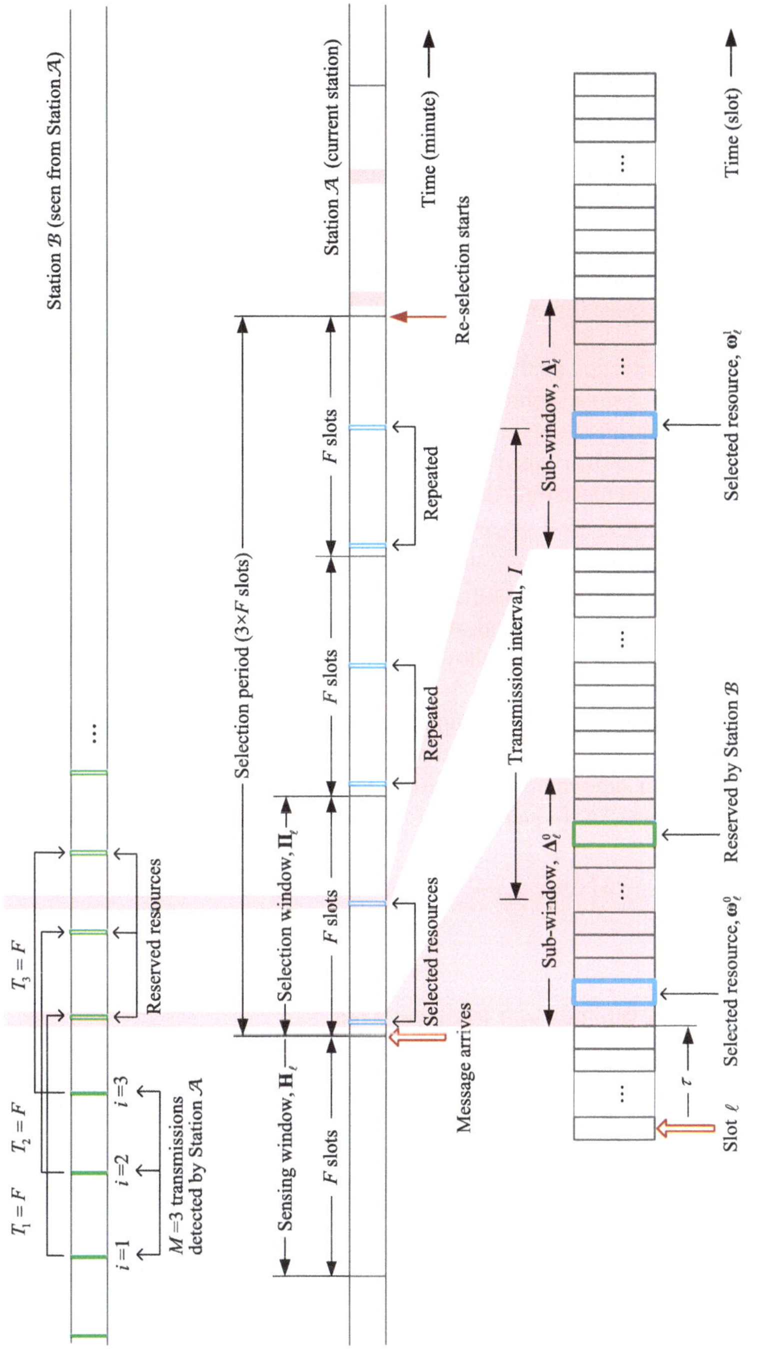

Fig. 7.4 Illustration of the self-organizing TDMA medium access mechanism, where $N=2$ and $R=3$ for illustration purpose

To this end, a *selection window* of length F slots,

$$\boldsymbol{\Pi}_\ell \triangleq \boldsymbol{\theta}_\ell^F, \forall \ell, \tag{7.1}$$

is established as illustrated in Fig. 7.4, where

$$\boldsymbol{\theta}_\ell^F \triangleq \{\theta_\ell, \theta_{\ell+1}, \cdots, \theta_{\ell+F-1}\} \tag{7.2}$$

contains $\left|\boldsymbol{\theta}_\ell^F\right| = F$ consecutive slots starting from slot ℓ. Without loss of generality, we choose F to be 1 min, i.e., F=2250 slots, following the legacy AIS convention. Note that here, the selection window does not necessarily align with the UTC minute epoch.

The resource selection goal is to reserve $N = F/I$ transmission resources separated by an interval of I slots within the selection window, $\boldsymbol{\Pi}_\ell$. For the ship reporting example, N=2 transmissions per selection window, or once per I=1125 slots. Due to the random nature of distributed medium access control, strict periodicity is hard to be enforced. To provide certain degrees of freedom or flexibility during selection, a *sub-window*, $\boldsymbol{\Delta}_\ell^n$, $n \in \{0, 1, \cdots, N-1\}$, is used to reflect certain "variations" in periodicity, i.e., between $I - \delta/2$ and $I + \delta/2$. In other words, the width of $\boldsymbol{\Delta}_\ell^n$, i.e., $\left|\boldsymbol{\Delta}_\ell^n\right| = \delta$slots. For instance, in ship reporting, the reporting period may vary within 10 percent of the reporting period, meaning $\delta/2 = 0.1 \times I \approx 112$slots for I=1125. Nonetheless, the actual variance allowed may depend on the latency constraint from the service requirement.

Now, denoting τ the amount of time (slots) that a station needs to complete the selection algorithm, the earliest selection sub-window starts τ slots later after slot ℓ. Therefore, the first sub-window, $\boldsymbol{\Delta}_\ell^0$, within the selection window $\boldsymbol{\Pi}_\ell$ is established, as illustrated in Fig. 7.4. In general,

$$\boldsymbol{\Delta}_\ell^n \triangleq \boldsymbol{\theta}_{\ell+\tau+nI}^\delta \subset \boldsymbol{\Pi}_\ell, \quad n \in \{0, 1, \cdots, N-1\}. \tag{7.3}$$

7.1.3.2 Sensing Window and Resource Map

The purpose of channel sensing is to produce a set of resource allocation candidates selected from the selection window, $\boldsymbol{\Pi}_\ell$, according to the channel usage predicted through a sensing process conducted in a time window, referred to as the *sensing window*, $\mathbf{H}_\ell$.

This window traces backward from the start of the selection window over a period of the same length as the selection window, i.e.,

$$\mathbf{H}_\ell \triangleq \boldsymbol{\theta}_{\ell-F}^F, \forall \ell, \tag{7.4}$$

Table 7.1 The communication state field of AIS Message 1

Field	Bit width	Remarks
`Sync state`	2	Indicates the synchronization source stratum
`Slot timeout`	3	0–7 means that 0–7 min respectively are left until slot re-selection
`Sub-message`	14	The slot offset when slot timeout is 0;

where $|\mathbf{H}_\ell| = |\mathbf{\Pi}_\ell| = F, \forall \ell$. From this sensing window, the ongoing traffic patterns are derived to help the resource selection in the selection window so that the collision with the ongoing traffic (if any) can be maximally avoided.

Suppose M transmissions are detected in the sensing window, $\mathbf{H}_\ell$,

$$\boldsymbol{\theta}_{n_m}^{s_m} \subset \mathbf{H}_\ell, m = 1, 2, \cdots, M. \tag{7.5}$$

Each of the detected bursts in the sensing window, $\mathbf{H}_\ell$, contains the information that indicates a set of its associated resource parameters such as the resource size in terms of slots, s_m, and a `slot timeout counter` (see Table 7.1), r_m, indicating the number of left reservations still yet to be transmitted in its current selection period, which infers a particular resource reservation pattern.

Similarly, the resource reservation interval is fixed to the same length of the selection window, i.e., 1 min or $T_m = F$, $m = 1, 2, \cdots, M$. The slots that are reserved by this burst, m, is

$$\boldsymbol{\eta}_m \triangleq \bigcup_{r=1}^{r_m} \boldsymbol{\theta}_{n_m + r \cdot F}^{s_m}, m = 1, 2, \cdots, M. \tag{7.6}$$

After taking into account all reservation intentions from the derived reservation patterns of the detected traffic, the sensing station builds up a map of which slots are in use by which station, $\bigcup_{m=1}^{M} \boldsymbol{\eta}_m$, which helps in the slot selection process.

7.1.3.3 Slot Selection

Assuming the required resource size is L slots, then the candidate resources that are free of conflict with the reserved resources indicated by the detected traffic can be expressed as

$$\mathbf{O}_\ell^n = \left\{ \boldsymbol{\theta}_k^L \subseteq \boldsymbol{\Delta}_\ell^n \middle| \boldsymbol{\theta}_k^L \cap \bigcup_{m=1}^{M} \boldsymbol{\eta}_m = \varnothing, \forall k \right\}, \quad n \in \{0, 1, \cdots, N-1\}, \tag{7.7}$$

which is the set of resources of size L that can be utilized by the sensing station for its own transmissions. These selected slots avoid the slots known to be in use by other stations, preventing stations in range of one another from using the same slots. In this example, only one reserved slot (by Station `B`) falls into the first selection sub-window of Station `A`, and hence that slot is not available for selection.

The final selected slot(s), $\boldsymbol{\omega}_{\ell}^{n}$, on $\boldsymbol{\Delta}_{\ell}^{n}$ is randomly picked from the above candidate set, i.e., $\forall \boldsymbol{\omega}_{\ell}^{n} \subset \mathbf{O}_{\ell}^{n}$. Note that there could be more than one station that happens to be in the selection process competing for the slots, so the random selection helps reduce the probability of selecting the same slot that causes "deadlock" or continuous collisions. Apparently, the more concurrent selecting stations there are, the higher the collision probability. The same selection process is repeated within the next sub-window, e.g., $\boldsymbol{\Delta}_{\ell}^{n+1}$, to obtain $\forall \boldsymbol{\omega}_{\ell}^{n+1} \subset \mathbf{O}_{\ell}^{n+1}$.

7.1.3.4 Selection Period and Re-selection

Once selected, the station may keep this transmission pattern determined within this selection window and repeat R times before re-selection. This period, called the *selection period*, spends $R \times F$ slots, where R is randomly selected from, e.g., 3 to 7 ($R = 3$ in the example of Fig. 7.4). Again, the randomness of R helps desynchronize the re-selection processes among stations, thereby reducing the number of concurrent reselecting stations, ergo the collision probability. As such, the `slot timeout` counter, r, of a transmission burst is initialized with $R - 1$, and decrements by one after every selection window, i.e., 1 min. Once this counter counts down to zero, i.e., the selection period expires, the resource needs to be reselected, and the selection process is repeated. When a transmission is detected with the `slot timeout` counter equal to zero (i.e., this detected station is entering re-selection), the sensing station may lose track of the next "move" (transmission) of that station. Therefore, additional information is added to the MAC signaling PDU of the transmission burst to indicate the `offset` (in slots) between the newly selected resource for future transmissions (as a result of the re-selection) and the current resource.

It is tempting to simply make R arbitrary large to negate the need for re-selection, but the possibility exists that more than one station happens to select the same resources with the same transmission pattern. When that occurs, the collision could continue for an extended time for a large R. And yet, even though not currently happening, collisions are likely to occur down the road in a dynamic maritime environment. The mobile station encounters new stations with different slot allocations as it moves from one area to another, which necessitates the station to modify its slot allocation, leading to a dynamic and self-organizing system over time and space.

SO-TDMA also provides for dynamic and autonomous management of capacity in busy areas. Should a situation occur where all TDMA slots are occupied, the "busy" slots occupied by the stations most distant to the sensing station (with the lowest detected signal energy) may be included in the selection process.

7.2 AIS Air Interface

Under the maritime MTC protocol structure for proximity networking presented in Chap. 3, applications are decoupled from the air interface. The application traffic is encapsulated in the network traffic PDUs and transported to its relevant peer (s) through the service-oriented network layer protocols, hiding the physical delivery (by the data link and physical layers) from the application.

Unfortunately, the legacy AIS was not designed with this layered protocol structure in mind, resulting in a "flat" structure, as shown in Fig. 7.5. The only flexibility is provided through a message type (the so-called `Message ID` field, as we will see later), leaving little space for adaptability to new applications, not to mention flexible network control and MAC signaling, as will soon become evident. This section briefly reviews the AIS interface for compare-and-contrast to the new proximity air interface design to be presented in Sect. 7.3.

7.2.1 Transmission Burst

A transmission burst is structured to enable the transmission and reception of information over a physical radio channel of a given bandwidth at a radio frequency. An AIS burst requires a minimum 1-W transmit power and runs at a baud rate of 9600 symbols per second over a 25-kHz VHF frequency channel (see Chap. 6), employing the Gaussian-filtered MSK or GMSK (β=0.4) modulation for transmission. Hence, each slot provides 256 bits, with a symbol interval of 104μs.

Figure 7.5 shows an AIS transmission burst structure. The 8-symbol (832 - μs) `ramp-up` and 8-symbol `ramp-down` (included in the `buffer`) are used to improve the spectral shaping of the AIS burst waveform, i.e., to suppress the out-of-band emissions aroused by the on-off transitions. Additionally, the `ramp-up` helps the receiver automatic gain control (AGC) converge over a wide dynamic range of input signal levels – typical for radio signals.

A 24-symbol preamble, consisting of a bit sequence with alternating 0 s and 1 s, is used for a receiver to detect the presence of a transmission burst and align the timing

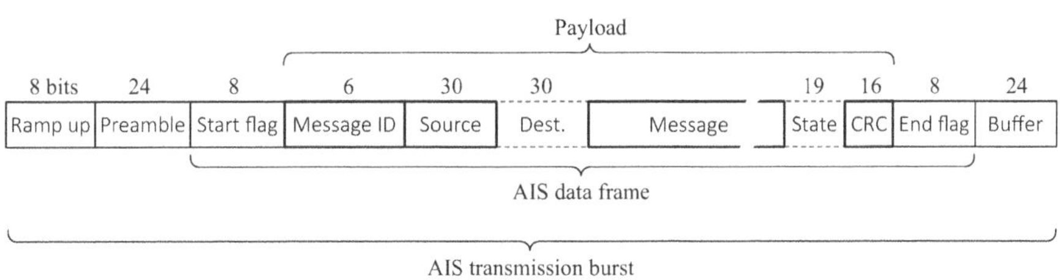

Fig. 7.5 AIS data frame and transmission burst structure, where the state field denotes the communication state field. The length or TTI of a burst is one slot up to five consecutive slots

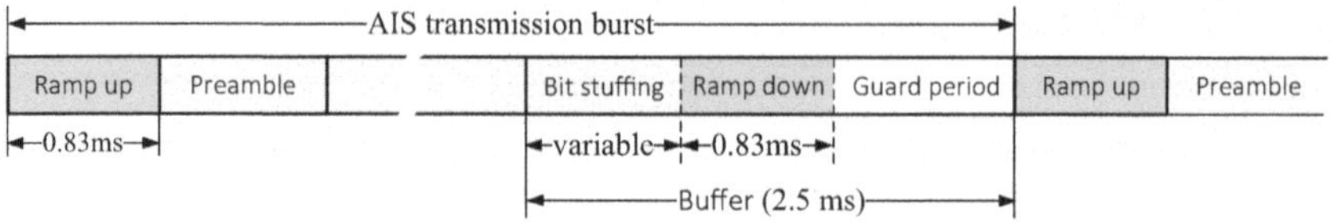

Fig. 7.6 Illustration of the buffer field of an AIS burst to show its effect in preventing the potential overlapping between two subsequent bursts from two transmitters

to the burst if present. In addition, for ad hoc communication, when GPS is not available, all transmissions by mobile stations are based on local oscillators and hence arrive at a receiver with potentially large carrier frequency offsets. The preamble is essential for a receiver to synchronize to the incoming burst. Estimation of the carrier phase is also necessary if coherent detection is employed.

At the end of the burst is a 24 symbol (2.5 ms) worth of time buffer, earmarked for bit stuffing, `ramp-down`, propagation delay, and timing jitter. Note that although the time slots are temporally non-overlapping, i.e., orthogonal; it is by no means a guarantee that the signals transmitted on separate slots from different mobile stations will not overlap at a receiving mobile station. In fact, overlap may happen at a receiver between consecutively transmitted bursts from different stations due to differences in propagation delays, and hence orthogonality may not be retained.

Assuming four symbols (416 μs) are normally consumed by bit stuffing (discussed next), 20 bits are left to guard the burst against overlapping, as illustrated in Fig. 7.6. After six symbols (624 μs) are deducted for timing jitters, this guard period is capable of absorbing a maximum propagation delay of at least 1.46 ms, corresponding to a communication range of at least 240 nautical miles (NM), noting that the protrusion of the ramp-down into the ramp-up field of the succeeding burst from another station does no harm to either burst.

7.2.2 Data Frame

The start and end of an AIS frame are "marked" by the start flag and the end flag. They are an 8-bit or one-byte "unique" bit sequence (i.e., `01111110`) for the division of a bitstream from the upper layer into frames, often found in a wired communication system (such as the High-level Data Link Control protocol and Ethernet) for framing. This technique is somehow adopted by AIS, a wireless system, to indicate the transmission burst boundary. In particular, the start flag breaks the bit pattern of the preamble, and together they compose a preamble that signals the start of the actual payload. It reduces the chance that a station that does not have UTC falsely locks onto a bit sequence that happens to look like the preamble. The `end flag` indicates the end of the payload since the transmission time interval or TTI of an AIS burst varies from one slot up to five slots, and a receiver needs to know the end of the burst.

The question that arises immediately is that the bit pattern used for the `start` and `end flags` could also be part of the original payload bitstream. This flag sequence is disambiguated by the scheme coined "bit stuffing"—the process of inserting non-information bits or delimiters into the information bit stream to break up the specific bit patterns to ensure the synchronous transmission of information. In this case, a bit of `0` is inserted after five consecutive `1`'s in the payload bitstream to avoid more than five consecutive `1`'s appearing in the payload. Consequently, the length of the bitstream after bit-stuffing is random, which is absorbed by the `buffer`. After seeing five consecutive `1`'s in the received payload at the receiving station, the following (sixth) 0 is stripped.

So far, nothing seems out of the ordinary until we notice that this scheme works only under the condition that bit errors are rare during transmission since such an error may destroy a flag or create a flag prematurely. This condition is likely true for communication protocols with a wired communication medium where bit errors are rare. However, the condition can hardly hold for those with a wireless communication medium since errors are the norm of a wireless channel, especially for the AIS channels with abundant co-channel interference due to the distributed medium access control mechanism and the inherent adjacent channel interference, as explained in Chap. 6. With that understood, AIS does look more like a wired communication interface like the Ethernet, as some may have already noticed.

After taking into account all the overhead, all that is left for traffic is the 184-bit payload. The payload has a *fixed* format: a 6-bit message identifier or `Message ID`, a 30-bit `source MMSI`, an *optional* `destination MMSI`, and the message, followed by an *optional* 19-bit `communication state` field for the distributed medium access control mechanism detailed in Sect. 7.1.3.

As shown in Table 7.1, the `communication state` field contains, among other things, a 3-bit `slot timeout` field, a 14-bit `offset` field, and a 2-bit `sync state` field. The `slot timeout` field indicates frames remaining until a new slot is selected (i.e., re-selection): The number `0` means that this was the last transmission in this slot; any number between `1` and `7` means that that number of minutes are left until re-selection, whereas offset indicates the reselected slot position `offset`, by the offset value, from the previously selected slot at the re-selection at timeout `0`. The `sync state` field is described in detail in Sect. 7.3.3.

The presence of the optional field is determined by the message. For example, the `communication state` is present for messages that employ the virtual channel sensing scheme, like SO-TDMA, such as `Message 1`, as shown in Table 7.2, where destination MMSI is absent since `Message 1` is a broadcast message.

Since it is often a necessity for the receiver to know whether the received payload bitstream is free of error, a 16-bit cyclic redundancy check (CRC) is appended for error detection to ensure the integrity of the message. The probability of failure to detect an error is about 2^{-16}or 0.0015%.

The message can be one of the 2^6=64 messages indicated by the 6-bit `Message ID`, predefined in the AIS air interface specification. The message size (`Message ID` inclusive) ranges from 168 bits for a one-slot TTI to 1192 bits for a five-slot TTI. These messages include anything from some most commonly used maritime services (e.g., ship position reporting and aids-to-navigation) to MAC controls (e.g., ARQ for safety-related messages).

Table 7.2 AIS `Message 1`

Field	Bit width	Remarks
`Message ID`	6	Message identifier: `000001`
`Source ID`	30	MMSI of transmitter
`Repeat indicator`	2	For relay operation
`Navigational status`	4	`0`: Under way using engine, `1`: At anchor, `2`: Not under command, and so on
`Rate of turn`	8	Turning right at up to 708°/min to left at up to 708°/min
`Speed over ground`	10	Speed over ground in 1/10 knot steps
`Position accuracy`	1	1: High ($\leq$10 m); 0: Low (>10 m)
`Longitude`	28	Longitude in 1/10000 min
`Latitude`	27	Latitude in 1/10000 min
`Course over ground`	12	Course over ground in 1/10 degree
`True heading`	9	Degrees (0–359)
`Time stamp`	6	UTC second (0–59 or 60); 61 if positioning system is in manual input mode
`Special maneuvers`	2	e.g., regional passing arrangement on Inland Waterway
`RAIM-flag`	1	RAIM (Receiver Autonomous Integrity Monitoring)
`Spare`	3	Not used. Should be set to zero. Reserved for future use
`Communication state`	19	See Table 7.1
`Total`	168	Total 21 bytes

For example, `Message 1` is commonly used by a ship to periodically broadcast its dynamic information for, e.g., collision avoidance, clearly an application-specific message. The message contains the ship's identification and information such as position (longitude and latitude), speed over ground, course over ground, and navigational status (e.g., underway using engine, at anchor, not under command), transmitted every a few seconds while underway and a few minutes when at anchor or moored.

Similarly, an aids-to-navigation device may use `Message 23` to provide precision piloting to passing ships in areas such as dangerous coastlines and channels/waterways and hazardous shoals and reefs, which is generally transmitted at a rate of once every 3 min. A SAR transmitter (i.e., life vests/rafts, maritime survivor locating devices, or man overboard devices) transmits a series of eight identical position report messages (four on Channel 2087 and four on 2088) per minute when active.

An ARQ protocol is needed for messages that require reliability. For each AIS message that has been successfully received, the destined station is expected to return an acknowledgment message in the same radio channel to negate retransmission. The ARQ message, i.e., `Message 7`, is used to acknowledge a unicast message, e.g., `Message 6`. This type of message requires an acknowledgment for each sent message for reliability. In a stop-and-wait ARQ, the sending station stops after every message and waits for the receiving station to acknowledge the receipt of the message (Table 7.3).

Table 7.3 AIS Message 6

Field	Bit width	Remarks
Message ID	6	Message identifier: 000110
Repeat indicator	2	For relay operation
Source	30	MMSI of transmitter
Destination	30	MMSI of destination
Sequence number	2	Sequence number for acknowledgement
Retransmit flag	1	0: New data; 1: Retransmission;
Data	Variable	Data area

Table 7.4 AIS Message 7

Field	Bit width	Remarks
Message ID	6	Message identifier: 000111
Repeat indicator	2	For relay operation
Source	30	MMSI of transmitter
Destination	30	Acknowledge destination MMSI
Sequence number	2	Acknowledgement sequence number (see Message 6)
Destination	30	Optional
Sequence number	2	
Destination	30	Optional
Sequence number	2	
Destination	30	Optional
Sequence number	2	
Spare	2	Reserved
Total	72–168	Total (variable): 9, 13, 17, or 21 bytes

The drawback is that it takes a whole burst, i.e., a slot worth of resources, to send a single acknowledgment since the AIS structure allows one message one transmission burst. There is no mechanism or flexibility for packing multiple AIS messages into one burst. To ameliorate this inefficiency, AIS exploits the sliding window protocol that allows a selected number of messages, the window, to be sent without waiting for an acknowledgment. Each message within the window includes a unique sequence number for identification, which allows a station to respond with a group of acknowledgments in a single message, also known as the *group acknowledgment*. The window slides along the stream of messages making up the transfer when all the messages within the window have been acknowledged. Message 7 thus contains a 2-bit sequence number field, and the sequence number is not to be reused until it has been acknowledged, or the maximum retransmission times (e.g., 3) have been reached, or the timeout counter (e.g., 4 s) has expired. Message 7 goes further to allow up to four acknowledgments to different senders to be addressed individually, as shown in Table 7.4.

There are specific AIS messages (not shown here) for broadcast or addressed safety-related messages. There are also other messages for a shore station to announce its position and UTC every 10 s identified by an MMSI. A station may also use a special message to request UTC and data from a station or to provide current UTC and data if available. A mobile station may transmit a safety-related message to a specified station by indicating the destination in the `destination MMSI` field or broadcast to all other stations (in the absence of the `destination MMSI` field). Also, some of the 64 messages are devoted to system control/management, and a shore station may use them to control the reporting rate of a mobile station. A shore station may also transmit a group assignment message to command the transmission behavior (e.g., reporting interval or transceiver mode) of a group of mobile stations in a geographical area.

Simply put, these 64 messages include everything that could be thought of at the time of AIS standardization – but still fall short when it comes to emerging maritime applications and services.

7.2.3 *Pitfalls*

The shortcomings and limitations of the legacy AIS are now evident. First and foremost, the lack of a layered protocol structure imposes a strong tie between the application and the air interface, fundamentally restricting the applicability of the air interface. The only flexibility is provided through the `Message ID` field, which allows up to 64 *predefined* messages that are literally carved into the air interface, thereby preventing it from providing communication services to emerging diverse maritime IoT applications that do not fit into the existing AIS message format. This flat structure also strips away the freedom to manage the burst payload efficiently, such as message multiplexing (i.e., multiplexing multiple messages, possibly from different applications, into one burst). In fact, there is no general way to multiplex different messages, such as MAC signaling messages and application messages, into one burst. For example, there is no way to multiplex the acknowledgment messages with other messages.

Second, AIS is intrinsically not service-oriented, rendering it unsuitable to proximity-based IoT applications. Since AIS does not have the notion of "application," it inherently lacks the capability of identifying applications. Different applications that send messages using the same message type will confuse receiving applications unless the sender sends the message addressed individually. However, it is not obvious how to identify all relevant stations in proximity in a dynamic environment. Even if the sender can, it creates an ambiguity issue when an AIS station hosts more than one application, possibly from different providers.

Third, the AIS air interface, adopted from a wired interface, essentially rests on the assumption that the AIS channel behaves as a wired channel; hence no error correction coding is provided, let alone any adaptive modulation and coding scheme necessary for efficient communication over wireless media. Without this layer of protection, it leaves the entire burst at the mercy of Mother Nature when traveling through the radio channel notoriously known to be unreliable, thereby making AIS extremely vulnerable to any bit error introduced by channel impairments. It only takes a single bit of error to kill the entire burst up to five slots. Moreover, as explained in Chap. 5, a lack of forward error correction coding nullifies the coding gain and consequently necessitates a high E_b/N_0 or energy per bit. And what makes things even worse is the lack of adaptability to varying channel conditions. Therefore, it is not surprising to see a minimum 1-W transmit power requirement by AIS and an excessive guard band for shielding the adjacent channel interference, producing a channel symbol rate of 9600 symbols per second over a 25-kHz channel – a less than 50% bandwidth usage, making AIS inherently energy and spectrally inefficient and unable to enforce QoS or support low-power and low-energy applications.

Another notable weakness of AIS is that its transmission burst structure itself provides minimal synchronization and tracking capability for a dynamic wireless communication environment characterized by the strong influence of interference and other channel impairments. Due to the random and bursty nature of ad hoc communication, the only opportunity for a receiver to time and frequency synchronize to the incoming AIS signal is the `preamble` at the beginning of the burst, making it susceptible to radio channel variability (e.g., phase ramping due to frequency error) during the burst, especially for long TTI bursts. Moreover, AIS has a weak preamble design: a 24-bit `preamble` sequence (plus an 8-bit `start flag`) leads to a significant residual frequency estimation error even in a benign channel condition, let alone in an ad hoc communication environment where co-channel interference is abundant.

Last, remember that the carrier phase is needed for coherent detection of the bitstream. Two factors may contribute to the variability of the received carrier phase: radio channel dynamics, mainly due to station mobility, and the frequency offset between the transmitter and receivers resulting from a synchronization error. Assume a carrier frequency offset, ς, remains between the transmitter and the receiver due to a synchronization error, the preamble-based "one-shot" estimation produces a residual error in the order from tens to hundreds of Hertz. Even with GPS to provide precision frequency reference for the transceivers (see Sect. 5.6), the motion of mobile stations can easily introduce a Doppler frequency shift in the range of tens of Hertz depending on the speed of a mobile station. From Chap. 5, the received baseband signal can be expressed as

$$s'(t) = s(t) \cdot e^{-j2\pi\varsigma \cdot t}. \tag{7.8}$$

This frequency offset *in effect* introduces a linear phase ramping factor,

$$\phi(t) \triangleq e^{-j2\pi\varsigma \cdot t}, \tag{7.9}$$

on the received baseband signal. Depending on the magnitude of ς and the elapsed time, t, the resultant time-varying phase offset, $\phi(t)$, may jeopardize the coherent detection of the bitstream.

Figure 7.7 plots $\phi(t)$ at a frequency offset of $\varsigma = 10$ Hz over a period up to five slots, where this frequency offset causes $\phi(t)$ to rotate over time, adding varying phase offsets on top of the received symbols. Nonetheless, AIS does not provide any

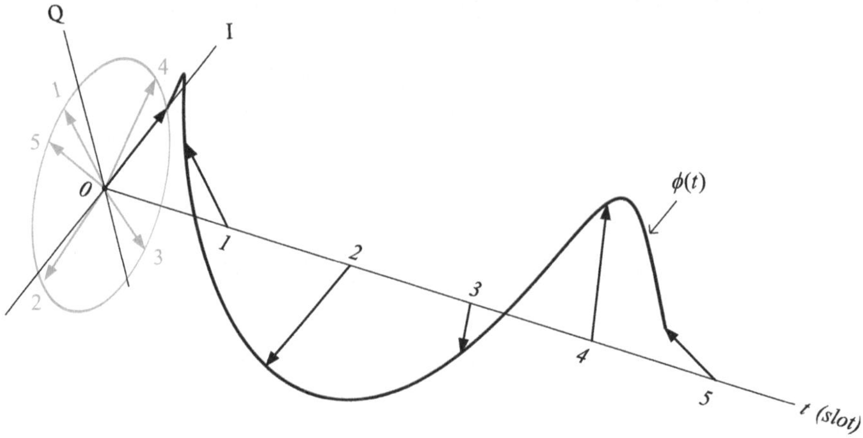

Fig. 7.7 Phase rotation of a received signal over the elapsed time as a result of a frequency offset, ς=10 Hz

pilot (see Chap. 5) for channel tracking during the burst to facilitate the detection and correction of the phase offsets at the receiver, rendering non-coherent detection or differential detection (non-optimal) the only viable options.

7.3 ASM Air Interface

So far, we have seen the inherent structural weakness within the legacy AIS, which now profoundly impacts its applicability to maritime IoT. Hence a new proximity communication air interface as a part of the maritime MTC system is much needed. In this section, we provide a design example of the ASM air interface. In order to avoid the same pitfalls as we have encountered in AIS, the design follows the protocol framework presented in Chap. 3 that decouples the application layer from the air interface for delivering more versatile maritime proximity-based applications and services. With well-defined interfaces between them, as shown in Figs. 7.8 and 7.9 under the framework from Chap. 3, an application does not need to know anything about the air interface, nor does the air interface protocol need to concern itself with specific applications. Under this framework, the ASM's sole job is to transfer the network layer PDUs that carry application traffic to the destined application peer(s). Therefore, it is expected of ASM to provide more flexibility for maritime proximity-based IoT. We also demonstrate a new physical layer design with significantly improved robustness and energy- and spectral efficiency over AIS.

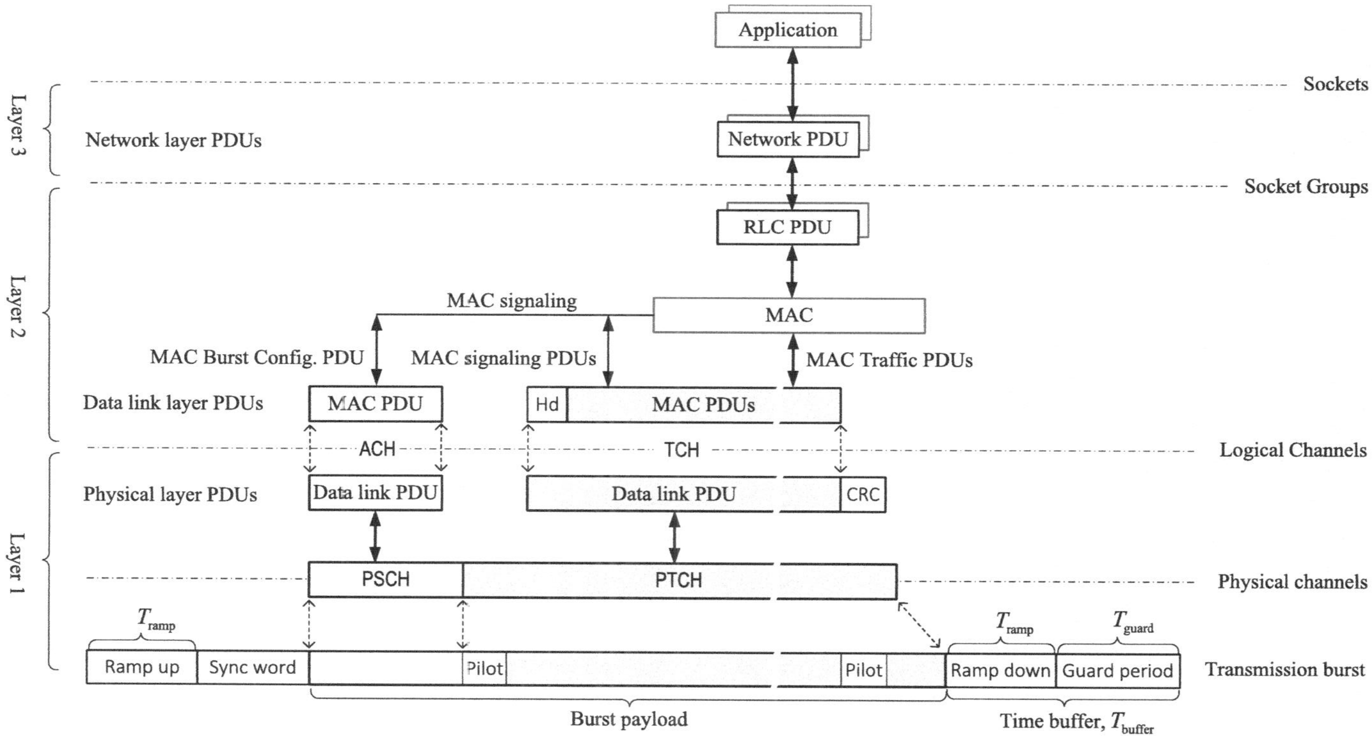

Fig. 7.8 ASM burst formation, where Hd stands for the TCH data link PDU header that includes a bit to indicate the presence of a Layer 2 source identifier (i.e., a source MMSI) in the current header: a four-byte header with MMSI, and one-byte without. A source identifier is needed when the data link PDU contains a MAC Traffic PDU that requires, e.g., ARQ when a peer-to-peer communication link is established per the application scenario. The data link PDU from ACH does not have a header

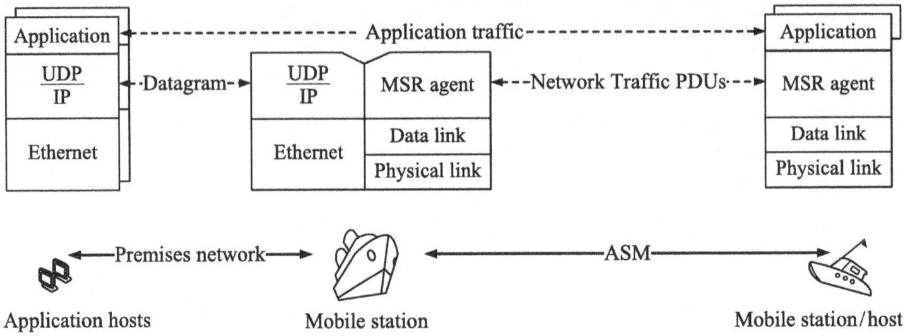

Fig. 7.9 Illustration of an exemplary ASM communication scenario: Applications running on an ASM terminal (the end host) aboard a yacht communicate with applications running on the end hosts on an on-premises network (e.g., an Ethernet network) aboard a ship equipped with an ASM terminal

7.3.1 Network Layer

As mentioned in Chap. 2, maritime proximity communication is a special type of MTC in that the target or destination of the communication is in the proximity. In this sense, it belongs to the short-range communication or direct communication between neighboring mobile stations. Since the destination is typically not the "station of interest" but the "service of interest," according to the service-centric networking principle, the network layer is no longer responsible for interconnecting stations inside the maritime MTC network. Instead, only the services of interest are monitored, and the messages are delivered to their destined applications through the socket channels. Under the ASM context, the Maritime Service Registry or MSR in the network layer is responsible for internetwork adaptation (e.g., Layer-3 address resolution), socket channel creation, maintenance, and mapping to the socket group per service profile maintained by the MSR of the Maritime Application Server, as outlined in Chap. 4. From Chap. 3, a socket is a logical construct that identifies a specific service or application, represented by the service identifier or `Service ID`, a Layer-3 identifier maintained by the MSR, whereas a socket group defines specific QoS constraints through abstraction of specific metrics from the service requirements, including rate, latency, and jitter. The local MSR agent decouples the air interface from the application by transporting application data to the data link layer through socket groups in the form of the network Traffic PDU, as described in Chap. 4.

A network layer Traffic PDU header for proximity MTC contains a `destination identifier` field set to the `Service ID` associated with the service. For each socket channel configured at the transmitting station, there needs to be a peer socket at each receiving station signed up for the service. Now a receiving station uses this list of socket channels to screen out the incoming irrelevant messages, i.e., messages that do not match the specifications of the socket channels, to avoid further processing of these messages. As an agent of the central network resource control or NRC, defined in the network structure in Chap. 4, the local mobile station NRC is in charge of the radio resource management involved in proximity communication over ASM, such as the frequency channels available for ASM. From Chap. 6, the default frequency channels are Channels 2027 and 2028.

Thus, the network layer (NRC and MSR agents) stays in touch with the network controller and the maritime application server for network and service updates via a backhaul, e.g., VDE-TER or VDE-SAT or other wide-area communication means.

7.3.2 Data Link Layer

With socket groups that define what type of data is transported, the data link layer uses logical channels to define how the respective PDU is transported to its peer through the physical channel with the required QoS. For that, the data link layer relies on a set of RLC and MAC protocols. In general, for identification at the receiving peer, the RLC and MAC PDU header formats outlined in Chap. 3 must be followed in creating the RLC and MAC PDUs. In this section, we place the focus on the most relevant MAC functions/protocols and PDUs under the ASM context.

Table 7.5 defines the MAC Traffic PDU headers, whereas Tables 7.6 and 7.7 define some MAC Signaling PDUs. A MAC PDU header devotes a type field to

Table 7.5 MAC traffic PDU header

Field	Bit width	Value	Remarks
`Type`	1	`0`	Traffic PDU
`RLC instance`	4	`0000 – 1111`	16 RLC instances
`SDU size`	6	0–63	64 SDU sizes
`ARQ`	1	`0`	No ARQ
`Reserved`	4		
`Type`	1	`0`	Traffic PDU
`RLC instance`	4	`0000 – 1111`	16 RLC instances
`SDU size`	6	0–63	64 SDU sizes
`ARQ`	1	`1`	ACK required
`Destination`	30	0 – 1,073,741,823	Layer 2 address (MMSI)
`Sequence number`	6	0–63	ACK sequence

Table 7.6 MAC signaling PDU: Feedback PDU (for ARQ)

Field	Bit width	Value	Remarks
`Type`	1	`1`	Signaling PDU
`Control function ID`	7	`0000011`	ARQ feedback, i.e., ACK
`Destination`	30	0 – 1,073,741,823	Layer 2 address (MMSI)
`Sequence number`	6	0–63	ACK sequence
Reserved	4		

Table 7.7 MAC signaling PDU: Slot assignment PDU (for SO-TDMA)

Field	Bit width	Value	Remarks
`Type`	1	1	Signaling PDU
`Control function ID`	7	`0000001`	SO-TDMA
`Slot timeout`	4	0–15	SO-TDMA signals (see Sect. 7.1.3)
`Offset`	8	0–255	
Reserved	4		

distinguish a Traffic PDU from a Signaling PDU. The MAC Traffic PDU header also contains an SDU size field to indicate the variable MAC `SDU size`. As such, multiple MAC PDUs (traffic or signaling) can be multiplexed into one data link PDU.

7.3.2.1 Burst Configuration

As shown in Fig. 3.8, the MAC PDUs are queued and multiplexed into the data link PDU, based on their priorities or RLC instances and the MAC scheduling policies, recalling that the data link layer utilizes the RLC instances to handle the socket group's QoS (cf. Chap. 3).

The MAC resource allocation function scheduler employs an adaptive modulation and coding scheme or MCS for configuring the Physical Traffic Channel or PTCH, which transports the data link PDU from the Logical Traffic Channel or TCH. This protocol enables the adaptation of the MCS to the channel condition and the QoS requirement of the respective RLC PDU. A benign radio channel and a low-priority RLC PDU instance favor an aggressive or weak MCS, i.e., (m, r), with a high modulation order, m, and a high code rate, r (i.e., a low processing gain, $\mathcal{g}$) to achieve high spectral efficiency. On the contrary, a poor channel condition or a high-priority RLC PDU suggests the opposite, i.e., a conservative or strong MCS, at the expense of spectral efficiency.

As explained in Chap. 5, the processing gain is the ability of a physical channel to reject noise, achieved by the added structure in the transmitted signal, like coding and repetition, so that a receiver can leverage the embedded structure to suppress the noise. A high processing gain also means high resistance to interference, hence an effective tool for a MAC scheduler to combat interference at the expense of the radio resources. When achieved through time spreading, the processing gain reaps an additional benefit of boosted signal energy as a result of increased transmission time at a given transmit power. Thus, it is also an effective means for MAC to deal with power-limited mobile stations by extending a physical channel's TTI to energy-boost the physical channel (in particular, E_b/N_0) to close the radio link (see (Eq. 5.123) and (Eq. 5.124)) – essentially trading resources for energy.

Specifically, a set of burst configurations is predefined, each indicated by an identifier coined `Link ID`. Table 7.8 lists examples of a few typical burst configurations. A burst configuration is associated with a unique combination of the

Table 7.8 Link ID examples

`Link ID` (6 bits)	PDU size	Pilot pattern	Burst TTI	Modulation	Code rate
…	…		…	…	…
2	21 bytes	(2, 32)	1 slot	QPSK	0.50
…	…		…	…	…
9	65 bytes	(1, 16)	1 slot	16-QAM	0.78
…	…		…	…	…
35	250 bytes	(1, 32)	3 slots	16-QAM	0.74
…	…		…	…	…

specific physical channel and burst parameters, including the PTCH PDU or SDU size and MCS and the burst TTI and pilot pattern (discussed in Sect. 7.3.3). For instance, `Link 2` corresponds to a conservative MCS with QPSK-modulation ($m = 2$) and a code rate of 1/2 ($r = 0.5$), whereas Link 9 is associated with an aggressive MCS with 16-QAM ($m = 4$) and a code rate of about 3/4 ($r = 0.78$).

The MAC burst configuration PDU includes a 6-bit `Link ID` that accommodates up to 64 burst configurations to facilitate adaptive channel modulation and coding schemes. The configuration is prescribed by the MAC scheduler, transported via the Logical Assignment Channel or ACH, and transmitted on the Physical Signaling Channel or PSCH. It contains the information necessary for the receiver to retrieve the data link PDU, carrying MAC Traffic PDUs and other MAC Signaling PDUs, transported via TCH and transmitted over a PTCH, as shown in Fig. 7.8.

Due to its critical role, the MAC burst configuration PDU is designated with a dedicated logical channel (i.e., ACH) and physical channel (i.e., PSCH) to be detected independently of PTCH by a receiver. Yet, its MAC PDU header can be omitted for signaling efficiency without ambiguity with other signaling PDUs.

7.3.2.2 Automatic Retransmission Request (ARQ)

The MAC Feedback PDU for MAC ARQ is defined in Table 7.6. Since it is a Layer 2 ARQ, this PDU contains information regarding the target Layer 2 identifier (MMSI) and ACK sequence number. Without a dedicated logical channel, a MAC Feedback PDU is transported via TCH, encapsulated (alone or piggybacked with other MAC PDUs) in a data link PDU, and transmitted over a PTCH.

7.3.2.3 Distributed Medium Access Control

The slot acquisition/assignment for a pending burst under ASM is through a distributed medium access mechanism (treated in Sect. 7.1) employed by the MAC resource allocation function. Depending on the specific scheme, information exchange between MAC peers may be needed and is done through one of the MAC

Signaling PDUs, i.e., the Slot Assignment PDU, transported through TCH over a PTCH, along with other MAC PDUs, such as MAC Traffic PDUs. The SO-TDMA based medium access scheme is an example whose PDU is defined in Table 7.7.

7.3.3 *Physical Layer*

With the upper layers taken care of, we are ready to proceed with the physical layer design. According to Fig. 7.8, the physical layer is the layer that converts logical channels into physical channels for physical transmission over the air in a burst and recovered back to the logical channel at the receiver. Although the transmission burst is an RF waveform in order to travel over the air or space, the design is focused on the baseband in the current and following chapters. The conversion between the baseband and passband waveforms is covered in Chap. 5.

7.3.3.1 Physical Channels

For a physical layer PDU (a bit sequence) to be transmitted over the air, it is transformed into a physical channel through a process comprising forward error-correction coding, rate-matching, channel-interleaving, and constellation-mapping. Each process is detailed in Chap. 5. The resultant constellation symbols are then constructed into a burst for being physically transmitted over the air, shown as a flowchart in Fig. 7.10.

The Physical Traffic Channel, PTCH, relies on the MAC scheduler to deal with various channel conditions (i.e., η) that a station experiences, different application QoS', and different category mobile stations. The MCS information, i.e., the choice of (m, r), carried by MAC Burst Configuration PDU, is conveyed to the receiver via ACH over the Physical Signaling Channel or PSCH. Without it, the receiver will not be able to decode the PTCH. Apparently, PSCH itself does not have the luxury of adaptive MCS. In fact, its MCS must be predefined, known to all stations, and designed with sufficient robustness and reliability since its transmit range determines the PTCH range.

PSCH

A PSCH carries the physical layer PDU that contains a 6-bit MAC Burst Configuration PDU from ACH. No CRC is provided by this physical layer PDU. Note that an *undetected* burst configuration PDU error (i.e., a `Link ID` error) causes no extra damage to PTCH decoding due to the fact that a `Link ID` error will fail the decoding of PTCH regardless of whether the error is detected or not and that the integrity of the PDU that the PTCH carries is protected by a powerful 24-bit CRC.

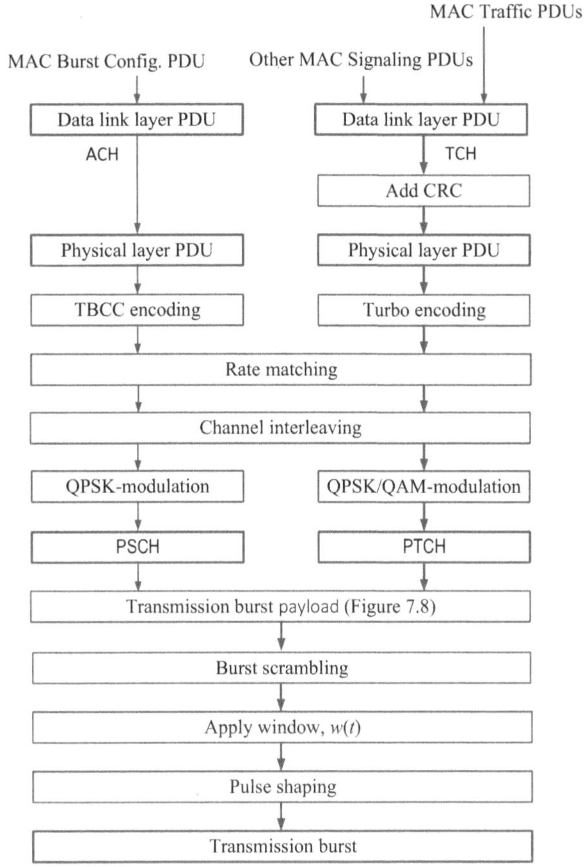

Fig. 7.10 ASM MAC PDUs and transmission burst flowchart, where $w(t)$ is defined in (Eq. 7.19)

Hence, this 6-bit burst configuration PDU is encoded with a code rate 1/3 tail-biting convolutional code or TBCC (see Chap. 5), and the 18-bit code block is rate matched, i.e., repeated three times, onto 27 QPSK symbols to synthesize the PSCH (refer to Fig. 7.8). The repetition provides extra processing gain to redouble the PSCH reliability. PSCH thus provides a processing gain of g=9.5 dB. TBCC is chosen over the turbo code for forward error correction because of its superior performance over the turbo code for small PDUs, as has been shown in Chap. 5 (see Fig. 5.33).

To determine ASM burst coverage, we simply find out PSCH transmission range. Since the required E_b/N_0 for decoding a 6-bit information block, at a target error probability of one percent, is about E_b^+/N_0 =1.5 dB, the minimum receive SNR of a PSCH receiver can be determined from (Eq. 5.124) as

$$\begin{aligned} \eta^+ &= E_b^+/N_0 + m - g \text{ (dB)} \\ &= -5 \text{ dB}. \end{aligned} \tag{7.10}$$

Now consider a transmitting station with transmit power, P, the channel SNR at a receiving station can be expressed as

$$\eta = \frac{P \cdot G(\theta_{tx}) \cdot G(\theta_{rx})}{L \cdot N_0 T_{sym}^{-1}}, \tag{7.11}$$

where $G(\theta_{tx})$ and $G(\theta_{rx})$ are the transmit and receive station antenna gains (including RF coupling losses) and L the propagation loss.

Letting $\eta \geq \eta^{+}$, the propagation loss must satisfy

$$L \leq \frac{P \cdot G(\theta_{tx}) \cdot G(\theta_{rx})}{\eta^{+} \cdot N_0 T_{sym}^{-1}}.. \tag{7.12}$$

where $N_0^{mobile} = \kappa T_{mobile}$, κ is the Boltzmann's constant, and T_{mobile} is the noise temperature of a mobile station receiver, with a default value of 30 dBK, plus an extra 5 dB to account for adjacent channel interference (e.g., AIS in close vicinity) as well as co-channel interference due to distributed medium access and potential incumbent land systems. Further assuming a minimum transmit power of P=5 mW and an average antenna gain of −3 dB for Category 1 mobile stations, we obtain the maximum propagation loss of

$$\widehat{L} = 130\,\text{dB}. \tag{7.13}$$

Factoring in an additional 20 dB loss on top of the free space propagation loss [1], (Eq. 7.13) corresponds to a transmit range of 48 km or 26 NM.

Higher transmit power (e.g., 100 mW and 1 W) extends the range. However, the communication range at sea is ultimately limited by the horizon, which is pertinent to the antenna height. A pair of transceiver antennas that measure 100 meters tall, i.e., $h_{tx} = h_{rx}$=100 m, covers up to a communication range of

$$\begin{aligned} d_{horizon} &= \left(h_{tx}^2 + 2R_{earth}h_{tx}\right)^{\frac{1}{2}} + \left(h_{rx}^2 + 2R_{earth}h_{rx}\right)^{\frac{1}{2}} \\ &\approx 73\text{ km or }40\text{ NM}. \end{aligned} \tag{7.14}$$

PTCH

A PTCH carries the physical layer PDU (i.e., the physical layer SDU plus a trailer – a 24-bit CRC) associated with TCH. The SDU integrity is protected by this reliable CRC for error detection with a failure probability of roughly $2^{-24} \approx 0.000006\%$. The PTCH is constructed by first turbo-encoding the physical layer PDU. The resultant code bits are then stored in a circular buffer and consumed in the similar manner outlined in Sect. 5.3.3 of Chap. 5. The number of bits drawn from this buffer to be transmitted over PTCH is determined through rate matching, based on the payload, a function of the burst configuration prescribed by the MAC resource assignment protocol or the scheduler, to achieve the desired effective code rate. The burst configuration is indicated by `Link ID` through the MAC Burst Configuration PDU and conveyed to the receiving peer. After channel-interleaving, these bits are mapped to the constellation symbols to fill up the PTCH.

7.3.3.2 Transmission Burst

An ASM transmission burst is constructed to aid the transmission and reception of the physical channels, a PSCH, concatenated with a PTCH, as diagrammed in Fig. 7.8.

So far as the transmitter is concerned, the emission power spectrum and the energy and spectral efficiency are the primary issues. Although the GMSK waveform has a power spectrum with superior frequency localization and a PAPR that allows the use of a high-efficiency nonlinear power amplifier, it lacks a simple way for an extension to more spectrally efficient higher-order modulations and, hence, precludes the incorporation of an adaptive MCS into the medium access protocol. So far as the receiver is concerned, the inter-symbol interference issue of GMSK complicates the demodulator, recalling that GMSK is not a Nyquist pulse. Taking all into account, QAM is a favored choice.

From Chap. 5, the QAM modulation symbol waveform can be represented as

$$s_k(t) = \boldsymbol{c}_k \cdot p\big(t - kT_{\text{sym}}\big), \tag{7.15}$$

where $\boldsymbol{c}_k$ is the data-dependent constellation symbol of the kth modulation symbol and $p(t)$ is a square-root raised cosine pulse defined in (Eq. 5.37). The choice of this pulse is to ensure that the station on the opposite side of the aisle receives $\boldsymbol{c}_k$ free of inter-symbol interference.

According to Sect. 5.2.1 in Chap. 5, ideally (i.e., with *infinite* pulse duration), the power spectrum of $p(t)$ is *strictly* confined within $(1+\sigma)T_{\text{sym}}^{-1}$, where σ is the roll-off factor. Given a radio channel bandwidth of $|\mathbf{V}^{\text{ASM}}|$=25 kHz per Fig. 7.1, we wish a maximum symbol rate, T_{sym}^{-1}, subject to

$$(1+\sigma)T_{\text{sym}}^{-1} \leq \left|\mathbf{V}^{\text{ASM}}\right|, \sigma \geq 0. \tag{7.16}$$

From Chap. 5, the maximum rate is the well-known Nyquist rate, i.e., $T_{\text{sym}}^{-1} = \left|\mathbf{V}^{\text{ASM}}\right|$ symbols/sec, implying $\sigma = 0$..

The problem here is that a practical pulse must be time-bounded, and hence $p(t)$ is truncated to become

$$p_{\text{trunc}}(t) \triangleq p(t), \quad -\frac{1}{2}T_{\text{pulse}} \leq t < \frac{1}{2}T_{\text{pulse}}, \tag{7.17}$$

with a finite duration of $T_{\text{pulse}} < \infty$ after truncation. The immediate consequence is the spectrum regrowth and that $p_{\text{trunc}}(t)$ in (Eq. 7.17) is no longer frequency-bounded and (Eq. 7.16) fails to hold, regardless of the choice of T_{sym}^{-1}..

To see the effect of truncation on the emission power spectrum, we need to consider the transmission burst as a whole since the on-off transition nature of the burst also affects the spectrum. We represent the burst as

$$s_{\text{burst}}(t) = w(t) \cdot \sum_{k=0}^{T_{\text{burst}} T_{\text{sym}}^{-1} - 1} c_k \cdot p_{\text{trunc}}(t - kT_{\text{sym}}), 0 \le t < T_{\text{burst}}, \tag{7.18}$$

where T_{burst} is the burst length in second and $w(t)$ is a window function that reflects the ramp-up and ramp-down of the burst for spectrum-shaping, defined as

$$w(t) = \begin{cases} T_{\text{ramp}}^{-1} \cdot t & 0 < t < T_{\text{ramp}}, \\ 1 & T_{\text{ramp}} \le t \le T_{\text{burst}} - T_{\text{ramp}}, \\ T_{\text{ramp}}^{-1} \cdot (T_{\text{burst}} - t) & T_{\text{burst}} - T_{\text{ramp}} < t < T_{\text{burst}}. \end{cases} \tag{7.19}$$

The power spectral density of the burst is plotted in Fig. 7.11 with various values of σ, given a ramp duration of $T_{\text{ramp}} = 4T_{\text{sym}}$.

As expected, Fig. 7.11 demonstrates that truncation gives rise to spectral regrowth, invalidating (Eq. 7.16), and larger values of σ incurs less out-of-band emissions. From Chap. 5, a larger σ also benefits time localization, ergo less potential inter-symbol interference. Therefore, we want to leverage σ to maximally

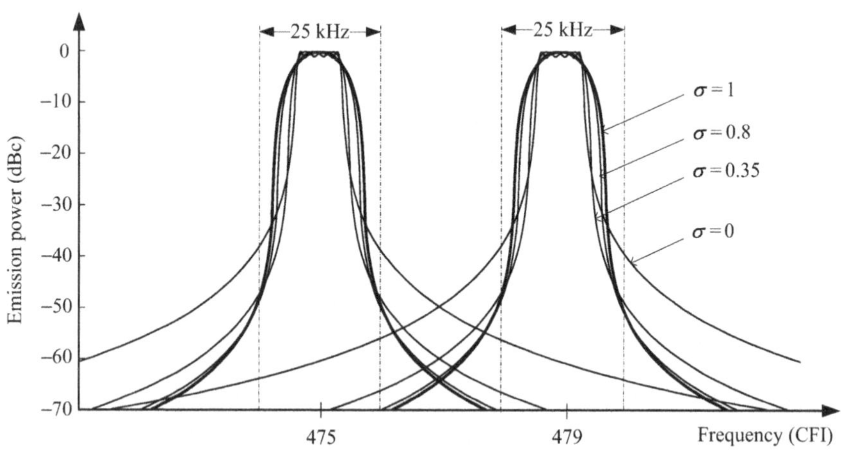

Fig. 7.11 Power spectral density of the ASM burst (one slot and $T_{\text{ramp}} = 4T_{\text{sym}}$) with square-root raised cosine pulses of various roll-off factors, where the channel symbol rate is chosen to be the same as AIS, i.e., 9600 symbols/sec, and $T_{\text{pulse}} = 10T_{\text{sym}}$ and the constellation symbols are randomly selected. CFI, i.e., the center frequency index, is defined in Fig. 7.1

suppress the out-of-band emissions for protection of the legacy AIS station from being interfered with by nearby ASM stations, knowing that AIS is vulnerable to adjacent channel interference due to lack of channel coding. Indeed, with the frequency channels assigned shoulder to shoulder, ASM does not have much choice but to run at a much lower channel symbol rate (than the Nyquist rate) to leave room for a larger σ.

Therefore, the ASM burst is designed to transmit at the same symbol rate as AIS, i.e., $T_{\rm sym}^{-1}$ =9600 symbols per second or 256 symbols per slot, with a roll-off factor of $\sigma = 1$ to ensure minimal interference to AIS.

Finally, just like the `buffer` of an AIS burst for avoiding the inter-burst interference from different transmitters at a receiver, a `guard period` may be needed at the end of the burst to provide a time buffer for absorption of propagation delays and timing errors. The propagation delay is upper-limited by the horizon, i.e., $d_{\rm horizon}$=73 km given by (Eq. 7.14), corresponding to a $\delta_{\rm MAX}\approx$250 μs one-way delay.

1. A station with a UTC source may synchronize directly to UTC using, e.g., the 1-pulse-per-second (or 1-PPS) signal from a GPS receiver. The UTC timing error can be assumed to be less than 200μs, i.e., $\left|\varepsilon_{\rm mobile}^{\rm UTC}\right| < 200$μsfor a mobile station. We thus require a time buffer of length

$$\begin{aligned} T_{\rm buffer} &> \delta_{\rm MAX} + 2\left|\varepsilon_{\rm mobile}^{\rm UTC}\right| \\ &\approx 650\ \mu\text{s}, \end{aligned} \tag{7.20}$$

about seven symbols, to prevent a receiver from seeing a burst from a transmitter protruding into the next slot used by another transmitter.

2. In the event that information from the dedicated GPS is interrupted, a mobile station may acquire UTC timing from the VDES network, i.e., the control station (either a VDE-SAT space station or a VDE-TER shore station). The acquisition procedure for network timing is described in Sect. 9.3 of Chap. 9, from which the accuracy of slot timing provided by a control station, i.e., $|\varepsilon_{\rm TA}| + \left|\varepsilon_{\rm sat}^{\rm UTC}\right|$, is at least 72 μs (with respect to UTC), excluding the error from mobility. After taking everything into account, a time buffer of

$$\begin{aligned} T_{\rm buffer} &> \delta_{\rm MAX} + 2\left(|\varepsilon_{\rm TA}| + \left|\varepsilon_{\rm sat}^{\rm UTC}\right|\right) \\ &\approx 400\ \mu\text{s}, \end{aligned} \tag{7.21}$$

i.e., four symbols, is needed for containing the propagation delay and timing errors. Details regarding the network timing protocol and associated errors are developed in Chap. 9.

Table 7.9 MAC signaling PDU: Sync state PDU

Type	Control Function ID	Sync stratum	Reserved
1	0000010	2 bits	6 bits

3. In case when neither GPS nor VDES network is present, a mobile station always tries to synchronize with the most accurate timing source available to it. Hence, a mobile station needs to be aware of the synchronization hierarchy in the proximity and correspondingly determines its timing stratum regularly. A mobile station thus looks for mobile stations in proximity and synchronizes with the highest synchronization stratum. This behavior allows a mobile station to stay as close to the UTC as possible. The mobile station in sync with GPS has the highest stratum, the one with derived UTC the second, and the one with neither the lowest. If more than one station is found to have the same highest stratum in its neighborhood, the mobile station synchronizes to the lowest MMSI. This rule prevents forming an endless synchronization loop. To this end, a station needs to indicate its synchronization stratum. The MAC Sync State PDU shown in Table 7.9 serves the purpose.

   Note that timing acquired from the ASM signal of a neighboring mobile station inherits a one-way propagation delay. Therefore, the time buffer should be able to absorb a two-way propagation delay, which requires a time buffer of

$$\begin{aligned} T_{\text{buffer}} &> 2\left(\delta_{\text{MAX}} + \left|\varepsilon_{\text{mobile}}^{\text{UTC}}\right| + \left|\varepsilon_{\text{detect}}^{\text{ASM}}\right|\right) \\ &\approx 1 \text{ ms}, \end{aligned} \tag{7.22}$$

   or 10 symbols, where $\left|\varepsilon_{\text{detect}}^{\text{ASM}}\right| \le T_{\text{sym}}/2 = 52\mu\text{s}$ is the timing detection error at the mobile station. This is clearly the "worst" scenario of the three, and a time buffer of T_{buffer}=10 symbols is needed to containing all three cases. It leaves 600 μs for buffering the mobility-induced error in the second scenario.

As aforementioned, the `ramp-up` and `ramp-down` of a burst provide smooth power-on and power-off transitions to shape the burst spectrum. At the receiver, they, in effect, provide an extra time buffer up to T_{ramp} (416 μs or four symbols) for collision protection between bursts caused by propagation delays as well as timing errors. Exploiting the `ramp-up` and `ramp-down`, a `guard period` of

$$\begin{aligned} T_{\text{guard}} &= T_{\text{buffer}} - T_{\text{ramp}} \\ &\approx 600\ \mu\text{s}, \end{aligned} \tag{7.23}$$

or six symbols, is included at the end of a burst to provide the burst with a 10-symbol worth of time buffer.

Another primary concern from a receiver's perspective is synchronization with the incoming signal, as noted in Chap. 5. Following the `ramp-up` is the `sync word` (see Fig. 7.8), a predefined waveform for use by a receiver to time-

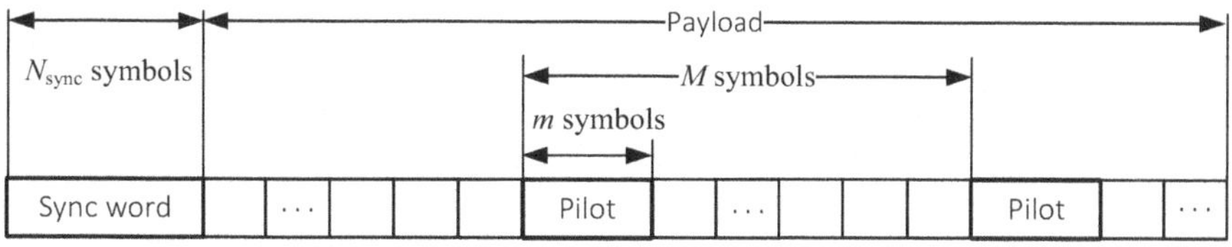

Fig. 7.12 Pilot pattern definition, where the `pilot` duty cycle, denoted as (m, M), is configurable through `Link ID` of the MAC Burst Configuration PDU

and-frequency synchronize to the incoming ASM burst and estimate the channel for coherent demodulation of the subsequent PSCH and PTCH. The sync word occupies 64 channel symbols, initialized with the constellation symbol, $e^{j\pi/4}$(see Fig. 5.11).

Note that the frequency error or offset between the incoming signal and the receiver may include the potential Doppler frequency shift that cannot be eliminated or reduced by a more accurate frequency source, e.g., GPS, and hence needs to be estimated and corrected through sync word. As explained in the previous section, residual frequency estimation errors cause the received signal phase to ramp, thereby jeopardizing the coherent demodulation of the modulation symbols of the burst payload. Moreover, channel dynamics may also introduce variations during the burst, which can only be estimated and compensated on the fly. Auxiliary synchronization signals or `pilots` (initialized with the constellation symbol, $e^{j\pi/4}$) are thus pre-inserted onto the burst to assist a receiver in tracking the radio channel variability and reducing synchronization errors. The pilot pattern is depicted in Fig. 7.12, where a smaller M, i.e., denser pilots, is suitable for higher dynamic radio channels, whereas a larger m is better in dealing with low SNR channels. A sequence of `pilots` with $M = 16$ is suitable for a residual frequency error up to 300 Hz. Pilot-based channel estimation has been extensively studied. Various algorithms are available in the literature, such as [2] and references therein.

In the absence of `pilots`, the maximum payload of a one-slot burst is 178 channel symbols for carrying the physical channels, of which PSCH takes 27 symbols. The actual payload left for PTCH depends on the pilot pattern, (m, M).

Finally, as shown in Fig. 7.10, the burst is scrambled. The scrambling process and the significance of scrambling are explained in Sect. 8.4.3 of Chap. 8.

References

1. *Method for point-to-area predictions for terrestrial services in the frequency range 30 MHz to 3000 MHz*, Rec. ITU-R P.1546-5, ITU, Geneva, Switzerland, Sep. 2013
2. T. Liu, W. Mao, A minimum complexity high-performance channel estimator for MIMO-OFDM communications. IEEE Trans. Veh. Technol. **59**(9), 4634–4639 (2010)

Chapter 8
Terrestrial Air Interface Design

Abstract Both AIS and ASM operate under a distributed medium access control mechanism through ad hoc networking. The advantage is the great flexibility that enables direct communication without the presence of a control station; nevertheless, the disadvantage is also the lack of supervision of control stations that gives rise to frequent collisions among transmitting stations, especially in the high traffic area, such as ports, harbors, and waterways. Collisions cause not only poor overall system efficiency that limits the system capacity but the instability of the system. Both VDE-TER and VDE-SAT are thus designed to alleviate this problem using a centralized resource management mechanism with the benefit of the infrastructure, i.e., the control stations. This chapter provides a practical design of such an air interface that includes a unified network layer, data link layer, and physical layer protocols for operating on the VDES spectrum within the maritime MTC framework outlined in the previous chapters. A unified air interface framework solution for both satellite and terrestrial segments (both under centralized resource management) is one of the important goals of the design. Therefore, although VDE-TER is the primary focus of this chapter, the air interface framework derived in this chapter applies to VDE-SAT, which differs only in configuration parameters by design. A unified framework reduces the complexity and cost of a VDES transceiver and, more importantly, eases the seamless integration of VDES segments, accentuated by spectrum sharing and interference coordination in meeting the unique maritime MTC requirements under the peculiar VDES spectrum assignments as discussed in Chaps. 1 and 6.

Keywords Terrestrial air interface · Centralized medium access control · Maritime machine-type communication · Maritime Internet of Things · Resource frame · Resource block · Interference mitigation · Power control · Coordinated transmission and reception · VDE-TER · VDES

M. M. Wang, J. Zhang, *Machine-Type Communication for Maritime Internet-of-Things*, Wireless Networks, https://doi.org/10.1007/978-3-030-77908-5_8

8.1 VDE-TER for Terrestrial MTC

VDE-TER, as a terrestrial segment of VDES, is an effective means to address the highly non-uniform characteristic of maritime traffic. It is structured via connected cells to alleviate traffic congestion near shore, using the highly efficient centralized medium access control strategy. More importantly, the VDE-TER infrastructure provides connectivity to the maritime information and service network or maritime cloud, allowing information exchange between a mobile station and the maritime service providers. As noted in Chap. 2, under centralized medium access control, all communications are through the control stations in a star network topology, consisting of a "central node," i.e., the control station (in this case, the shore station) of the VDE-TER infrastructure, to which all other nodes, i.e., the mobile stations, are connected. Thus the shore station provides a common connection point, in particular, an access point to the maritime network for all mobile stations (see Fig. 2.1 in Chap. 2). The mobile-originated message is communicated over the uplink to the shore station. The uplink is thus a many-to-one communication link, hence a *multiple access link*. In the current design based on the assigned VDE-TER spectrum, multiple mobile stations share the link mainly in a TDMA fashion, following the slotted transmission timeline depicted in Fig. 7.2 in Chap. 7, however, under a centralized medium access control by the shore station. In the opposite direction, the network-originated message is communicated through a shore station over the downlink to the mobile station(s) following the same slotted TDM timeline. Thus, the downlink is a one-to-many link, essentially a *broadcast* or *multicast link*.

8.2 Network Layer

As outlined in Chap. 4, the maritime MTC network layer consists of two functional elements, network resource control or NRC and maritime service registry or MSR. We have seen in Chap. 4 how they work in tandem for service-centric networking. This section focuses on the radio spectrum and the air interface configuration and provisioning as well as control station coordination and interference management for terrestrial MTC networks, although the same applies to the satellite network.

8.2.1 Radio Resource and Interference Management

The default radio resources globally available for the maritime MTC terrestrial segment are assumed to be the VDE-TER spectrum, as replotted in Fig. 8.1 from Chap. 6. It certainly does not exclude the possible spectrum geographically available to it.

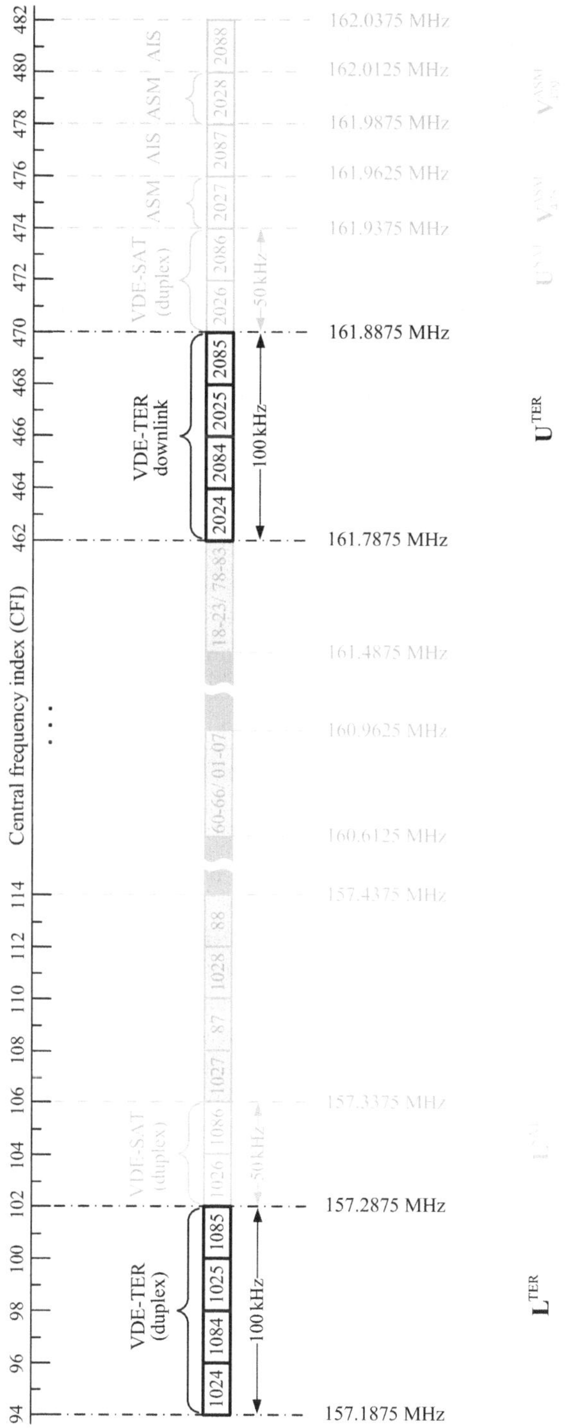

Fig. 8.1 VDE-TER radio spectrum assignments on the VHF maritime mobile band. (Replotted as FDD from Fig. 6.1 in Chap. 6)

At the network level, control stations or cells share the same spectrum for spatial spectrum reuse, and inter-cell interference is a major issue. NRC is responsible for coordination among control stations to mitigate interference, employing techniques such as coordinated transmission and reception. With the coordinated transmission and reception, the downlink data to a mobile station are simultaneously transmitted from a group of control stations to improve the received signal quality at the target mobile station, some of which may simply choose to mute (no transmission) on the specific resource. Joint reception means that the signals received at multiple control stations on this specific resource are jointly processed for enhanced reception performance at the network.

These operations orchestrated by the NRC are transparent to the mobile stations to simplify the mobile station operation and minimize the over-the-air control messages. A detailed example is given in Sect. 8.4.3. Nonetheless, such coordination typically involves heavy message exchanges and timely orchestration at the network level, complicating the network layer design, but agrees with the design principle of maximal offloading of the mobile station complexity to the infrastructure. At the same time, the design emphasizes the simpler and more robust interference suppression technique at the data link layer that leverages the processing gain. It is a data link layer technique but involves coordinating scrambling codes – a special "resource" in interference management at the network layer. Since these topics involve the physical layer scrambling sequence design, we defer them to the end of this chapter.

8.2.2 Air Interface Configuration

Another important responsibility of NRC is related to the configuration and provisioning of an air interface for specific maritime service requirements, given the specific resources in a specific geographic area. NRC works with MSR allocating and configuring network resources to provide appropriate QoS for applications, recalling that MSR holds application portfolios and service profiles and performs transport level QoS marking (via socket groups) and enforcement.

This air interface configuration information is conveyed to the mobile station through a System Bulletin Board (BB) network layer message. A System Bulletin Board message includes the primary system information, such as network and air interface configuration, necessary for mobile stations of different categories to access the system. The Bulletin Board PDU is exemplified in Table 8.1. The BB is broadcast over the PBBCH (see Chap. 3). The PBBCH burst has a synchronization preamble at the beginning of the burst for a mobile station to achieve time and frequency synchronization with the system. The PBBCH burst is also used as a beacon to mark a VDE-TER cell's no-share zone in the spectrum sharing scheme, as described in Chap. 6, for restricting a mobile station from transmitting VDE-SAT signals on the VDE-TER uplink spectrum, $\mathbf{L}^{\mathrm{TER}}$. We have a detailed design on PBBCH later in this chapter.

Table 8.1 Network layer System Bulletin Board PDU

Field	Bit width	Logical channel	Physical channel
`Control station identifier`	30	BBCH	PBBCH
`Bulletin Board version validity`	16		
`System Version`	6		
`Frame configuration`	9		
`Transmit power,` P_{shore}	6		
`Target receive power,` p^{+}_{shore}	6		
`Neighbor list`	4×8		
`System information block`	9		
`Reserved`	78		
Total	24 bytes		

The `frame configuration` field in the Bulletin Board allows 512 predefined radio resource configurations for centralized radio resource management and adaptation to diverse communication environments and application traffic. In particular, by controlling the resource configuration, the network can balance coverage, transmission latency, and system overhead/spectral efficiency as needed, including the configuration of downlink signaling channels, like PDSCH (such as its resources and pilot pattern), choice of bandwidths for different channel conditions, as well as FDD and TDD transmission modes. For TDD, by partitioning the uplink and downlink resources, the network can accommodate asymmetric uplink and downlink traffic loads.

The `control station identifier` is for identifying the control station at the network level. It is a 30-bit field. In the current design, MMSI is used as the identifier.

The `bulletin board version validity` field indicates the lifetime of the current Bulletin Board message, which can be up to 2^{16} minutes or 45 days. Furthermore, the `system version` represents the release version number of the VDES currently in use. If a newer version is detected, a mobile station updates the current version with the latest one.

The `transmit power` and `target receive power` are the default transmit power by the control station on the downlink, and the expected received power by the control station on the uplink. They are used for power control during initial random access, which is treated in Sect. 8.4.2.5.

The `neighbor list` contains eight indices to the scramble sequence used by the eight neighboring control stations for PBBCH scrambling, as discussed in Sect. 8.4.3, for monitoring the interference from neighboring control stations. If necessary, e.g., significant interference from a nearby control station is identified, interference and channel measurements are reported to the NRC via a network layer control PDU for potential coordinated transmissions with interfering cells and, possibly, handover.

Finally, a valid (i.e., non-zero) `system information block` pointer points to the resource block that contains additional system information or announcements, if so needed.

A control station broadcasts the Bulletin Board message every minute via designated Socket Group 0 (see Chap. 3). The latest valid System Bulletin Board message received is instigated in the frame immediately following the frame in which it is received. Unlike the other network layer PDUs (control or application) that include a header, the System Bulletin Board does not have a network PDU header since it has designated layer channels, i.e., Socket Group 0, BBCH, and PBBCH, hence no ambiguity there when traversing from the control station to its mobile station peers.

8.3 Data Link Layer

We now turn our attention to the data link layer and focus on the MAC sublayer design. This section provides a detailed design of the radio resource control framework pertaining to the centralized medium access control under the radio spectrum constraints and the TDMA slot structure.

8.3.1 Resource Block and Frame

Remember that under the distributed medium access control employed in ASM for proximity communication, the time slot resource structure is sufficient for TDMA-based medium access. However, under centralized medium access control, a more sophisticated structural framework is needed for efficient resource management at the data link layer. From the data link layer's perspective in the current design, the radio resource is structured through "framing," on top of the time slots, to enable centralized TDMA-based medium access. Under this framework, the radio resources are further hierarchically structured into radio resource blocks and frames, facilitating various levels of data link layer functional association and control for ease of implementations of diverse MAC functions and efficient centralized radio resource management.

8.3.1.1 Resource Block

Based on the logical and physical channels defined in Sects. 3.3.3 and 3.3.4, a unified resource organization framework is established for centralized medium access, using the *radio resource block* concept – the building block for air interfaces with centralized medium access control. In other words, the resource block represents the base structure and time scale of MAC control. It is introduced for ease of

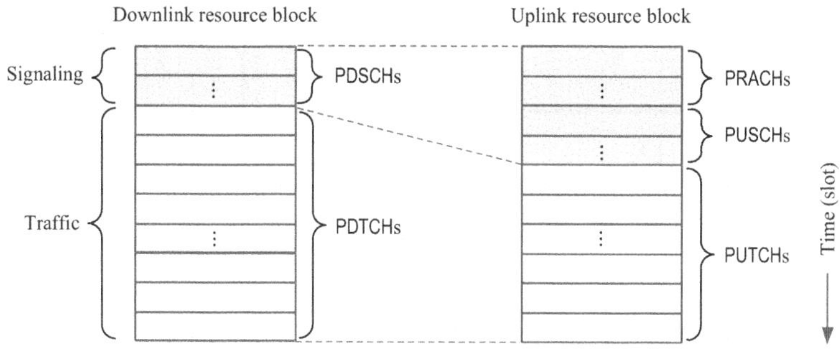

Fig. 8.2 Transmission resource block for centralized resource control

adaptation to various radio channel allocations for scalability and, more importantly, for ease of adaptation to various types of deployments, including terrestrial and satellite deployments as well as FDD and TDD deployments.

As illustrated in Fig. 8.2, a resource block consists of a group of sequential time slots divided into a *signaling zone* and a *traffic zone*. The downlink signaling zone contains a PDSCH, and the traffic zone includes a plurality of PDTCHs. The uplink resource block contains a plurality of PRACHs and PUSCHs in the signaling zone shared among multiple mobile stations in a *code-division multiplexing* (CDM) fashion and multiple PUTCHs in the traffic zone *in a time-division multiplexing* (TDM) fashion. CDM relies on three techniques, power control, processing gain, and scrambling sequence, to suppress interference, which will be discussed later in this chapter. Although both types of resources are shared among multiple mobile stations, they are shared differently. The PRACH resource is shared among multiple random mobile stations (unknown to the control station), contending for the resource in an unscheduled hence contention-based fashion under *open*-loop power and timing control; the PUSCH resource, in contrast, is shared among the mobile stations that are expected to send acknowledgments on these resources in a scheduled and hence non-contentious fashion under *closed*-loop power and timing control. Timing adjustment is deferred to Chap. 9. The reason to use CDM to share the resource is that the SDU that the PUSCH carries is small, as we will see, and a slot is simply too large of a resource for one PUSCH. CDM is a simple and "clean" solution to this resource granularity issue.

The TTI (in units of slots) of these physical channel bursts is configurable by MAC. The MAC may leverage the TTI to provide adaptive processing gains to varying channel conditions. For the uplink, TTI may also be used by MAC to accommodate various categories of mobile stations with different transmit power limitations. A longer TTI helps power-limited mobile stations boost the signal energy.

The resource block configuration, i.e., the actual size of the resource block and the TTIs of the respective signaling channels, i.e., PDSCH, PUSCH, and PRACH, are inferred from the `frame configuration` field of the Bulletin Board. The TTI of PUSCHs can be dynamically altered via CCCH by datalink layer MAC. The traffic channels, i.e., PDTCH and PUTCH, are dynamically configured through ACH.

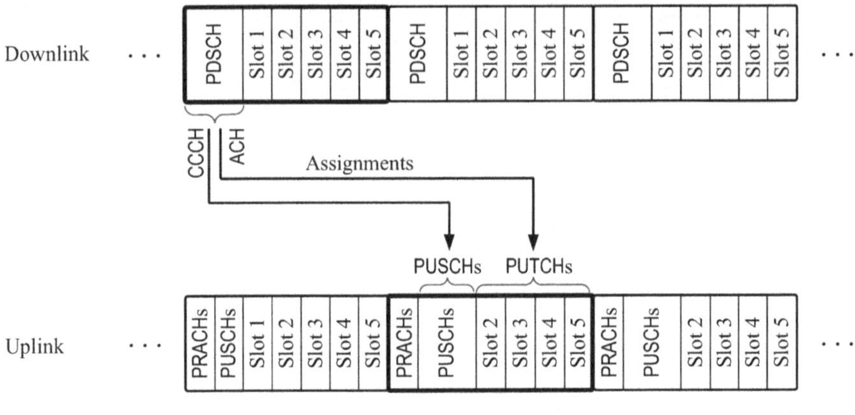

(a) Uplink UTCH assignments and PRACH and PUSCH configurations

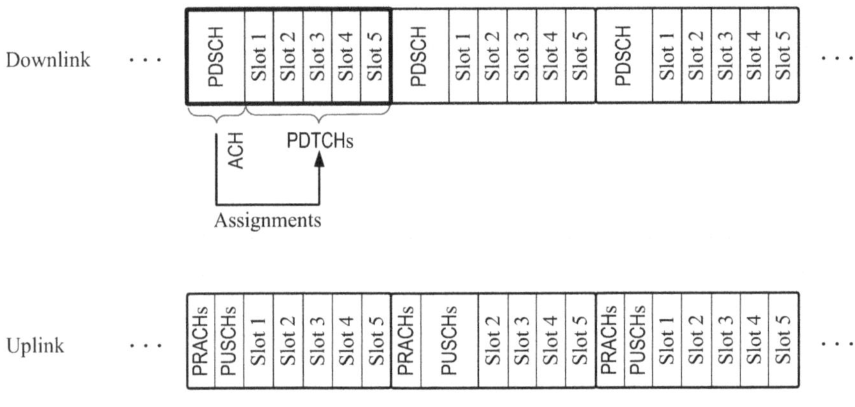

(b) Downlink PDTCH assignments

Fig. 8.3 Resource block pairing for centralized resource assignments under FDD: (**a**) uplink resource assignments for PUTCHs via ACH, and PUSCH resource configurations (e.g., TTIs) via CCCH; (**b**) downlink resource assignment of PDTCHs via ACH

For ease of centralized resource management and transmission error control, a downlink resource block and an uplink resource block are implicitly paired in a way such that the logical association or interaction between those channels involved in a transmission can be clearly defined. This logical pairing creates cohesion between the logical channels, as depicted in Figs. 8.3 and 8.4. Sufficient time for propagation delays and physical layer processing at the receiving end (e.g., filtering, demodulation, and decoding) is reserved between the associated channels. For instance, one resource block worth of time is available for propagation delay and the mobile station to process the PDSCH in terrestrial and LEO satellite communications.

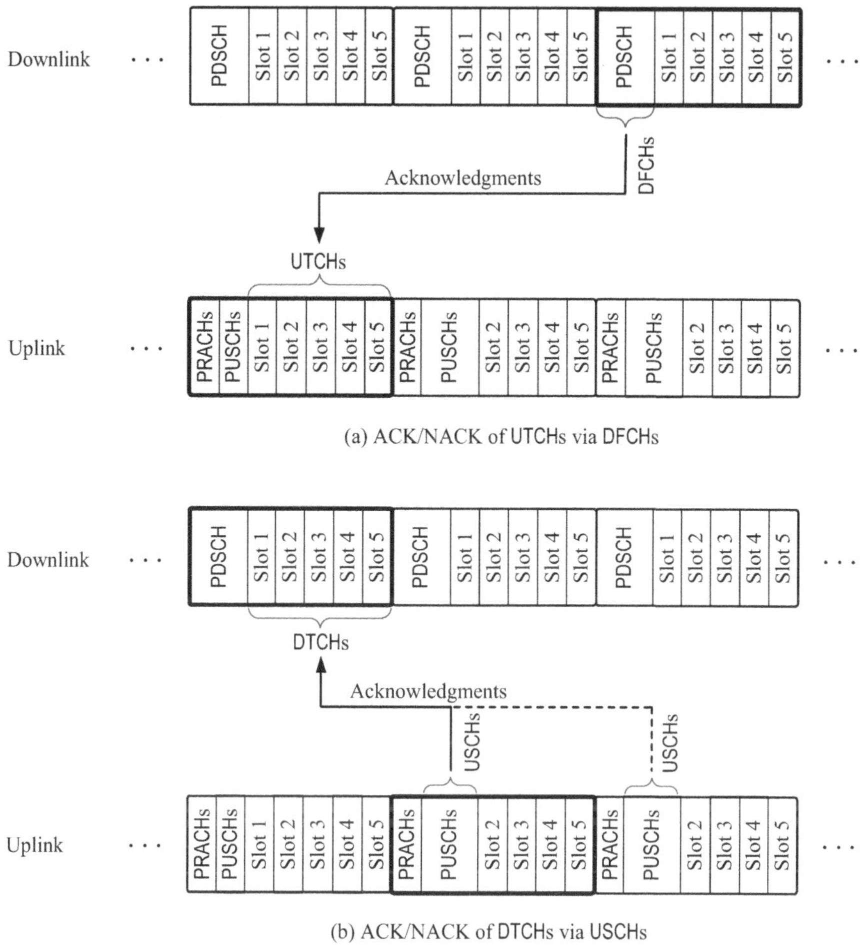

Fig. 8.4 Resource block pairing for ARQ under FDD: ACK/NACK feedbacks of the respective (**a**) UTCHs via DFCH; (**b**) DTCHs via USCHs, where at least one slot (or one block if longer delay is needed) worth of time is available for propagation delay and for mobile station to process the PDTCH and prepare to transmit ACK/NACK on PUSCH

8.3.1.2 Resource Frame

A resource frame in the current design is one minute long, containing exactly 2250 slots although a shorter frame (e.g., 30 seconds) is possible to allow more dynamics at the cost of more system overhead. These one-minute worth of slots are time-aligned with the UTC minute boundary for ease of global frame synchronization among stations. A resource frame contains a plurality of resource blocks for MAC *scheduling*, error control, random access, and other functions, which will become clear as we get into the details of these functions in the following sections. Figure 8.5 depicts such a frame configuration under FDD.

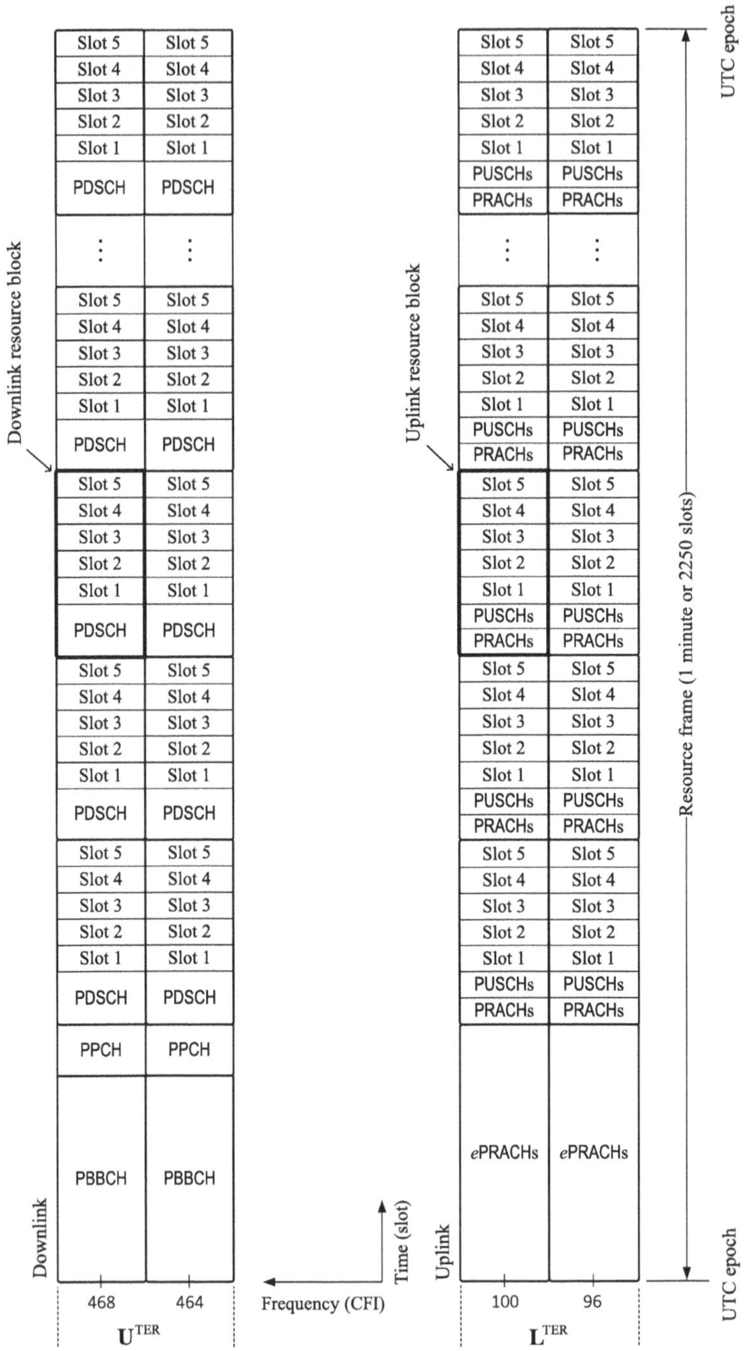

Fig. 8.5 An example resource frame configuration, embodied in an FDD setting, for centralized radio resource control

The frame structure allows dynamic control of these resource blocks, on a frame-by-frame basis, at the network level through the System Bulletin Board PDU for adaptation to application traffic characteristics and deployment environments. The only thing that is kept unconfigurable by design is the PBBCH for ease of system acquisition. However, additional system information, e.g., additional VDE-TER bands to allow scalability for future growth, can be broadcast as needed via a pointer embedded in the System Bulletin Board (see Table 8.1). It always appears at the beginning of a frame (i.e., at the top of the minute) and at the fixed carrier frequency with fixed duration and bandwidth (i.e., 50 kHz), regardless of the system bandwidth, under FDD or TDD. The resource is dedicated, meaning not sharable between different air interfaces.

8.3.2 MAC Protocols and PDUs

As described in Chap. 3, the network layer PDUs (network Control PDUs and Traffic PDUs outlined in Chap. 4) are sorted by the socket group and prepared by corresponding RLC instances of the data link layer for QoS marking and transmission through the physical layer to the receiving network peer. For that, a suite of MAC control or signaling protocols must be in place to ensure reliable and timely delivery. The RLC PDU is carried by the MAC Traffic PDU, and the MAC control and signaling messages for downlink or uplink resource allocations and other operational aspects are conveyed by the MAC signaling PDUs.

The MAC Traffic PDU and formation into the data link PDU have been defined in Chap. 3. As will become clear, the Layer-2 identifier, i.e., the station's MMSI, is implicitly indicated by scrambling the physical channel signals (Sect. 8.4.3) or explicitly indicated in the resource assignment (next). Hence, the data link header is saved without causing ambiguity.

This section describes the implementation of the MAC signaling protocols and PDUs based on the resource block and the resource frame, under the framework of MAC protocols and PDUs defined in Chap. 3. Without loss of generality, we use the frame configuration in Fig. 8.5 in the following discussions.

8.3.2.1 Resource Assignment

To enable centralized resource scheduling, the MAC resource allocation or scheduling function uses Resource Assignment PDU, via ACH transmitted on PDSCH, to commute the PTCH (PDTCH or PUTCH) resource assignments to the mobile peers. One way of doing that is the use of the *assignment element*. At the finest granularity level, slots in the traffic zone of the resource block pair are *individually* assigned to mobile stations as the resource for PDTCH or PUTCH, indicated by an assignment element included in the MAC Resource Assignment PDU shown in Table 8.2. In general, the assignment granularity can be more than one slot and is inferred through the frame configuration field of the System Bulletin Board. Finer granularity favors short burst data services but requires more signaling overhead.

The definition of the assignment element is shown in Table 8.2a. An assignment element includes an MMSI, a PDU size, a power adjustment (PA), a timing adjustment (TA), and a new data indicator (ND).

The MMSI field contains the MMSI of the destined mobile station of the TCH (UTCH or DTCH). As we recall from Sect. 2.1 in Chap. 2, a valid ITU MMSI ranges from 0 to 999,999,999; an assignment element with an assignee MMSI value of 1,000,000,000 is dedicated to a broadcasting TCH. Similarly, an assignment element with an MMSI value of 1,000,000,001 represents a *void* assignment element that can be used as a place holder.

The new data indicator indicates whether the TCH that the PTCH carries is a retransmission. The PDU size field contains an index into a predefined PDU size table that stores up to 256 PDU sizes in ascending order. A 4-bit field, representing the four least significant bits (LSBs) of the table index, can thus index the 16 smallest PDU sizes in the table.

The power adjustment field is meant for the MAC power control function to instruct the mobile station to adjust its transmit power for its next uplink transmission, such as acknowledging the PDTCH transmitted on PUSCH, as we shall see shortly.

Finally, the timing adjustment field is for uplink transmission timing adjustment, which is elaborated in Sect. 9.3 of Chap. 9.

Each of these individual slots can be a PTCH (PDTCH or PUTCH) or part of a larger PTCH that consists of more than one slot. For example, the UTCH of a mobile station is assigned a 2-slot PUTCH when receiving two assignment elements destined for it. A continuation bit is included in the first assignment element to flag this scenario. If true, i.e., "1", the subsequent assignment element is a continuation of the current assignment element, which contains the four most significant bits (MSBs) of the PDU size, as shown in Table 8.2b, thereby extending the PDU size to an 8-bit index capable of indexing the entire PDU size table. Hence, a larger PTCH can be configured using the continuation bit.

Given the assigned resource and the PDU size from the assignment elements, the spectral efficiency of the PTCH is then known. From the relationship between the spectral efficiency and the modulation order, as shown in Fig. 8.6, the

Table 8.2a Assignment element Format a

Field	Assignee (MMSI)	PDU size	ND	PA	TA	Continuation
Bit width	30	4 LSBs	1	2	2	1

Table 8.2b Assignment element Format b

Field	Assignee (MMSI)	PDU size	Reserved	Continuation
Bit width	30	4 MSBs	5	1

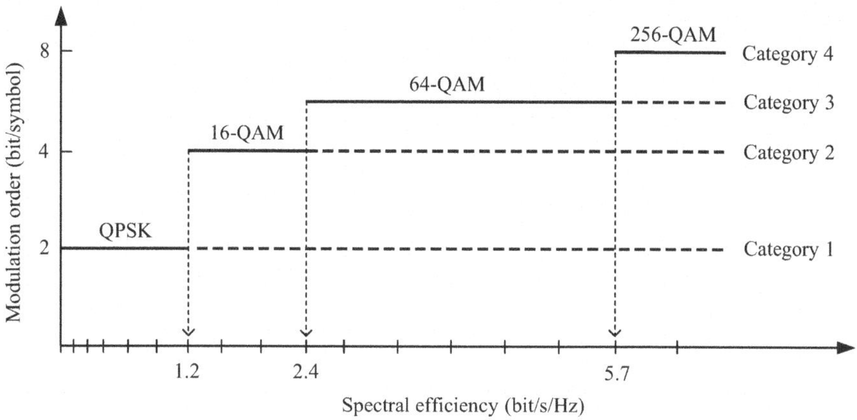

Fig. 8.6 Modulation order vs spectral efficiency and the upper limits set by the mobile station categories (see Chap. 5)

Table 8.2 MAC Resource Assignment PDU (for the frame configuration in Fig. 8.5)

Field	Bit width	Remarks
`Assignment element`	40	Downlink: Slot 1
`Assignment element`	40	Downlink: Slot 2
`Assignment element`	40	Downlink: Slot 3
`Assignment element`	40	Downlink: Slot 4
`Assignment element`	40	Downlink: Slot 5
`Assignment element`	40	Uplink: Slot 1
`Assignment element`	40	Uplink: Slot 2
`Assignment element`	40	Uplink: Slot 3
`Assignment element`	40	Uplink: Slot 4
`Assignment element`	40	Uplink: Slot 5
Total	400	Total 50 bytes

modulation order can be derived at the mobile station (and hence no explicit indication in the assignment is necessary). The highest order is constrained by the mobile station category. For example, the highest order for Category 2 is 16-QAM (Table 8.2).

8.3.2.2 Transmission Error Control

As aforementioned, unlike wired communications, wireless channels are notoriously error-prone. Hence, ARQ must be performed at the lower layer, i.e., the data link layer controlled physical layer ARQ discussed in Chap. 5, rather than the network layer (see Sect. 5.5 of Chap. 5) for more efficient error control. In the current design, the transmission error control follows a synchronous acknowledgment process, i.e.,

Table 8.3 MAC Feedback PDUs

Field	Bit width	Remarks
`ACK/NACK of UTCH on uplink Slot 1`	1	MAC Downlink Feedback PDU (under the frame configuration in Fig. 8.5)
`ACK/NACK of UTCH on uplink Slot 2`	1	
`ACK/NACK of UTCH on uplink Slot 3`	1	
`ACK/NACK of UTCH on uplink Slot 4`	1	
`ACK/NACK of UTCH on uplink Slot 5`	1	
`ACK/NACK on the respective DTCH`	1	MAC Uplink Feedback PDU
`Downlink channel quality indicator`	4	
`Reserved`	1	

with a deterministic timing embedded in the resource block pairing framework, as shown in Fig. 8.4, and is managed by a MAC ARQ function/protocol.

As tabulated in Table 8.3, acknowledging the receipt of the DTCH, transmitted on the PDTCH of the corresponding downlink resource block, is via the MAC Feedback PDU through the USCH, transmitted on the PUSCH of the uplink resource block. Similarly, the response to the receipt of the respective UTCH, transmitted on the PUTCH of the corresponding uplink resource block, is controlled by the MAC ARQ function via the MAC Feedback PDU (see Table 8.3) via DFCH, transmitted on the PDSCH. The DFCHs are transmitted along with the ACH and CCCH on the corresponding PDSCH.

For the pairing scheme depicted in Fig. 8.4, the MAC ARQ protocol has at least one slot worth of time (80/3 ms) for buffering the propagation delay as well as the transceiver processing delay which includes PDTCH/PUTCH reception and PUSCH/ PDSCH transmission preparation. This time buffer is sufficient for VDE-TER but may not be so for VDE-SAT due to a much longer propagation delay in satellite communication. Depending on the satellite orbit (detailed in Chap. 9), a longer time buffer may be necessary to absorb the excess delay (e.g., an extra one resource block worth of time, as indicated by the dashed line in Fig. 8.4). A detailed analysis of this issue is provided in the next chapter, where the impact of the propagation delay on the transmission timing is analyzed.

8.3.2.3 Random Access

The MAC random access function enables a mobile station to access the network's radio resource without being scheduled first. The first slot(s) of each uplink resource block is reserved for PRACHs for this purpose and is shared among mobile stations in a contention-based fashion. However, the PRACH for random access is not logically linked to any other channels in the resource block pair, and the actual TTI is configured by the network through `frame configuration`.

The MAC control PDU for random access is transported via RACH. Table 8.4 is the format of such a Random Access PDU. It supports three types of MAC signaling messages: one for the *uplink resource request*, one for the *paging response* used by

Table 8.4 MAC Random Access PDUs

Field	Bit width	Remarks
`MMSI of mobile station`	30	Uplink Resource Request
`Pending PDU size (value: 1-256)`	8	
`RLC instance`	4	
`Downlink channel quality indicator`	4	
`Mobile station category`	2	
`MMSI of mobile station`	30	Paging Response / slot timing request
`Pending PDU size (value: 0)`	8	
`RLC instance`	4	
`Downlink channel quality indicator`	4	
`Mobile station category`	2	
Total	48	6 bytes

Categories 2 and 3 mobile stations, and one for the explicit *network slot timing request*. All have the same PDU format: the `MMSI` of the requesting mobile station, the `pending PDU size` and `RLC instance` for the requested uplink transmission on UTCH, and the `downlink channel quality indicator` for future downlink transmissions on PDTCH if needed.

The `pending PDU size` indicates the MAC PDU size that the accessing mobile station is requesting for transmission on UTCH. A `pending PDU size` of zero indicates a response to the paging (discussed in the following section) or a network slot timing request when the response is sent by a mobile station that is not on the paging list.

As mentioned in Chap. 2, the MAC resource assignment function or scheduler at the control station determines the modulation and coding scheme according to the spectral efficiency that the radio channel can support (as well as other factors related to the quality of service and scheduling policy) and allocates the corresponding resource for the PTCH per the pending PDU size. The radio channel quality that a receiver sees is often measured by the SNR. However, even for the same offered SNR, the spectral efficiency that a receiver can deliver may vary depending on the receiver's capability. Not surprisingly, a higher-end receiver has superior performance over a lower-end receiver and hence is capable of delivering higher spectral efficiency for the same offered SNR, which people pay the big bucks for. Therefore, a mobile station is expected to translate the radio channel SNR it sees into a corresponding spectral efficiency value (according to its receiver performance) and reports it to the control station via the 4-bit `downlink channel quality indicator` (see Table 8.4).

As also shown in Fig. 8.5, an *extended* PRACH with a 10-slot TTI is allocated at the beginning of an uplink frame, denoted as *e*PRACH, providing an extra 10-dB processing gain. Its primary purpose is energy boost for Category 1 mobile stations with limited transmit power, especially the 5-mW stations. In VDE-SAT (Chap. 9), *e*PRACH has an additional function for absorbing excess propagation delays before an accurate uplink timing is acquired.

8.3.2.4 Paging

As noted in Chap. 3, the data link layer MAC has two means to notify the mobile stations of the pending downlink data, either via ACH or via PCH. The former has the lowest latency at the cost of higher energy consumption since the mobile station needs to listen in on the ACH, broadcast on the PDSCH at the beginning of every resource block, whereas the latter is broadcast through a paging channel PCH broadcast on PPCH, at the beginning of every resource frame. Hence, paging in the current design is for Category 2 and 3 mobile stations (cf. Sect. 5.7 in Chap. 5). These mobile stations receive the network-originated message with an energy-efficiency priority over downlink latency. Nevertheless, the most energy-efficient mobile stations belong to Category 1 mobile stations (cf. Chap. 5), which check ACH for pending downlink messages only during the wake-up for a mobile-originated uplink transmission from, e.g., the mobile application, hence the largest downlink latency. On the other side of the aisle are the Category 4 mobile stations that check every ACH for pending downlink message notifications, ergo the lowest downlink latency.

The MAC Paging PDU contains a list of MMSIs of the mobile stations currently being paged for pending downlink messages. This PDU is transported via PCH and broadcast on PPCH located at the beginning of a frame. Once found in the list, a mobile station (Category 2 and 3) responds with a MAC Random Access PDU (see Table 8.4 with `pending PDU size` = 0) through RACH and transmits it on one of the PRACHs randomly selected from the current frame. It then monitors the ACH for possible downlink assignments on the following downlink resource blocks. Otherwise, the mobile station goes back to sleep until the next PPCH cycle.

8.3.2.5 Resource Block Configuration

A resource frame containing many resource blocks is configured as a whole by the network layer NRC to adapt to the overall traffic statistics through the `frame configuration` field of the System Bulletin Board PDU via BBCH broadcast on PBBCH. Within the so configured frame, adjustments can be made on a resource block basis, i.e., on a smaller time scale, to deal with individual groups of mobile stations. If necessary, the data link MAC may use the Resource Configuration PDU (Table 8.5) to dynamically alter some aspects of a specific resource block via CCCH transmitted on PDSCH. For instance, the MAC may alter a specific PUSCH TTI as a means to deal with, on a block-by-block basis, various categories of mobile stations with various transmit power limitations (see Fig. 8.5). A value 0 means an unchanged TTI, 1, 2, and 3 mean the scaled-down TTIs by factors of 2, 4, and 8, respectively, and 4, 5, and 6 mean scaled-up by 2, 4, and 8, respectively.

The data link MAC may also use the Resource Configuration PDU to specify the `pilot` pattern for PTCH (PDTCH or PUTCH) and PUSCH to dynamically adapt to the specific radio channels that some mobile stations are experiencing. The 3-bit `pilot pattern` field can indicate up to eight predefined pilot patterns.

Table 8.5 MAC Resource Configuration PDU (for the frame configuration in Fig. 8.5)

Field	Bit width	Remarks
PUSCH TTI	3	PUSCH TTI scaling factor
Pilot pattern	3	Transmission burst pilot pattern (except for the PDSCH bursts)
Reserved	2	Reserved for future use
Total	8	One byte

The pilot pattern for the downlink signaling channel, i.e., PDSCH, is dependent on the frame configuration, indicated through the frame configuration field of the Bulletin Board PDU from the network resource controller or NRC for adaptation to different radio channel conditions of deployment environments. The pilot pattern is defined in Fig. 7.12 of Chap. 7.

8.3.3 Transmission Procedure

The logical associations of the logical channels between two resource blocks shown in Figs. 8.3 and 8.4 enclose a complete schedule-transmission-acknowledgment cycle. However, the actual transmission time depends on the origin – the timing of the incoming data call and the number of retransmissions. We use the examples in Fig. 8.7 to further illustrate this process.

In the network-originated example, a network PDU arrives at the data link layer through a socket group and becomes a data link PDU of the shore station at Slot 11, which just misses the scheduling opportunity. It then has to wait for the next PDSCH, i.e., the PDSCH on Slots 17 and 18, on which the pending PDU is scheduled to be transmitted on PDTCH on downlink Slot 19. The data link PDU, now the physical layer PDU via DTCH, is transmitted on the PDTCH on Slot 19 as scheduled, and the mobile station receives PDSCH and decodes the PDTCH sometime during Slot 20. It takes about nine slots or ~0.24 s for the PDU to be transmitted. If successfully received, an ACK is sent on the PUSCHs (Slot 25) of the next resource block pair, which completes the transmission. Otherwise, a NACK is sent instead, which triggers the retransmission process. After a NACK is detected, the shore station schedules the retransmission on Slot 33, announces it on the PDSCH on Slot 31, and the same transmission process repeats until it is successfully received or the maximum number of retransmissions is reached.

In the example of a mobile station originated call, a network PDU arrives at the data link layer of the mobile station at Slot 11, just missing the PRACH opportunity for PUTCH resource request. The request is then sent later on the PRACH on Slot 17 and scheduled in PDSCH on Slot 24 with a PUTCH grant on uplink Slot 33. The elapsed time from the arrival to the transmission and reception of the data is about 22 slots or 0.6 s. Upon successfully decoding the PUTCH, the shore station sends an ACK on the PDSCH on downlink Slot 45 of the next resource block pair.

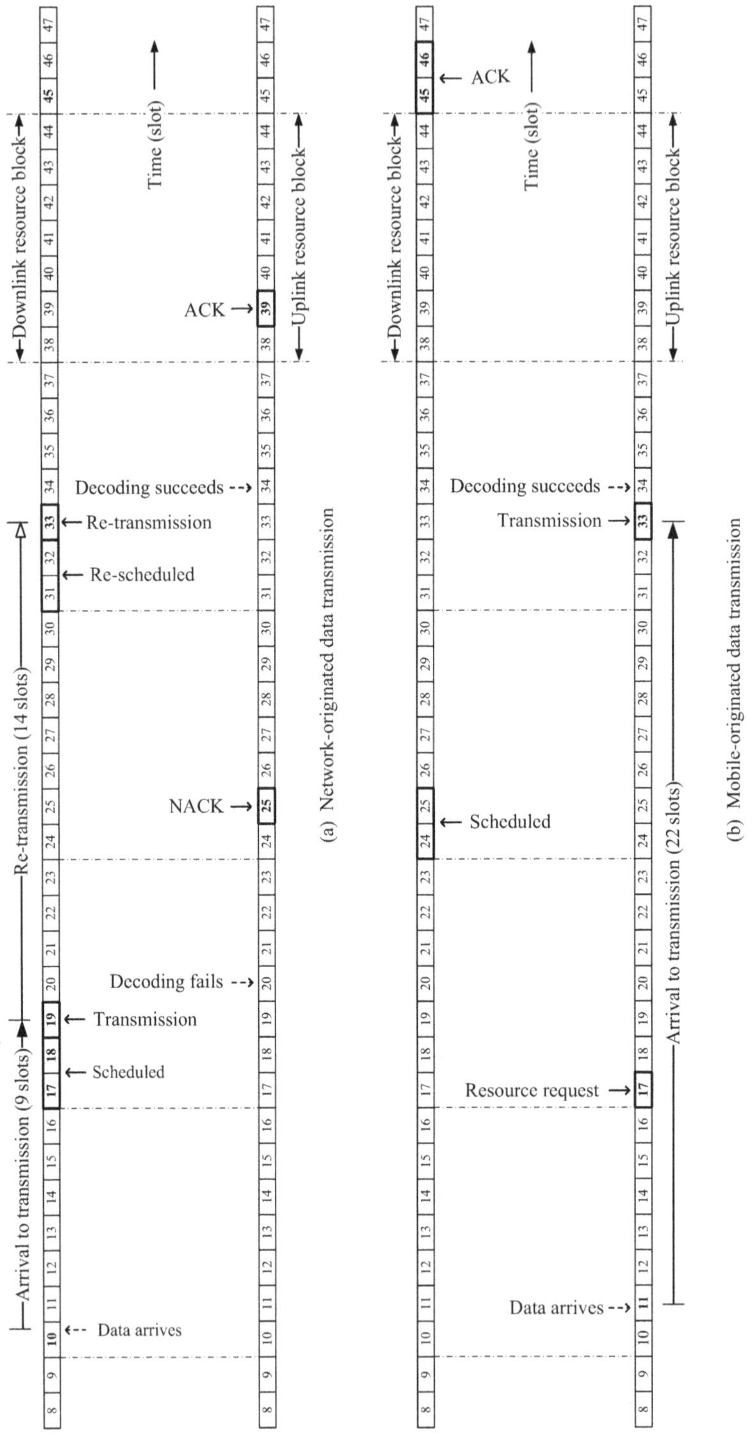

Fig. 8.7 Illustration of (**a**) network-originated, and (**b**) mobile-originated data transmission

8.4 Physical Layer

Built on the general resource framework presented above, the physical layer design strives to achieve the following goals: (1) high adaptability, (2) simplicity and cost-effectiveness, (3) low latency, (4) high energy efficiency, and (5) high overall system spectral efficiency.

However, these goals are generally contradictory since the enhancement of one goal undermines the others. For example, simplicity, cost-effectiveness, and low latency typically come at the cost of overall system spectral efficiency. Hence trade-offs have to be made, and dynamical adjustment and optimization through frame configuration adapted to traffic characteristics at the network level and dynamic resource assignment customized for individual stations at the data link level help ameliorate this dilemma.

Without overstressing the system and causing instability, for the given VDE-TER spectrum (Chap. 6) and the maximum transmit power constraint, the following design example targets a 50-NM (~93 km) radius of coverage and less than 1-second over-the-air transmission latency (from the arrival at the data link layer to the reception at the data link peer of the receiving end) with minimized overhead to maintain overall sustainable system spectral efficiency. Since the maximum coverage of a shore station is upper-bounded by the radio horizon, here, we assume the shore station antenna altitude measures up to 250 m, and the mobile station antenna (e.g., aboard a ship) is up to 100 m.

The physical layer in the current design is structured per the format in Figs. 8.8 and 8.9, following the framework laid out in Sect. 3.3 of Chap. 3.

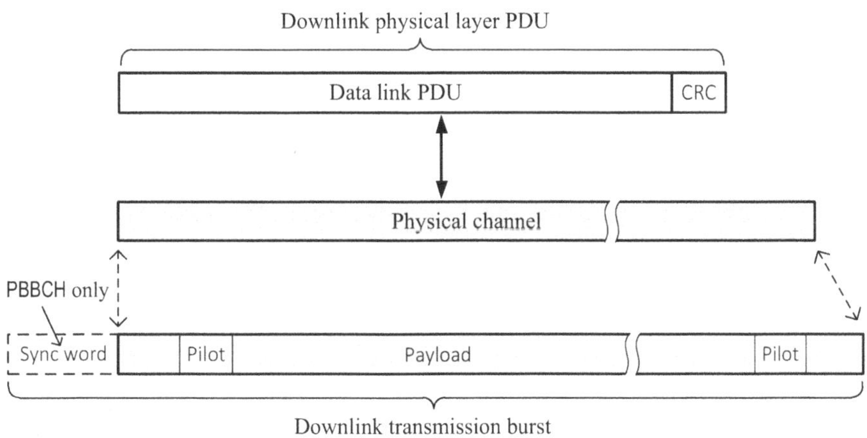

Fig. 8.8 VDE-TER and VDE-SAT transmission burst structure for centralized medium access control. The reason that the downlink burst (except the PBBCH) does not have a `sync word` is explained later in Sect. 8.4.2

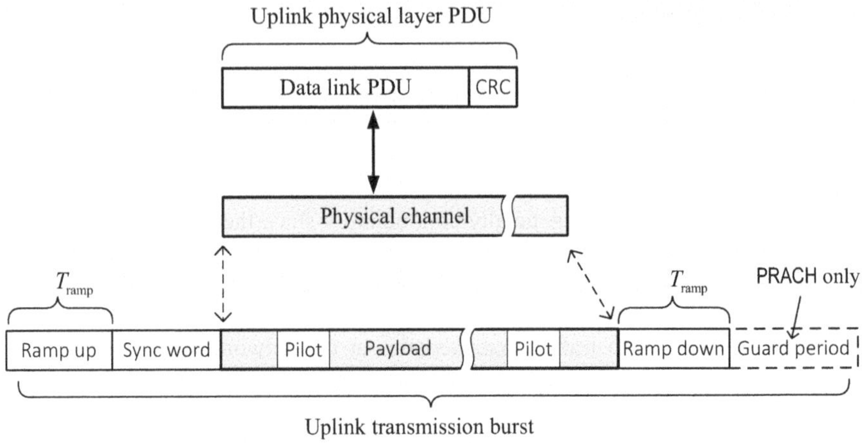

Fig. 8.9 VDE-TER and VDE-SAT uplink transmission burst structure for centralized medium access control. The reason that only the PRACH burst has a `guard period` is explained later in Sect. 8.4.2

8.4.1 Physical Layer PDUs

The data link layer PDU, via the physical layer SDU, is carried by a physical layer PDU. As explained previously, a data link PDU under the current design does not have a header. Therefore, for discussion convenience, the terms "MAC PDU" and "data link PDU" are interchangeable unless a data link PDU includes multiple MAC PDUs, such as MAC Signaling PDU(s) piggybacked with MAC Traffic PDU(s).

A physical layer PDU does not have a header either, but does have a trailer for an error detection code, i.e., CRC. All downlink physical layer PDU provides a 24-bit CRC for the integrity of the SDU that contains the data link PDU. For the uplink, the CRC length of the physical layer PDU depends on the type of MAC PDU it carries. For the physical layer PDU that carries MAC Traffic PDUs, the same 24-bit CRC is used. For the PDU that carries the MAC Feedback PDU, a 6-bit CRC is used, and for MAC Random Access PDU, a 12-bit CRC is the default. The reason that the latter two PDUs have shorter CRCs is due to the fact that they have smaller PDUs (see Tables 8.3 and 8.4).

8.4.2 Physical Channels and Bursts

A physical layer PDU is transformed into a physical channel through a process comprising forward error-correction encoding, rate matching, channel interleaving, and modulation, as detailed in Chap. 5. The resultant constellation symbols are then constructed into a transmission burst that can be physically transmitted over the air, as shown in Fig. 8.10, much like how an ASM burst is constructed, but only that here

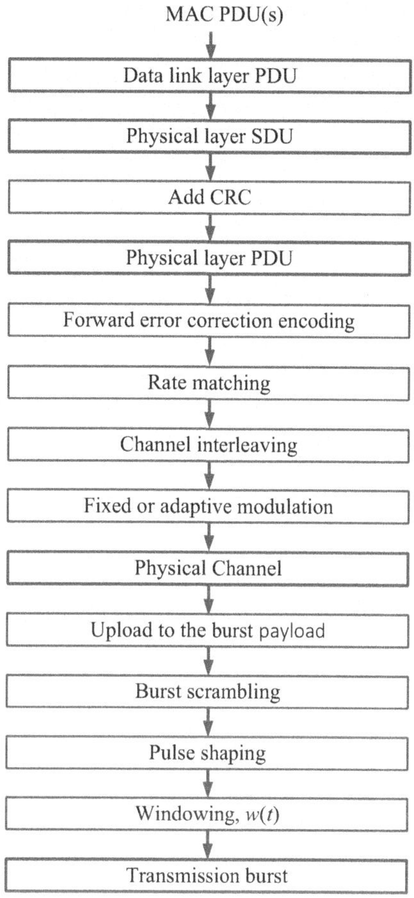

Fig. 8.10 Physical layer PDU, physical channel, and transmission burst generation, where `sync word` and `ramp-up/down`, if present, are always QPSK-modulated, independent of the physical channel's modulation scheme

a burst carries only one physical channel, either a signaling channel or a traffic channel (see Figs. 8.8 and 8.9). This is because, under distributed medium access control, an ASM burst is a standalone burst. It must include a physical signaling channel with a known configuration that carries a MAC Burst Configuration PDU to indicate the current burst configuration. Under centralized medium access control, they are transmitted on separate bursts by taking advantage of the a logical association between the physical signaling channel bursts and the physical traffic channel bursts, under the resource frame and resource block pair framework.

Special techniques to suppress interference are needed for these bursts mainly because of the frequency reuse among cells. In the current design, we rely on the processing gain and scrambling technique to suppress the interference. Processing gain is an effective and robust means of combating interference, as described in Chap. 5. Scrambling will be treated in Sect. 8.4.3 at the end of this chapter.

8.4.2.1 Waveform Selection

Every air interface design problem eventually boils down to the waveform selection. When it comes to waveform selection, the main crossroads one faces is between single-carrier and multi-carrier. Unfortunately, there is no general answer to this "dilemma" since each design problem has its own interrelated requirements and constraints involving the radio spectrum and deployment environments. In this chapter, we continue to follow the design principle that prioritizes mobile stations over control stations in the sense that we shift the complexity and energy burden off the mobile station towards the network infrastructure (e.g., the control station) whenever possible.

Single-Carrier vs. Multi-carrier

Firstly, a typical multi-carrier waveform, particularly an OFDM waveform, transforms a wideband frequency-selective channel into multiple narrowband frequency-flat subchannels negating the need for multi-tap channel equalizers and enabling efficient FFT-based baseband processing. An OFDM-based waveform is thus well-suited for wideband communication systems. The drawback of a single-carrier waveform is that its transceiver is complicated by the need for multi-tap channel equalization in a wideband channel that is susceptive to frequency-selectivity and could become costly depending on the dynamics of the channel.

Secondly, the multi-carrier modulation waveform is the sum of multiple independent subcarrier signals, resulting in a large variation in amplitude, approaching a Gaussian distribution as the number of subcarriers grows, hence poor peak-to-average power ratio or PAPR. On the other hand, a single carrier waveform allows for more relaxed front-end requirements, such as power amplifier linearity, synchronization accuracy, and out-of-band emissions. The relaxed linearity requirement is due to the improved PAPR for single-carrier waveforms, meaning that the power amplifier of a single-carrier transmitter requires a less dynamic range to support a given average output power because of less peak power back-off.

Thirdly, the OFDM time-and-frequency resource grid (each represents one degree of freedom) provides fine granularity in radio resource allocation, particularly in the frequency domain, ergo great flexibility for centralized resource management and scheduling for multiple access. The single carrier waveform falls short in frequency granularity compared to OFDM.

Finally, as pointed out in Chap. 5, a rectangular time-domain window is implied by an OFDM symbol, equivalent to a sinc pulse in the frequency domain, to maintain Nyquist between subcarriers. The high spectral sidelobes entail inter-carrier interference that is sensitive to frequency errors. Moreover, a considerable amount of guard band at two sides of the band is often necessary to suppress them by filtering – an acceptable overhead for wideband systems, however, not for narrowband systems. The increased symbol length of an OFDM symbol (compared to that of a single-carrier waveform of the same overall bandwidth) leads to less sensitivity to symbol timing errors. However, timing errors compromise a cyclic prefix's ability to absorb the multipath delay spread. A cyclic prefix needs to be present at every OFDM symbol – a considerable overhead.

Nevertheless, a single-carrier waveform is preferred to multi-carrier ones for mobile stations in the current design, mainly due to the following considerations:

First, a multi-carrier like an OFDM waveform is an effective solution for combating the multipath effects typical in a wideband channel. However, for line-of-sight predominant narrowband communication at sea, especially via satellites, the single-carrier waveform is more suitable than multi-carrier due to the advantage in transceiver simplicity, consequently, a simpler and cheaper transceiver than a comparable OFDM transceiver, which is critical for low-cost mobile stations.

Second, the poor frequency localization due to excess out-of-band emissions of the OFDM waveform makes it harder for adjacent-channel interference management, especially on the peculiar VDES spectrum, which we have discussed at great lengths in Chap. 6.

Third, a multi-carrier waveform like OFDM is inherently more sensitive to frequency errors than a single-carrier waveform, and synchronization is one of the most resource and energy-expensive components in wireless mobile communications, especially in a high mobility environment where there is no shortage of frequency and time uncertainty. For instance, the Doppler effect may introduce a frequency shift or spread, depending on the traveling speed and direction of a mobile station relative to the control station. High-precision synchronization and tracking is a challenge to low-cost MTC in the highly dynamic maritime IoT environment and, hence, is not well aligned with the maritime IoT principle.

Last but not least is the unification of air interfaces among ASM, VDE-SAT, and VDE-TER, even though multi-carrier may have some advantages for VDE-TER deployment environments, as will be seen next.

Bandwidth

Now that the waveform type is settled, the next natural question is the waveform bandwidth, which is particularly crucial for a single-carrier waveform. Although the VDE-TER spectrum comprises multiple 25 kHz frequency channels, it does not necessarily mean that it is the best bandwidth for the single-carrier waveform, nor does it mean that it should take the whole assigned bandwidth.

On the one hand, the waveform bandwidth for each physical channel should be as large as possible (e.g., the maximum 100 kHz) to avoid splitting the band into multiple 25-kHz sub-bands to save the guard bands that could become a significant overhead. Recall that the orthogonality between adjacent single carriers is protected by the guard band for suppressing inter-carrier interference when transmitting single-carrier waveforms on individual sub-bands.

On the other hand, a smaller bandwidth favors finer resource granularity (given the time granularity, i.e., one slot), allowing scheduling more PUTCHs for more

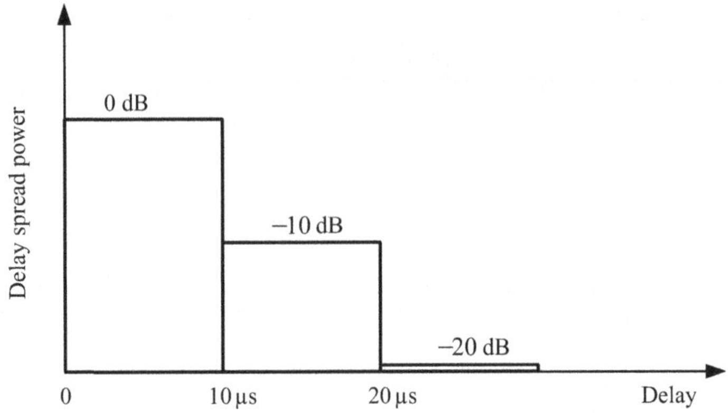

Fig. 8.11 Typical delay spread power profile of a radio channel in a VDE-TER environment

mobile stations on the uplink (a many-to-one multiple access link), useful for short burst traffic. More importantly, a narrower band single-carrier waveform is also more likely to stay frequency-flat, i.e., more resilient to multipath delay spread effects without resort to multi-tap equalizers. In simple terms, assuming a 100-kHz bandwidth, the Nyquist sample width is 10μs. That means, for such radio channel to stay frequency-flat, it requires the delay spread to be confined within 10μs, a condition that could be difficult to be met since VDE-TER is mainly deployed near ports or waterways, typically surrounded by tall buildings and high-rise structures and hills. Significant delay spread over 10μs is not uncommon, as shown in Fig. 8.11. As such, the single-carrier waveform bandwidth should be kept less than 100 kHz, e.g., 50 kHz, at least for the downlink to avoid multi-tap equalizers at the mobile station receiver, at the expense of degraded power amplifier efficiency of the shore station since a single-carrier waveform can no longer be retained on the downlink – less of an issue for a shore station, nonetheless.

As for the uplink, since the transmit power on an uplink physical channel is ultimately limited by the maximum transmit power of a mobile station, dividing the uplink spectrum into smaller sub-bands (e.g., minimum 25 kHz) for more PUTCHs allows more simultaneous transmitting mobile stations, meaning higher overall transmit power on the uplink channel and ergo higher overall spectral efficiency – especially true for the low-power class (Category 1) mobile stations.

Therefore, the default transmission bandwidth is selected 50 kHz for both VDE-TER uplink and downlink – a balance among the above factors. Nonetheless, the bandwidth is configurable through BBCH (Chap. 3) for a particular deployment.

The square root raised cosine pulse and the modulation and coding scheme or MCS, similar to ASM, are employed but supports higher-order modulations of 64-QAM (for Category 3 mobile stations) and 256-QAM (for Category 4) in a better-controlled environment for higher spectral efficiency. The roll-off factor of $\sigma = 0.3$ for the square root raised cosine pulse is selected to satisfy the spectrum mask (as the one shown in Fig. 5.7 and replotted in Fig. 8.12) and at the same time

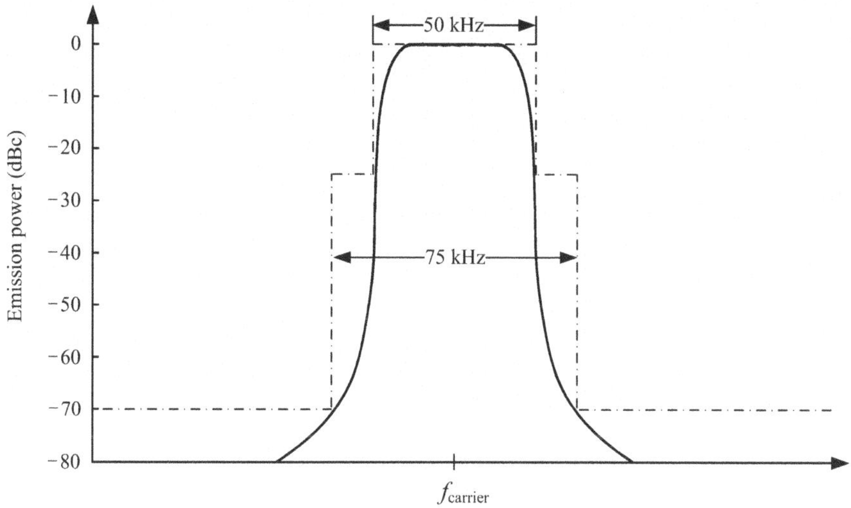

Fig. 8.12 Spectral emission mask and power spectral density of a transmission burst, where burst TTI = 1 slot, and pulse roll-off factor is 0.3

maximize the symbol rate, i.e., $T_{sym}^{-1} = 38{,}400$ channel symbols per second with a symbol interval of $T_{sym} = 26\mu s$, providing 1024 channel symbols per slot.

8.4.2.2 Burst Structure

The physical channel is transformed into a transmission burst to be physically transmitted over the air/space on an assigned radio resource. The burst structure is different for downlink (Fig. 8.8) and uplink (Fig. 8.9) to reflect the following factors: The downlink channel is a broadcast or one-to-many transmission, remembering that the downlink channel is exclusively used for the control station transmissions within a cell. In contrast, the uplink channel is a multiple access or many-to-one channel, shared among multiple mobile stations within the cell, much like ASM, only that sharing is scheduled by the MAC resource assignment protocol in VDE-TER and VDE-SAT, with assignments communicated to the mobile station peers through ACHs.

Downlink and Uplink Bursts

Since the downlink is dedicated to control station transmissions, the downlink has a single continuous transmission source, i.e., the shore station, and hence neither a `guard period` nor `ramp-up/down` is needed in the downlink transmission burst. A `sync word` is only present in the PBBCH burst at the beginning of a frame for frame detection and synchronization. Nevertheless, `pilots` may exist in a burst for channel tracking as well as time and frequency tracking, especially for long bursts and dynamic radio channels. The pilot pattern is configurable per resource block by the MAC resource block configuration function via CCCH, except for the PDSCH burst whose pilot pattern is updated per fame, implied by the `frame configuration` in the System Bulletin Board message through BBCH.

In contrast, the uplink is shared among multiple mobile stations, much like in ASM where stations transmit on the same frequency. As such, an uplink transmission burst by a mobile station contains a smooth power `ramp-up` preamble and a `ramp-down` postamble, each including 16 channel symbols (~0.41-ms), for spectral shaping to reduce spectral spillage. A `sync word` is also present in an uplink burst for a control station to detect and synchronize with the incoming burst from a mobile station. It occupies 64 channel symbols, initialized with a default QPSK constellation, $e^{j\pi/4}$. Like the downlink, `pilots` may be present in an uplink burst whose pattern is configured via CCCH. The uplink bursts received by the shore station are from multiple mobile stations with varying propagation delays depending on the mobile station's distance to the shore station. This discrepancy necessitates a time buffer on the uplink to avoid overlapping between successive transmission bursts from different mobile stations at the control station receiver. In the current design, a `guard period` is saved through uplink transmission timing alignment (detailed in Sect. 9.4 of Chap. 9), a mechanism to align uplink bursts with the network timing at the control station through the MAC timing control protocol. The only exception is for PRACH bursts, at which time the timing adjustment is not possible (the communication between the mobile station and the control station is not yet established).

Both uplink and downlink bursts undergo a scrambling process, for reasons detailed later in Sect. 8.4.3.

Traffic and Signaling Bursts

Depending on a mobile station's position in the cell (i.e., how close to the control station), the radio link between the mobile station and control station experiences different propagation loss, hence different link conditions or receive SNRs. The worst SNR location is the cell edge. Moreover, since all cells share the same spectrum (i.e., 100% frequency reuse) to exploit the spatial spectrum reuse, radio links at different cell positions see different inter-cell interference. The maximum interference within a cell happens at the cell edge, hence the worst SINR – 0 dB at best. Here SINR denotes signal-to-noise-plus interference ratio.

From the previous section (Sect. 8.3.2), we know that physical traffic channels (i.e., PDTCH and PUTCH) rely on the MAC resource allocation protocol (or scheduler) that leverages *adaptive* MCS to deal with various channel conditions as well as different QoS requirements and mobile categories. They also depend on the MAC ARQ protocol to ensure the transmission reliability through retransmissions, i.e., "a second chance," when the intended receiver fails to decode the PDU, as a result of, e.g., a mismatch between the MCS used in the transmission and the actual channel (due to, e.g., inaccuracy in channel prediction).

However, unlike the traffic channels, the physical channels for MAC signaling (i.e., the downlink PBBCH, PPCH, and PDSCH, and the uplink PRACH and PUSCH) do not have the luxury of adaptive MCS or ARQ, which itself involves MAC signaling between transmit and receive peers, implying that the receiver only has one chance to decode the signaling PDU. Therefore, transmission bursts that carry the physical signaling channels must be designed with sufficient reliability and robustness so that the MAC Signaling PDU is reliably received by all category mobile stations, anywhere within the required coverage, with high probability (e.g., 99%). It warrants a careful calculation and design through the link budget analysis.

Furthermore, different signaling channels have distinctive characteristics and operating conditions. Specifically, PDSCH and PPCH are downlink signaling channels expected to be received *within* the cell coverage; that is to say, the transmission range of the signaling channels essentially determines the cell radius. Moreover, their link budgets need to not only cover the propagation loss but absorb the inter-cell interference at the cell edge.

PBBCH is a downlink physical signaling channel typically required to be "visible" *beyond* the current cell, i.e., visible from neighboring cells for, e.g., cell reselection or handover between cells, where SINR can be much worse than 0 dB. For VDE-TER, PBBCH also serves as the beacon signal under the spectrum sharing strategy described in Sect. 6.5 of Chap. 6, requiring an extended transmission range beyond the cell boundary.

The PUSCH resource is shared among mobile stations not only in different cells (full frequency reuse like other physical channels) but also within a cell in a CDM manner (unlike other channels) and hence is typically interference dominated, with an SINR much lower than 0 dB regardless of the mobile station's position in the cell. And the same goes for PRACH.

Therefore, these signaling channels are the focus of the following link budget analysis. Link budget analysis determines the amount of resources needed for the signal to be received over the radio link with an adequate SINR, given the transceiver RF parameters.

8.4.2.3 Downlink Link Budget

The construction of a physical channel is similar to the PTCH in ASM, described in Sect. 7.3.3 of Chap. 7.

For the traffic channel, PDTCH relies on the adaptive MCS, based on the QoS of the MAC Traffic PDUs, and the mobile station's radio channel condition (including link quality and RF capability) for spectral efficiency. It also depends on ARQ to achieve reliability, as detailed in Sect. 5.5 of Chap. 5. The TTI of the burst is dynamically assigned by the MAC resource assignment function on a per resource basis, alone with the MCS, based on the MAC Traffic PDU size, channel condition, mobile category, and ultimately the scheduling policy, to achieve a balanced goal

Table 8.6 Burst configurations of the VDE-TER downlink bursts (T_{sym}^{-1} =38,400 symbols/sec)

Physical signaling burst	PBBCH	PPCH	PDSCH	PDTCH
TTI (slot)	8	2	2	Variable
Number of Symbols, $N_{\mathrm{sym}}^{\mathrm{burst}}$	8192	2048	2048	1024 × TTI
Ramp up/down (symbols)	None	None	None	None
Sync word (symbols)	256	None	None	None
Guard period (symbols)[a]	None	None	None	None
Pilots pattern	(16, 4)	Configurable (via BBCH)		
Payload (symbols)	5952	≤2048	≤2048	≤1024 × TTI

[a]Detailed in Sect. 9.3

among spectral efficiency, QoS, and fairness, as noted in Chap. 2. The transmission of a PDTCH burst is thus dependent on the RLC instance of the PDU (i.e., the QoS mark inherited from the socket group) and controlled by the MAC scheduler and transmission error control protocol. The MCS is inferred from the PDU size and the assigned TTI, i.e., the assignment, by the MAC Resource Assignment PDU via ACH. A one-slot downlink burst provides a payload of a maximum of 1024 channel symbols for PDTCH. The actual number depends on the pilot pattern dynamically configured by the MAC resource block configuration function via CCCH. The coordinated interference mitigation scheme, such as the coordinated transmission and reception, is supported on these PDTCH bursts and is transparent to the mobile station.

A signaling channel employs the lowest code rate, i.e., the base code rate 1/3, and the lowest modulation order, i.e., QPSK modulation, for the highest reliability and lowest RF requirements and utilizes low effective-code rates ($r' < 1/3$), to achieve high processing gains for interference rejection.

Table 8.6 summarizes the burst configuration parameters for downlink physical channels. In the following sections, we evaluate the link budget for these signaling channels under such burst configurations.

Since the physical channel performance depends heavily on the link condition affected mainly by the propagation distance and interference, let us take a look at the lowest SNR that a physical channel may experience under the required operating condition inside a cell. For now, we ignore the VDE-TER inter-cell interference, which will be considered when evaluating the individual physical channels in the subsequent sections.

Here, we adopt the same notation used in Chap. 6, where ϕ denotes the control station's off-nadir angle and θ mobile station's elevation angle. The relationship between ϕ and θ is specified in (6.27) of Chap. 6. With that understood, the shore station EIRP on the VDE-TER downlink can be represented by

$$\Lambda_{\mathrm{shore}}(\phi) = P_{\mathrm{shore}} \cdot G_{\mathrm{shore}}(\phi), \tag{8.1}$$

where $P_{\text{shore}} \geq 5$ W/50 kHz is the transmit power of a shore station and $G_{\text{shore}}(\phi)$, $\phi \in [0^\circ, 90^\circ]$, is the antenna gain of the shore station. The minimum receive power or sensitivity at a mobile station is then

$$p_{\text{mobile}} = \Lambda_{\text{shore}}(\phi) \cdot B_{\text{TER}}^{\text{DL}} \cdot G_{\text{mobile}}(\theta)/L^{\text{TER}}(\theta), \tag{8.2}$$

where $G_{\text{mobile}}(\theta)$ is the mobile station antenna gain, $B_{\text{TER}}^{\text{DL}}$ the VDE-TER downlink bandwidth, and $L^{\text{TER}}(\theta)$ the propagation loss.

For VDE-TER, we assume a transmission range of 50-NM to the horizon, and $\phi \approx 90^\circ$ and $\theta \approx 0^\circ$. At the VHF band, a terrestrial radio link also experiences a 24 dB additional loss on top of the free space propagation loss of $\ell = 116$ dB, which gives rise to a maximum total propagation loss of $L^{\text{TER}}(0^\circ) = 140$ dB. The assumed average antenna gain of a Category 1 mobile station is $G_{\text{mobile}}(\theta) \approx -3$dBi, $\forall \theta \in [0^\circ, 90^\circ]$, the mobile station with the minimum RF capability. $B_{\text{TER}}^{\text{DL}}$ is 50 kHz for the frame configuration in Fig. 8.5. The shore station antenna gain at $\phi \approx 90^\circ$ is $G_{\text{shore}}(90^\circ) = 8$ dBi and $P_{\text{shore}} = 5$ W/50 kHz. The receive sensitivity is $p_{\text{mobile}} = -98$ dBm/50 Hz.

As discussed in Sect. 6.5, the VDE-TER spectrum is shared with the VDE-SAT downlink constrained by the PFD mask defined in Chap. 6. This interference has to be factored into the VDE-TER downlink budget. From the PFD mask definition, the PFD from the VDE-SAT space station impinging on the surface of the earth, $\widehat{\Psi}^{\text{SAT}}(\phi)$, shall satisfy the PFD mask derived in equation (6.25) as plotted in Fig. 6.2. The corresponding co-channel interference power from VDE-SAT collected by a mobile station's VDE-TER receiver is thus

$$I_{\text{mobile}}^{\text{SAT}}(\theta') = (4\pi)^{-1}\lambda^2 \cdot \widehat{\Psi}^{\text{SAT}}(\phi') \cdot G_{\text{mobile}}(\theta'), \tag{8.3}$$

where θ' and ϕ' are the respective elevation angle and off-nadir angle between the mobile station and the interfering space station. The maximum interference from a VDE-SAT space station (the aggressor), seen by a VDE-TER mobile receiver, is

$$\begin{aligned} \widehat{I}_{\text{mobile}}^{\text{SAT}} &\triangleq \max_{\theta' \in [0^\circ, 90^\circ]} I_{\text{mobile}}^{\text{SAT}}(\theta') \\ &= -102 \text{ dBm/50kHz or } -149 \text{ dBm/Hz}. \end{aligned} \tag{8.4}$$

Given the receiver sensitivity in (8.2) and the channel symbol rate of $T_{\text{sym}}^{-1} =$ 38,400 symbols per second under the default 50-kHz channel bandwidth for both uplink and downlink, the corresponding VDE-TER downlink SNR, i.e., the matched filter output SNR or the symbol sample SNR, is

$$\eta_{\text{mobile}} = \frac{p_{\text{mobile}}}{\left(N_0^{\text{mobile}} + \widehat{I}_{\text{mobile}}^{\text{SAT}}\right) \cdot T_{\text{sym}}^{-1}}, \tag{8.5}$$

where $N_0^{\text{mobile}} = \kappa T_{\text{mobile}}$, κ is the Boltzmann's constant (−229 dBJ/K), and T_{mobile}= 35 dBK is the noise temperature of a mobile station receiver nearshore. Similar to the treatment in proximity communication, an extra 5 dB is added to the noise temperature, assuming a nearshore mobile station receiver is 5-dB hotter than an offshore one to account for the interference from adjacent channel activities in close proximity and the interference from co-channel incumbent land systems. The corresponding minimum downlink SNR from (8.5) is

$$\eta_{\text{mobile}} \geq 5 \text{ dB}. \tag{8.6}$$

However, remember that unlike other signaling channels, the PBBCH is supposed to be visible by a mobile station deep inside a neighboring cell beyond the cell boundary. Besides carrying BBCH, the PBBCH of VDE-TER in particular also serves as the beacon signal for spectrum sharing by creating a no-share zone to prevent mobile stations from VDE-SAT uplink transmissions on the VDE-TER uplink spectrum, $\mathbf{L}^{\text{TER}}$, within the no share-zone. To this end, PBBCH is protected from interference posed by the VDE-SAT downlink transmissions by design such that the interference is nulled, i.e., $\widehat{I}_{\text{mobile}}^{\text{SAT}} = 0$, on the PBBCH resource (refer forwards to Fig. 9.4 in Chap. 9). Its corresponding SNR is hence

$$\begin{aligned} \eta_{\text{mobile}}^{\text{PBBCH}} &= \frac{p_{\text{mobile}}}{N_0^{\text{mobile}} \cdot T_{\text{sym}}^{-1}} \\ &\geq 20 \text{ dB}. \end{aligned} \tag{8.7}$$

We will use (8.6) and (8.7) to determine the link budgets of the downlink signaling channels in the presence of inter-cell interference next.

PBBCH

The System Bulletin Board PDU from the network layer is transported via the dedicated BBCH, carried by a physical layer PDU, and transmitted on the dedicated PBBCH at the beginning of every resource frame, as shown in Fig. 8.5.

As a physical signaling channel, the configuration is relatively static and equipped with the default lowest code rate and modulation order. For PBBCH that carries the top-level system configuration information, its configuration is fixed for the sake of easy acquisition and protected with the lowest effective code rate (or the highest processing gain) to extend its coverage beyond cell boundary. As enumerated in Table 8.7, the System Bulletin Board is 24-bytes long. Two PBBCH bursts transmit simultaneously on two 50-kHz channels at CFIs 468 and 464, over an eight-slot TTI positioned at the very beginning of a downlink frame.

Table 8.7 VDE-TER downlink budgets under the burst configuration in Table 8.6

Physical channel	PBBCH	PPCH	PDSCH	PDTCH
SDU (bits)	192	420	413	Variable
CRC (bits)	24	24	24	24
Code rate, r	TC: 1/3	TC: 1/3	TC: 1/3	TC: $\geq$1/3
Modulation	QPSK ($m = 2$)	QPSK	QPSK	Adaptive
1%-PER[a] $E_b/N_0, E_b^+/N_0$	1.8 dB	1.5 dB	1.5 dB	Variable
TTI	8 slots	2 slots	2 slots	Variable
Effective code rate, r' (g)	0.018 (17 dB)	0.108 (10 dB)	0.107 (10 dB)	Variable
Transmit power, P_{shore}	5 W/12.5 W (per 50 kHz)			
Noise temperature, T_{mobile}	35 dBK			
Receive SNR, η_{mobile}	20 dB/24 dB	5 dB/9 dB	5 dB/9 dB	5 dB/9 dB
Receive SINR, γ_{mobile}	0 dB	-1 dB	-1 dB	Variable

[a]*PER* physical layer PDU error rate

PBBCH is the only downlink physical channel transmitted in a burst with a `sync word` for frame detection and initial synchronization to a control station. The `sync word` occupies 256 channel symbols, initialized with 256 default QPSK constellation symbols, i.e., $e^{j\pi/4}$. The `pilot` pattern with a (4, 16) duty cycle, shown in Fig. 7.12 of Chap. 7, is employed for robust channel tracking and time and frequency tracking. After deducting the `sync word` and `pilot` overhead from the burst, a payload of 5952 channel symbols is left for PBBCH.

From (8.7), the PBBCH SNR at a cell edge is $\eta_{\text{mobile}}^{\text{PBBCH}} = 20$ dB, which is obtained without considering inter-cell interference. After taking into account the inter-cell interference at a cell edge, where the PBBCH power received from the current cell, $p_{\text{mobile}}^{\text{PBBCH}}$, and from the interfering cell, $I_{\text{mobile}}^{\text{PBBCH}}$, are assumed the same, the SINR of PBBCH at a cell edge becomes

$$\begin{aligned}\gamma_{\text{mobile}}^{\text{PBBCH}} &= \frac{p_{\text{mobile}}^{\text{PBBCH}}}{N_0^{\text{mobile}} \cdot T_{\text{sym}}^{-1} + I_{\text{mobile}}^{\text{PBBCH}}} \\ &= \left(\frac{1}{\eta_{\text{mobile}}^{\text{PBBCH}}} + 1\right)^{-1} \\ &\approx 0 \text{ dB}.\end{aligned} \tag{8.8}$$

As summarized in Table 8.7, for the System Bulletin Board PDU size of 192 bits (from Table 8.1), the required E_b/N_0at a code rate of 1/3 for a one-percent PDU error rate (PER), denoted as E_b^+/N_0, is about 1.8 dB. An effective code rate of $r' = 0.018$ or a processing gain of g= 17.4 dB produces an

$$E_b/N_0 = \gamma_{\text{mobile}}^{\text{PBBCH}} - m + g \\ = 14.4\ (\text{dB}), \tag{8.9}$$

at an SINR of $\gamma_{\text{mobile}}^{\text{PBBCH}} = 0$ dB, according to (5.124). Thus, the PBBCH burst boasts an

$$\Delta E_b/N_0 \triangleq E_b/N_0 - E_b^{+}/N_0 \\ = 13\ (\text{dB}) \tag{8.10}$$

extra processing gain for additional interference suppression, needed to penetrate neighboring cells or create a no-share zone for uplink spectrum sharing.

Since two PBBCH bursts are transmitted shoulder-to-shoulder (see Fig. 8.4), an additional gain of up to 3 dB can be obtained through combining if necessary.

PDSCH

PDSCH carries a physical layer PDU that contains the MAC signaling PDUs, transported via CCCH, ACH, and DFCHs. As a signaling channel, it is relatively static but configurable through the `frame configuration` field of the System Bulletin Board. A two-slot burst provides a payload of 2048 channel symbols for PDSCH. The SNR at the cell edge is 5 dB from (8.6). Using the same treatment as for PBBCH, the SINR at the cell edge is

$$\gamma_{\text{mobile}}^{\text{PDSCH}} \approx -1\ \text{dB}. \tag{8.11}$$

Since the E_b/N_0 at a code rate of 1/3 for a one-percent PDU error rate is about 1.5 dB, PDSCH enjoys an extra 4-dB processing gain at the cell edge under the two-slot TTI configuration in Fig. 8.5.

PPCH

PPCH is designated to PCH for transporting the MAC Paging PDU. As shown in Fig. 8.5, a two-slot TTI configuration provides 2048 channel symbols that allow the MAC Paging PDU to include 14 mobile station MMSIs. The actual number of mobile stations currently being paged may vary depending on the MAC traffic control protocol. Here, the *void* MMSI (Chap. 2) comes in handy for a place holder to represent the vacancy in the list. With a similar PDU size and the same TTI configuration as PDSCH (Table 8.7), PPCH has a 4-dB processing gain margin as PDSCH.

8.4.2.4 Uplink Link Budget

Unlike the downlink burst, an uplink burst must possess a transmission structure for multiple access on the uplink. It has a `ramp-up` preamble and a `ramp-down` postamble, each with a duration of T_{ramp}= 0.41 ms (16 symbols) at the end of a burst, as depicted in Fig. 8.9 and summarized in Table 8.8. Each burst is also equipped with a `sync word` that occupies 64 channel symbols (initialized with the default QPSK constellation of $e^{j\pi/4}$) to enable a control station to detect and synchronize to the incoming burst on the uplink.

A one-slot uplink burst provides a maximum payload of up to 928 channel symbols except for the PRACH burst with a `guard period` (see Fig. 8.9). The actual number depends on the pilot pattern that is dynamically configured by the MAC resource block configuration function via CCCH and by the `frame configuration` via BBCH, respectively.

Like PDTCH, the MCS of PUTCH and the TTI of the burst is controlled by the MAC resource assignment function or the scheduler, announced via ACH on the PDSCH, based on the channel condition and the RLC instance (i.e., priority) of the MAC Traffic PDU that it carries, as well as the MAC scheduling policy.

Unlike the downlink that is dedicated to the control station, the uplink is shared among multiple mobile stations in the same cell in a TDM fashion among the physical traffic channels (i.e., PUTCHs) and in a CDM fashion among the physical signaling channels (i.e., PUSCHs and PRACHs), as illustrated in Fig. 8.5. As noted earlier, CDM relies on three techniques, processing gain, power control, and scrambling codes to work, which are covered next.

A mobile station's transmit power P_{mobile} on the uplink normally ranges from 100 mW to more than 1 W, depending on its category, regardless of the transmission bandwidth. The uplink receiver sensitivity is then

$$\begin{aligned} p_{shore} &= \Lambda_{mobile}(\theta) \cdot G_{shore}(\phi)/L^{TER}(\theta) \\ &\geq -115 \text{ dBm}, \end{aligned} \tag{8.12}$$

Table 8.8 VDE-TER uplink burst configurations (T_{sym}^{-1} = 38,400 symbols/sec)

Physical signaling burst	PRACH	ePRACH	PUSCH	PUTCH
TTI (slot)	1	10	1	Variable
Number of symbols, N_{sym}^{burst}	1024	10240	1024	1024 × TTI
`Ramp up/down` (symbol)	16/16	16/16	16/16	16/16
`Sync word` (symbol)	64	640	64	64
`Guard period` (symbol)[a]	24	24	None	None
`Pilot` pattern	Configurable (via BBCH)		Adaptive (via CCCH)	
Maximum payload[b] (symbol)	904	9544	928	1024 × TTI-96

[a]Detailed in Sect. 9.3
[b]`Pilots` not present

where $\Lambda_{\text{mobile}}(\theta) = P_{\text{mobile}} \cdot G_{\text{mobile}}(\theta)$, $\theta \approx 0^\circ$, $\phi \approx 90^\circ$, and $G_{\text{shore}}(90^\circ) = 8$ dBi. The average antenna gain of a Category 1 mobile station is assumed $G_{\text{mobile}} \approx -3$ dBi.

Given the typical noise temperature of $T_{\text{shore}} = 30$ dBK, with the benefit of the directivity of the antenna system of a shore station that shields itself from the land system interference, the corresponding uplink SNR inside a VDE-TER cell is then

$$\begin{aligned} \eta_{\text{shore}} &= \frac{p_{\text{shore}}}{N_0^{\text{shore}} \cdot T_{\text{sym}}^{-1}} \\ &\geq 8 \text{ dB}, \end{aligned} \tag{8.13}$$

where $N_0^{\text{shore}} = \kappa T_{\text{shore}}$ is the noise power spectral density at a shore station.

For some Category 1 mobile stations, the transmit power can be as low as 5 mW. The corresponding receive SNR then becomes

$$\eta_{\text{shore}} \geq -5 \text{ dB}. \tag{8.14}$$

The equality holds at cell edges, and let us use it to evaluate the uplink signaling channel link budget next.

PRACH

Under the centralized medium access control, the transmit timing of a mobile station is also supervised by the control station so that the `guard period` is saved to minimize the overhead through the MAC timing control function, detailed in Sect. 9.4 of Chap. 9. The only exception is the transmissions on the PRACH resource, at which time a connection with the control station is not yet established. Therefore, PRACH is the only physical channel with a `guard period` (refer back to Fig. 8.9). To determine the PRACH link budget, we need to know how many resources or channel symbols the `guard period` occupies.

The analysis of the required time buffer is similar to VDE-SAT and is deferred to Chap. 9, where the determination of time buffer is accentuated by the excessive propagation delay ensued from the prolonged communication range and space station traveling speed in satellite communication. Using the same argument and treatment in determining the `guard period` of VDE-SAT PRACH, a time buffer for both PRACH and *e*PRACH of

$$T_{\text{buffer}}^{\text{PRACH}} = T_{\text{buffer}}^{e\text{PRACH}} = 1.026 \text{ ms} \tag{8.15}$$

is derived. Deducting the `ramp-down` period from (8.15) leads to a `guard period` of

$$\begin{aligned}\Delta_{\mathrm{GP}}^{\mathsf{PRACH}} &= T_{\mathrm{buffer}} - T_{\mathrm{ramp}} \\ &= 0.61 \text{ ms}\end{aligned} \tag{8.16}$$

or 24 symbols at the end of the PRACH burst. A one-slot PRACH burst thus provides a payload of up to 904 channel symbols and a two-slot burst up to 1928 channel symbols.

Since the PRACH resources are shared among accessing mobile stations in a CDM and contention-based manner (i.e., scheduling not required), an accessing mobile station randomly selects a scrambling code from the predefined scrambling codes (detailed in the next section). A control station first determines which scrambling code is transmitted by detecting the `sync word` from the incoming PRACH bursts on the PRACH resource and uses the detected scrambling code to descramble the burst payload that follows. Assuming ideal power control, the incoming PRACH powers from all accessing mobile stations are the same, i.e., $p_{\mathrm{shore}}^{\mathsf{PRACH}}$, at the control station receiver.

After taking into account the co-channel interference from $L-1$ contending mobile stations (including the ones potentially from the neighboring cell), the channel SINR of a mobile station PRACH at the control station receiver is

$$\begin{aligned}\gamma_{\mathrm{shore}}^{\mathsf{PRACH}} &= \frac{p_{\mathrm{shore}}^{\mathsf{PRACH}}}{N_0^{\mathrm{shore}} \cdot T_{\mathrm{sym}}^{-1} + \sum_{l=1}^{L-1} p_{\mathrm{shore}}^{\mathsf{PRACH}}} \\ &= \left(1/\eta_{\mathrm{shore}}^{\mathsf{PRACH}} + L - 1\right)^{-1},\end{aligned} \tag{8.17}$$

where $\eta_{\mathrm{shore}}^{\mathsf{PRACH}} \triangleq p_{\mathrm{shore}}^{\mathsf{PRACH}} / N_0^{\mathrm{shore}} \cdot T_{\mathrm{sym}}^{-1}$, a function of the mobile station transmit power.

Figure 8.13 plots (8.17) in terms of $\Delta E_{\mathrm{b}}/N_0 = E_{\mathrm{b}}/N_0 - E_b^{+}/N_0$ (dB), where$E_{\mathrm{b}}/N_0 = \gamma_{\mathrm{shore}}^{\mathsf{PRACH}} - m + g$ from (5.124) and $E_{\mathrm{h}}^{+}/N_0 = 2.6$ dB from Table 8.9. Clearly,

$$\begin{aligned}E_{\mathrm{b}}/N_0 &\rightarrow (L-1)^{-1} - m + g \\ &\approx 5.8 \text{ dB},\end{aligned} \tag{8.18}$$

and therefore,

$$\Delta E_{\mathrm{b}}/N_0 \rightarrow 3.2\text{dB}, \tag{8.19}$$

as $\eta_{\mathrm{shore}}^{\mathsf{PRACH}} \rightarrow \infty$.

For a 100-mW mobile station at a cell edge, the receive SNR at the control station is $\eta_{\mathrm{shore}}^{\mathsf{PRACH}} = 8$ dB from (8.13). It is observed that a PRACH has a 3-dB extra processing gain essential for withstanding the additional inter-mobile interference due to, e.g., imperfect power control, as we will see later in this section, and inter-cell interference from neighboring cells.

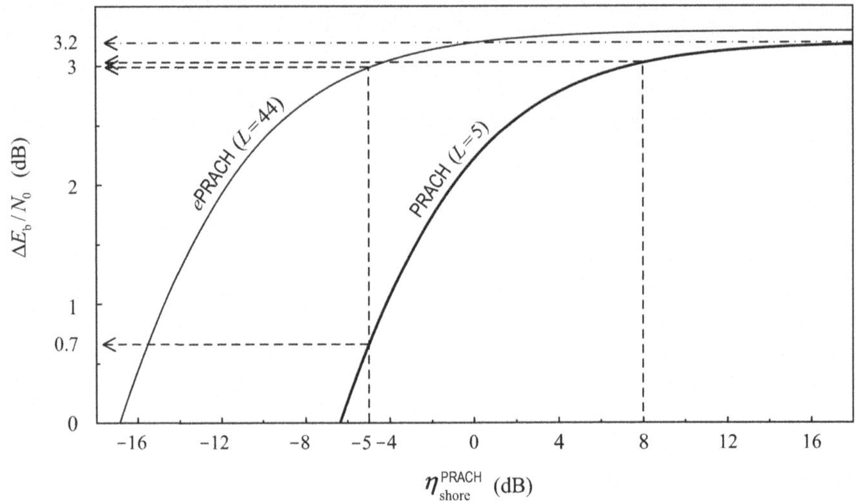

Fig. 8.13 Plot of (8.17) as an E_b/N_0 margin with respect to $E_b^+/N_0 = 2.6$ dB for $L = 5$ and 44

Table 8.9 VDE-TER uplink budgets under the burst configuration in Table 8.8

Physical channel	PRACH	*e*PRACH	PUSCH	PUTCH
SDU (bits)	48	48	6	Variable
CRC (bits)	12	12	6	24
Code rate, r	TC: 1/3	TC: 1/3	TBCC: 1/3	TC: $\geq$1/3
Modulation	QPSK ($m = 2$)	QPSK	QPSK	Variable
1%-PER $E_b/N_0, E_b^+/N_0$	2.6 dB	2.6 dB	1.7 dB	Variable
TTI	1 slot	10 slot	1 slot	Variable
Effective code rate, r' ($\mathcal{g}$)	0.033 (15 dB)	0.003 (25 dB)	0.006 (22 dB)	Variable
Transmit power, P_{mobile}	Power controlled			5 mW / 1W
Noise temperature, T_{shore}	30 dBK			
Receive SNR, η_{shore}	8 dB	-5 dB	8 dB	-5 dB / 18 dB
Receive SINR, γ_{shore}	-6 dB[a]	-16 dB[b]	-6 dB[a]	Variable

[a]In the case of 5 power-controlled PRACH transmissions
[b]44 PRACH transmissions

However, for a 5-mW station, $\eta_{\text{shore}}^{\text{PRACH}} = -5$ dB from (8.14), leaving less than 1 dB of processing gain margin (Fig. 8.13). Sharing the PRACH resources with higher-power stations restricts their transmit power and hence strips away the extra processing gain. In this case, the extended PRACH or *e*PRACH (see Fig. 8.5) comes in handy for the low-power stations. The 10-slot resource decreases the effective code rate to 0.003, an instant 10-dB boost in processing gain, capable of accommodating $L = 44$ simultaneous accessing mobile stations, plus an extra 3-dB processing gain.

PUSCH

Similar to PRACH, the PUSCH resources are shared among mobile stations. However, for PRACH transmissions, sharing resources designated to all accessing mobile stations is because the transmissions are unscheduled, and resource reservation for all potential mobile stations is obviously impractical. On the contrary, sharing PUSCH resources is only among the mobile stations acknowledging the receipt of the corresponding DTCHs with pre-assigned scrambling codes (covered later in Sect. 8.4.3) known to the control station and hence is neither random nor contention-based. A slot (the minimum schedulable resource element) is too wasteful for a PUSCH since the PDU that it carries contains only 12 bits, a 6-bit MAC Feedback PDU (see Table 8.3), plus a 6-bit CRC (Table 8.9). Sharing a resource element (slot) in a CDM fashion among multiple mobile stations solves the resource granularity issue, with an added complexity of power control to minimize the co-channel interference due to the near-far effect inherent to CDM.

Also, unlike other physical channels that employ the turbo code for forward error correction, PUSCH is TBCC-encoded due to its superiority over the turbo code for small PDUs noted in Sect. 5.3.3 of Chap. 5. Using the resource configuration in Fig. 8.4b, up to five mobile stations share the PUSCH resource; the PUSCH SINR that a control station sees under five mobile stations on the one-slot PUSCH resource is -6 dB, similar to the derivation for the PRACH. For the one-percent target E_b/N_0 of 1.7 dB from Table 8.9, PUSCH enjoys an 11-dB extra processing gain for a 100-mW mobile station, more than 8-dB for a 5-mW, useful for absorbing the inter-mobile station interference due to imperfect closed-loop power control and potential inter-cell interference at a cell edge.

8.4.2.5 Transmit Power Control

Albeit for different reasons, both PUSCH and PRACH share transmission resources in a CDM manner, rendering transmission power control crucial for keeping inter-mobile station interference under control.

PUSCH

When transmitting on the shared PUSCH resource, the mobile station is assumed to follow the MAC `power adjustment (PA)` command piggybacked in the `slot assignment element` (Table 8.2a) to minimize the near-far effect among the PUSCH bursts that are transmitted on the same PUSCH slot(s) from different mobile stations. As discussed in Sect. 5.1.3 of Chap. 5, the main idea to reduce the near-far effect is to keep PUSCHs all received at the same power level just enough to decode the USCHs. To that end, the MAC power control protocol of the control station keeps an eye on the received power of the previous transmission from each mobile station, based on which it instructs the mobile station to adjust its transmit power following the `power adjustment` command.

The PRACH power control is based on the estimate of the downlink propagation loss and, therefore, is an open-loop control with low accuracy. After a successful PRACH (i.e., successfully detected by the control station), a closed-loop power control is then established with improved accuracy for PUSCH via the feedback from the control station: The control station measures the received power and makes corrections/adjustments accordingly through the `power adjustment` command of the `assignment element` field of the MAC Resource Assignment PDU (see Table 8.2a). It is communicated via ACH over PDSCH to the mobile station.

PRACH

Like the PUSCH resource, the PRACH resource is shared among mobile stations. However, interference control is more challenging than the PUSCH since there is no `power control command` or the like from the control station for the obvious reason: PRACH is meant for the mobile station's first attempt to communicate with the control station, and hence it is either unexpected (Uplink Resource Request) or partially unexpected (Paging Response) by the control station. So, the burden is on the accessing mobile station (well, almost).

To figure out the appropriate transmit power for PRACH, the MAC random access function of the mobile station reckons that the ultimate goal of power control is for *all* PRACHs to reach the control station at the *same* power level, say p^{+}_{shore}, just sufficient for it to decode the PRACH, at, e.g., a one-percent PER. As such, after taking into account the extra power to compensate for the propagation loss of the radio channel, $\mathcal{L}^{\text{UL}}$, including the radio link and transceiver RF paths (see Fig. 5.1 in Chap. 5), the corresponding transmit power is

$$P_{\text{mobile}} = p^{+}_{\text{shore}} + \mathcal{L}^{\text{UL}} \ \ (\text{dBm}). \tag{8.20}$$

Since PRACH is the first transmission attempt, the mobile station has no way of knowing its uplink radio channel loss, $\mathcal{L}^{\text{UL}}$. However, this information can be derived from its downlink counterpart, i.e., the downlink radio channel loss, $\mathcal{L}^{\text{DL}}$, which can be measured from the downlink signals, knowing that

$$P_{\text{shore}} = p_{\text{mobile}} + \mathcal{L}^{\text{DL}} \ (\text{dBm}), \tag{8.21}$$

where P_{shore} is the control station transmit power and p_{mobile} the receive power at the mobile station.

However, under FDD over the VDES spectrum, the radio channel pair for uplink and downlink are separated by almost 5 MHz plus different transceiver RF paths; they are not exactly the same, i.e., $\mathcal{L}^{\text{UL}} \neq \mathcal{L}^{\text{DL}}$, but something to begin with when there is nothing else to go on.

Secondly, the mobile station has no knowledge of p^{+}_{shore} and P_{shore} for solving (8.21). These parameters are only known to the control station and, therefore, made available to mobile stations via the System Bulletin Board PDU (see Sect. 8.2), broadcast through the BBCH on PBBCH. We thus have

$$\begin{aligned} P^{+}_{\text{mobile}} &\approx p^{+}_{\text{shore}} + \mathcal{L}^{\text{DL}} \\ &= p^{+}_{\text{shore}} + P_{\text{shore}} - p_{\text{mobile}} \text{ (dBm)}. \end{aligned} \tag{8.22}$$

From Sect. 8.4.2.4, $\eta^{\text{PRACH}}_{\text{shore}} = 8$ dB; p^{+}_{shore} is then $\eta^{\text{PRACH}}_{\text{shore}} + N^{\text{shore}}_{0} \cdot T^{-1}_{\text{sym}} = -128$ dBm.

Moreover, depending on the radio channel and mobile station transmit power, (8.22) may not be satisfied as mobile stations are power-limited, especially for Category 1 mobile stations. The actual transmit power is upper-bounded by the maximum transmit power, $P^{\max}_{\text{mobile}}$, i.e.,

$$P_{\text{mobile}} = \min\left\{P^{\max}_{\text{mobile}}, p^{+}_{\text{shore}} + P_{\text{shore}} - p_{\text{mobile}}\right\} \text{ dBm}. \tag{8.23}$$

Therefore, as aforementioned, there are special PRACH channels by design in every frame reserved specifically for this scenario. The most transmit power-limited mobile stations (e.g., Category 1) or whoever finds itself with a power shortage (i.e., unable to satisfy (8.22), i.e., $P_{\text{mobile}} < P^{+}_{\text{mobile}}$) may take advantage of the extended TTI, sending PRACH on this extended PRACH or *e*PRACH resource for signal energy boosting. Once such an *e*PRACH is detected, the control station may assign a PUSCH with extended TTI to the accessing mobile station via CCCH, although it is unnecessary for the PUSCH under the current frame configuration (Fig. 8.5) that enjoys an 11-dB extra processing gain in link budget.

8.4.2.6 Transmit Timing Control

Transmit timing control is just as important as power control but is purposely deferred to the next chapter, where the propagation delay in satellite communication is more prominent, and the significance of timing control is more appreciable.

8.4.3 Interference Mitigation

As noted in Chap. 6, interference is present at every level of a wireless communication system. In a multi-mobile and multi-control station, multi-frequency channel, multi-air interface, and spectrum-sharing deployment environment like VDES, interference rejection and mitigation for overall system robustness and spectral efficiency is even more crucial and challenging, ergo the topic in focus for the remainder of this chapter.

8.4.3.1 Signal Ambiguity

Waveforms like the `sync word` and `pilot` are structured specifically as *reference signals* for burst detection, channel estimation, and time and frequency synchronization and tracking. Therefore, these signals are made universal and are used by all transmitters and known to all receivers for ease of detection. The caveat of this convenience is the ambiguity among the reference signals from different transmitters sharing the same radio spectrum.

To see this point more clearly, consider the example in Fig. 8.14, where full spectrum reuse is assumed. Ship $\mathcal{A}$ is receiving transmissions (Burst $\mathcal{A}$) from Station A, and Ship $\mathcal{B}$ is receiving transmissions (Burst $\mathcal{B}$) from Station B.

Assume the `sync word` and `pilot` signals on the downlink from both stations are identical, and both are heard by Ship $\mathcal{A}$ due to the broadcast nature of wireless transmissions, but through different radio channels on the same frequency band, $\boldsymbol{\alpha}_{\mathrm{A}\to\mathcal{A}}$and $\boldsymbol{\alpha}_{\mathrm{B}\to\mathcal{A}}$. What Ship $\mathcal{A}$ hears through the `sync word` or `pilot` signal is

$$\begin{aligned}\mathbf{y}_{\mathcal{A}}^{\mathtt{pilot}} &= \boldsymbol{\alpha}_{\mathrm{A}\to\mathcal{A}}\cdot\mathbf{c}_{\mathrm{A}\to\mathcal{A}}^{\mathtt{pilot}} + \underbrace{\boldsymbol{\alpha}_{\mathrm{B}\to\mathcal{A}}\cdot\mathbf{c}_{\mathrm{B}\to\mathcal{B}}^{\mathtt{pilot}}}_{\text{Interference}} \\ &= \underbrace{\left(\boldsymbol{\alpha}_{\mathrm{A}\to\mathcal{A}} + \boldsymbol{\alpha}_{\mathrm{B}\to\mathcal{A}}\right)}_{\text{Composite channel}}\cdot\mathbf{c}^{\mathtt{pilot}},\end{aligned} \tag{8.24}$$

where $\mathbf{c}_{\mathrm{A}\to\mathcal{A}}^{\mathtt{pilot}} \subset \mathbf{c}_{\mathrm{A}\to\mathcal{A}}$ is the `sync word` or `pilot` signal from Station A, embedded in the burst, $\mathbf{c}_{\mathrm{A}\to\mathcal{A}}$, for estimating $\boldsymbol{\alpha}_{\mathrm{A}\to\mathcal{A}}$ at Ship $\mathcal{A}$. For ease of

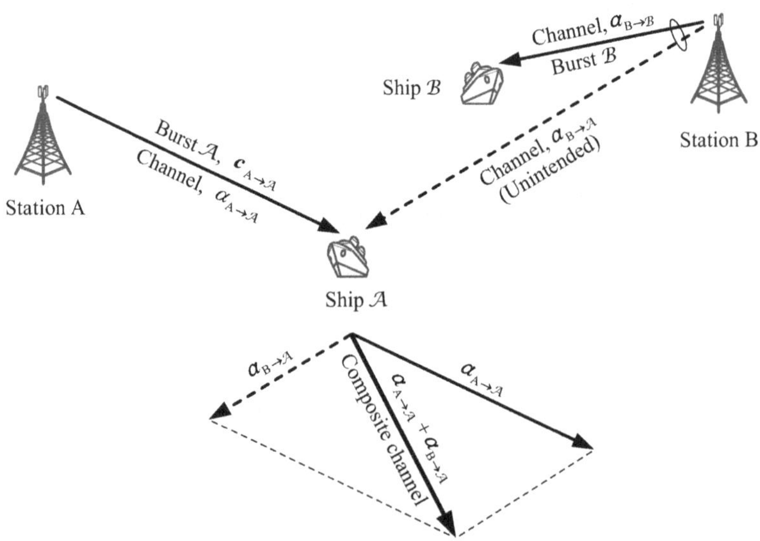

Fig. 8.14 Graphical illustration of a "composite channel," due to ambiguity of signals between transmitters sharing the same radio spectrum

explanation, they are represented as constellation symbol sequences rather than the actual waveforms. Since both pilots share the same waveform, ergo $\mathbf{c}_{\mathrm{A}}^{\texttt{pilot}} = \mathbf{c}_{\mathrm{B}}^{\texttt{pilot}} = \mathbf{c}^{\texttt{pilot}}$, Ship $\mathcal{A}$ will fail to reject the interference (i.e., $\boldsymbol{\alpha}_{\mathrm{B}\to\mathcal{A}} \cdot \mathbf{c}_{\mathrm{B}\to\mathcal{B}}^{\texttt{pilot}}$) when using `sync word` or `pilot` for channel estimation, resulting in a composite channel, i.e., $(\boldsymbol{\alpha}_{\mathrm{A}\to\mathcal{A}} + \boldsymbol{\alpha}_{\mathrm{B}\to\mathcal{A}})$.

Now for the `payload` part of the burst, what Ship $\mathcal{A}$ receives is

$$\mathbf{y}_{\mathcal{A}}^{\texttt{payload}} = \boldsymbol{\alpha}_{\mathrm{A}\to\mathcal{A}} \cdot \mathbf{c}_{\mathrm{A}\to\mathcal{A}}^{\texttt{payload}} + \underbrace{\boldsymbol{\alpha}_{\mathrm{B}\to\mathcal{A}} \cdot \mathbf{c}_{\mathrm{B}\to\mathcal{B}}^{\texttt{payload}}}_{\text{Interference/noise}}. \tag{8.25}$$

where $\mathbf{c}_{\mathrm{A}\to\mathcal{A}}^{\texttt{payload}} \subset \mathbf{c}_{\mathrm{A}\to\mathcal{A}}$ is the `payload` of Burst $\mathcal{A}$, destined to Ship $\mathcal{A}$, and $\mathbf{c}_{\mathrm{B}\to\mathcal{B}}^{\texttt{payload}} \subset \mathbf{c}_{\mathrm{B}\to\mathcal{B}}$ is the `payload` of Burst $\mathcal{B}$, meant for Ship $\mathcal{B}$ (unintended for Ship $\mathcal{A}$) from Station B.

Assuming the `payload` is purely random, $\boldsymbol{\alpha}_{\mathrm{B}\to\mathcal{A}} \cdot \mathbf{c}_{\mathrm{B}\to\mathcal{B}}^{\texttt{payload}}$ in (8.25) hence remains as noise to Ship $\mathcal{A}$ and can be suppressed by the processing gain, whereas the erroneous composite channel in (8.24) cannot be corrected since all reference signals share the same waveform and the receiver has no way of telling them apart. It is just like the human immune system that considers the cancer cell one of ours and fails to attack it.

Ship $\mathcal{A}$ will likely fail to decode $\mathbf{c}_{\mathrm{A}\to\mathcal{A}}^{\texttt{payload}}$ from Burst $\mathcal{A}$ since $(\boldsymbol{\alpha}_{\mathrm{A}\to\mathcal{A}} + \boldsymbol{\alpha}_{\mathrm{B}\to\mathcal{A}})$ is clearly not the actual channel $(\boldsymbol{\alpha}_{\mathrm{A}\to\mathcal{A}})$ that $\mathbf{c}_{\mathrm{A}\to\mathcal{A}}^{\texttt{payload}}$ goes through. The same happens to Ship $\mathcal{B}$, and it is vice versa for the uplink at Station A or Station B.

A similar situation occurs when multiple mobile stations *within* a cell share the same resource, such as PUSCH and PRACH. As we have already seen (Sect. 8.4.2.4), more than one mobile station transmits its PUSCH or PRACH on the shared PUSCH or PRACH resources, potentially causing the same ambiguity problem.

A solution to the signal ambiguity issue is through "discriminative scrambling" of the waveforms from different sources, similar to the modern medical techniques used in cancer treatment, where the cancer cells are biologically "painted" as foreign invaders to stimulate the immune system to unleash the natural killer cells and macrophages against the cancer cells. This approach is the topic to be discussed next.

8.4.3.2 Signal Whitening and Discrimination

If not avoidable, we wish the interference behaves like the additive white Gaussian noise or AWGN, uncorrelated with the signal of interest, just like the thermal noise, so that an optimal receiver is readily available to deal with it. Thus, from the perspective of Ship $\mathcal{A}$ in Fig. 8.14, we wish $\boldsymbol{\alpha}_{\mathrm{B}\to\mathcal{A}} \cdot \mathbf{c}_{\mathrm{B}\to\mathcal{B}}^{\texttt{pilot}}$ and $\boldsymbol{\alpha}_{\mathrm{B}\to\mathcal{A}} \cdot \mathbf{c}_{\mathrm{B}\to\mathcal{B}}^{\texttt{payload}}$ in (8.24) and (8.25) appear like AWGN or thermal noise to the unintended receiver (Ship $\mathcal{A}$) if unavoidable, i.e., $\boldsymbol{\alpha}_{\mathrm{B}\to\mathcal{A}} \neq 0$ on the resource or in the dimension that

$\boldsymbol{\alpha}_{\mathrm{A}\rightarrow\mathcal{A}}$ resides. Therefore, making $\mathbf{c}_{\mathrm{B}\rightarrow\mathcal{B}}^{\mathtt{pilot}}$ and $\mathbf{c}_{\mathrm{B}\rightarrow\mathcal{B}}^{\mathtt{payload}}$ AWGN is the "last line of defense" in the interference mitigation solution portfolio.

A real example of such interference is the VDE-SAT downlink transmission on the VDE-TER downlink spectrum when sharing $\mathbf{U}^{\mathrm{TER}}$. A simple spectrum sharing and interference mitigation technique has been discussed in Chap. 6, and its impact on the downlink link budget has been assessed earlier in this chapter, where the interference is treated as thermal noise, statistically modeled as uncorrelated AWGN, and hence can be absorbed by the processing gain. However, this assumption may not be true in all scenarios. Unlike the thermal noise without an embedded structure (hence manifesting a "white" spectrum), the interference is man-made and generally demonstrates a strong structure. It is especially true for waveforms like the `sync word` and `pilots` that are structured specifically as reference signals. Therefore, if these signals happen to have the same waveform, they, as interference from unintended transmitters, are 100-percent correlated with the signal from the intended transmitter, thereby invalidating the AWGN assumption and nullifying the processing gain that we rely on to suppress the interference, remembering that a processing gain is achieved through exploiting the structure enshrined in the signal buried in the structureless or uncorrelated AWGN.

A simple and robust solution to destroying the correlation between signals from different transmitters is the whitening technique through scrambling so that residual interference (i.e., unavoidable interference) appears close to AWGN and can be suppressed via processing gain as a final resort. This strategy requires special design at the physical waveform level and coordination at the network level.

First, as pointed out earlier, differentiating physical channel waveforms (e.g., `sync word` and `pilot` signals) across neighboring cells, air interfaces, and radio channels is necessary to avoid signal ambiguity that may hinder its ability to reject interference.

Second, `payload` whitening is also necessary to break up any potential structure present in the `payload`, such as a repetitive pattern, which may weaken the AWGN assumption. Moreover, a long streak of all ones or all zeros in the burst also induces DC offset or bias that is bad for the transceiver. Besides all this, scrambling or whitening helps shape the burst spectrum (see Fig. 7.10 in Chap. 7). Some may argue that it is an overdesign as the application data that the payload carries is most likely random and "white." The counter-argument is straight out of high school philosophy: a system's performance should never depend on a factor that is beyond control – which is the case for application traffic. Therefore, as the ice on the cake, for robustness and fully exploiting the processing gain, time variance must be enshrined *in* and *across* transmission bursts to *create* "interference diversity" over time, just like the AWGN that never repeats itself.

To this end, a scrambling technique that accomplishes both is employed in the current design in which *all* symbols in a transmission burst are scrambled with a pseudo-random sequence, as depicted in Fig. 8.15.

Assume a transmission burst (`ramp-up` and `ramp-down` inclusive) is comprised of a constellation symbol sequence of length $N_{\mathrm{sym}}^{\mathrm{burst}}$,

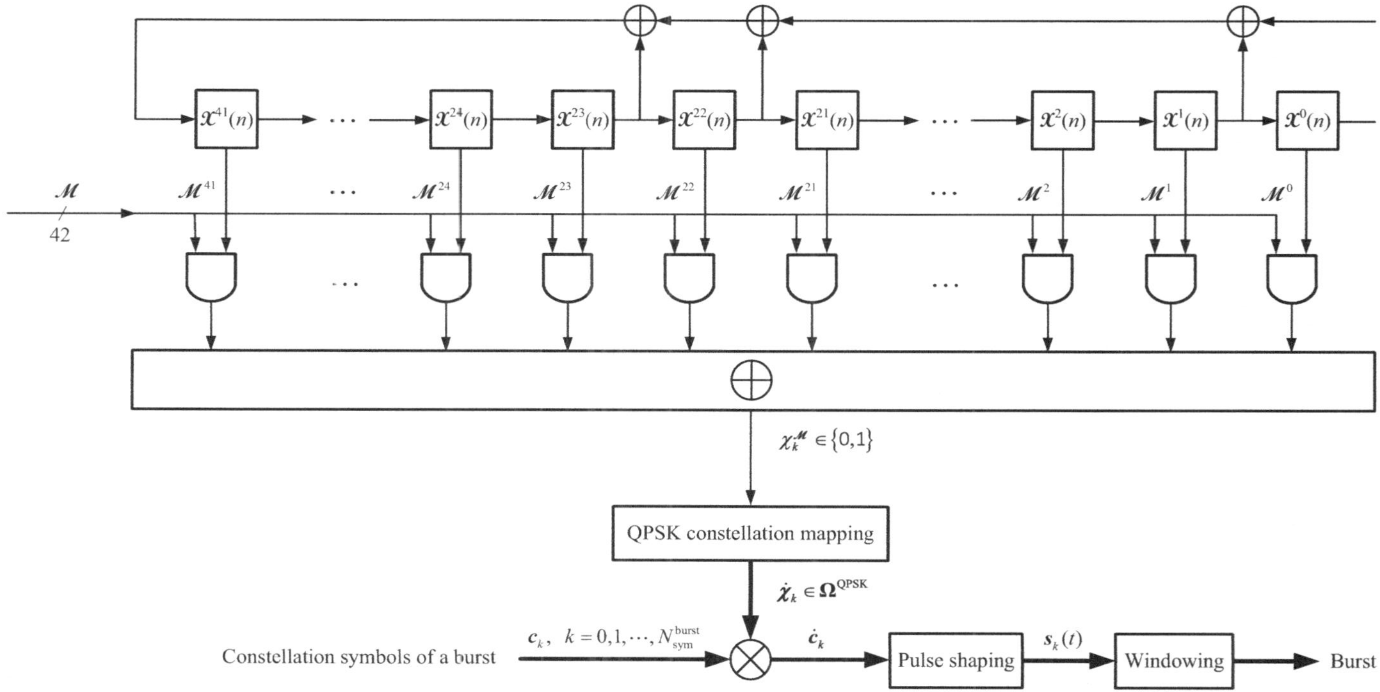

Fig. 8.15 Burst symbol scrambling for discrimination and interference suppression, employing a closed-loop shift register with 42 delay elements (see Sect. 5.3.2 of Chap. 5 for the definition of a delay element). Recall that the constellation symbols of a `sync word` and a `pilot` are intialized with default QPSK symbols, i.e., $e^{j\pi/4}$

$$\mathbf{c} \triangleq \left[c_0 \; c_1 \; \ldots \; c_{N_{\mathrm{sym}}^{\mathrm{burst}}-1} \right]. \tag{8.26}$$

Each constellation symbol is then operated upon according to

$$\dot{c}_k = c_k \cdot \dot{\chi}_k, k = 0, 1, \ldots, N_{\mathrm{sym}}^{\mathrm{burst}} - 1 \tag{8.27}$$

to produce a scrambled or randomized constellation symbol sequence,

$$\dot{\mathbf{c}} \triangleq \left[\dot{c}_0 \quad \dot{c}_1 \quad \ldots \quad \dot{c}_{N_{\mathrm{sym}}^{\mathrm{burst}}-1} \right], \tag{8.28}$$

where $\dot{\chi}_k \in \mathbf{\Omega}^{\mathrm{QPSK}}, k = 0, 1, \ldots, N_{\mathrm{sym}}^{\mathrm{burst}} - 1,$ is a QPSK symbol created from the pseudo-random sequence: $\left\{ \chi_k^{\mathcal{M}} \in \{0, 1\}, k = 0, 1, \cdots, 2N_{\mathrm{sym}}^{\mathrm{burst}} - 1 \right\}$.

This pseudo-random sequence is generated by the modulo-2 inner product of a 42-bit mask

$$\mathcal{M} \triangleq \left[\mathcal{M}^{41} \cdots \mathcal{M}^1 \; \mathcal{M}^0 \right], \mathcal{M}^i \in \{0, 1\}, \tag{8.29}$$

and the 42-bit state vector

$$\mathcal{X}(n) \triangleq \left[\mathcal{X}^{41}(n) \cdots \mathcal{X}^1(n) \; \mathcal{X}^0(n) \right], \mathcal{X}^i(n) \in \{0, 1\} \tag{8.30}$$

of a pseudo-random sequence generator, as depicted in Fig. 8.15.

The pseudo-random sequence satisfies the linear recursion specified by the characteristic polynomial of

$$F(x) = x^{42} + x^{23} + x^{22} + x + 1. \tag{8.31}$$

For creating differentiation among air interfaces, control stations, and carriers, the 42-bit mask $\mathcal{M}$ is specified by

$$\sum_{i=0}^{41} 2^i \cdot \mathcal{M}^i = n_{\mathrm{AI}} \cdot 2^{39} + n_{\mathrm{MMSI}} \cdot 2^9 + n_{\mathrm{CFI}}, \tag{8.32}$$

where n_{CFI} is the 9-bit CFI of the occupied channel, n_{MMSI} the 30-bit MMSI of the control station, and n_{AI} the air interface identifier: n_{AI}= `00` for VDE-TER uplink, `01` downlink, `10` for VDE-SAT uplink, and `11` for downlink.

For creating a time-variance for interference diversity, the pseudo-random sequence generator is initialized with $\mathcal{X}(0)$, determined by

$$\sum_{i=0}^{41} 2^i \cdot \mathcal{X}^i(0) = n_{\text{slot}}, \tag{8.33}$$

where n_{slot} is the frame slot index (0–2249) of the starting slot occupied by the transmission burst.

For PUSCH, n_{slot} is the frame slot index of the PDTCH with which the PUSCH is associated, thereby creating a one-to-one association between the USCH (that the PUSCH burst carries) and the DTCH (that the PDTCH burst carries).

8.4.3.3 Signal Disambiguation

Discriminative scrambling differentiates the signals from different sources with different scrambling codes, numerically painting the signals according to their sources so that the intended receiver can tell them apart. Reusing the example in Fig. 8.14, Ship $\mathcal{A}$ uses the respective scrambling code, $\dot{\boldsymbol{\chi}}_{\text{A}}$, of the intended signal, $\dot{\mathbf{c}}_{\text{A}\to\mathcal{A}}$, to descramble the received signal, $\mathbf{y}_{\mathcal{A}}$, by symbol-wise multiplying $\mathbf{y}_{\mathcal{A}}$ by $\dot{\boldsymbol{\chi}}_{\text{A}}^*$ to give

$$y_{\mathcal{A}} \cdot \dot{\chi}_{\text{A}}^* = \alpha_{\text{A}\to\mathcal{A}} \cdot \dot{c}_{\text{A}\to\mathcal{A}} \cdot \dot{\chi}_{\text{A}}^* + \alpha_{\text{B}\to\mathcal{A}} \cdot \dot{c}_{\text{B}\to\mathcal{B}} \cdot \dot{\chi}_{\text{A}}^*, \tag{8.34}$$

where $y_{\mathcal{A}} \in \mathbf{y}_{\mathcal{A}}$ is the kth received symbol of the burst, $\dot{c}_{\text{A}\to\mathcal{A}} \in \dot{\mathbf{c}}_{\text{A}\to\mathcal{A}}$ the kth constellation symbol, $\dot{\chi}_{\text{A}} \in \dot{\boldsymbol{\chi}}_{\text{A}}$ the kth scrambling code symbol used by Station A, and k is omitted for notational simplification. Substituting (8.27) into (8.34) gives

$$\begin{aligned} y_{\mathcal{A}} \cdot \dot{\chi}_{\text{A}}^* &= (\alpha_{\text{A}\to\mathcal{A}} \cdot c_{\text{A}\to\mathcal{A}} \cdot \dot{\chi}_{\text{A}}) \cdot \dot{\chi}_{\text{A}}^* + (\alpha_{\text{B}\to\mathcal{B}} \cdot c_{\text{B}\to\mathcal{B}} \cdot \dot{\chi}_{\text{B}}) \cdot \dot{\chi}_{\text{A}}^* \\ &= \alpha_{\text{A}\to\mathcal{A}} \cdot c_{\text{A}\to\mathcal{A}} + \underbrace{\alpha_{\text{B}\to\mathcal{B}} \cdot c_{\text{B}\to\mathcal{B}} \cdot \dot{\chi}_{\text{B}} \cdot \dot{\chi}_{\text{A}}^*}_{\text{Noise}}, \end{aligned} \tag{8.35}$$

where the second equality assumes $\dot{\chi}_{\text{A}} \cdot \dot{\chi}_{\text{A}}^* = |\dot{\chi}_{\text{A}}|^2 = 1$.

Therefore, the interference from Station B remains as noise, which includes the pilot $\left(\mathbf{c}_{\text{B}\to\mathcal{B}}^{\texttt{pilot}} \subset \mathbf{c}_{\text{B}\to\mathcal{B}}\right)$, and can be suppressed with the processing gain. Since there is more than one neighboring control station in a practical deployment, collectively, the noise is approximately Gaussian according to the central limit theorem.

8.4.3.4 Scrambling Code Coordination

Scrambling codes are network resources for interference mitigation. The underlying framework is based on (8.32). Nevertheless, practical issues may prevent it from being implemented on some signals, and special care must be taken.

PBBCH

The scrambling scheme prescribed in (8.32) relies on n_{MMSI}, i.e., the control station MMSI, to differentiate signals from different cells. A potential issue is that a mobile station looking for VDES services may not know the MMSI of the control station of which it is currently under coverage, not until BBCH is acquired. An exhaustive search for all MMSIs is almost certainly out of the question.

A mobile station is aware of the MMSIs of the neighboring cells from the serving station by reading the `neighbor list` off the System Bulletin Board PDU. When it is not the case, e.g., waking up in a new area, acquiring a new control station could be painful.

For ease of system acquisition without knowing the MMSI of a control station, the n_{MMSI} of a PBBCH burst is selected from one of the predefined numbers, as listed in Table 8.10, for identifying the PBBCH burst from other control stations. The assignment of n_{MMSI} to each control station in the network is managed by NRC to avoid "collisions" between neighboring cells. This constraint reduces the blind search complexity from millions of possibilities to just a few, at the cost of network planning of the n_{MMSI}, thereby off-loading the complexity from mobile stations to the network, perfectly in line with the maritime MTC requirement.

PRACH

PRACH has a similar "identity issue" since PRACH resources are shared among multiple accessing mobile stations. Hence, mobile stations need distinctive scrambling codes to differentiate themselves from the others to avoid "collision" during the random access attempt. It is not a problem when PRACH carries a MAC Paging Response PDU, where n_{MMSI} is simply the MMSI of the mobile station being paged, which the control station fully expects.

However, this strategy fails when PRACH carries a MAC Uplink Resource Request or Slot Timing Request since the control station does not know apriori which mobile station is requesting. An exhaustive search for all MMSIs is already deemed impractical. Here, we adopt the same strategy for PBBCH for this scenario. The n_{MMSI} for these two types of MAC requests (Uplink Resource Request and Slot Timing Request) is randomly selected from one of the 64 predefined 30-bit numbers. To reject PRACHs from mobile stations in other cells using the enshrined PRACH processing gain by design, each set is uniquely associated with one of the 16 n_{MMSI}values used by the PBBCH. Therefore, there are 16 sets of such numbers, one-to-one associated with the 16 PBBCH codes in Table 8.10. These 64×16 n_{MMSI}values range from 1,000,000,002 to 1,000,001,025 (omitted from Table 8.10). A control station blindly detects the 64 respective PRACH bursts on the PRACH resource, corresponding to the 64 scrambling codes associated with the current control station. Successful decoding (i.e., CRC checks) indicates a successful detection.

Table 8.10 Examples of mask $\mathcal{M}$ generator parameters

Air interface	Physical channel	n_{AI}	n_{CFI}	n_{MMSI}	
VDE-TER	PBBCH PPCH PDSCH PDTCH	00	464	1 2 3 … 15	0x12200C43 0x1A43B1A8 0x0C2B9AF9 … 0x0B66BB25
			468	1 2 3 … 15	0x24ED24AC 0x37D9E0BB 0x0CBD92D2 … 0x2A1B04A8
	PUSCH	01	96/100	MMSI of the mobile station	
	PRACH[a]	01	96/100	64 × 16 unused ITU MMSIs	
	PUTCH	01	96/100	MMSI of the serving shore station	
VDE-SAT	PBBCH	10	472	1	0x3B787680
	PPCH PDSCH PDTCH	10	468	2 3 … 15	0x0EBE3816 0x1B2A38F7 … 0x32D14918
	PUSCH	11	96/100/104	MMSI of the mobile station	
	PRACH[a]	11	96/100/104	64 × 16 unused ITU MMSIs	
	PUTCH	11	96/100/104	MMSI of the serving space station	
ASM	PSCH/PTCH	00	475	1 2 3 4	0x0EDC81FC 0x365D71D5 0x35BD9CD8 0x0C6F61AA
			479	1 2 3 4	0x0F36ABE5 0x1687F4A5 0x21A64872 0x06AE3C4E

[a]For PRACH carrying an Uplink Resource Request or Slot Timing Request PDU

ASM

As noted in Chap. 7, ASM is a proximity MTC air interface for proximity-based maritime IoT services. As a service-oriented network, the destination of an ASM burst is not the "station of interest" but the "service of interest." Naturally, only the service of interest is monitored by a station, not the MMSI of the source or destination station, unless a peer-to-peer link is established, and communication is destined to the mobile station identified by the Layer 2 address, i.e., MMSI. Therefore, it seems straightforward to let n_{MMSI} be a number associated with a maritime service, i.e., the `Service ID`, managed by MSR and universally known to all stations, except for the case of peer-to-peer communication where n_{MMSI} is set to the MMSI.

However, the caveat is that in distributed medium access control using the virtual channel sensing approach, such as SO-TDMA, as described in Sect. 7.1.3 of Chap. 7, a station needs to detect and decode all nearby ongoing traffic, not just the ones of interest. It is clearly impractical that a station has to exhaustively detect all possible services. An intuitive solution is to use a common n_{MMSI}, risking the potential signal ambiguity problem that we previously raised. A compromised solution is similar to PBBCH that restricts the number of n_{MMSI}'s. A station *randomly* selects from the four predefined candidate n_{MMSI}'s, listed in Table 8.10, for the ASM burst. More candidates to choose from certainly help reduce the collision probability, however, at the cost of increased detection complexity.

8.4.3.5 Coordinated Transmission and Reception

So far, we have seen that interference, if not avoidable, can be robustly suppressed through discriminative scrambling and leveraging the processing gain. As aforementioned, inter-cell interference coordination, i.e., coordinated multi-station transmission and reception, through corresponding network layer functions, is another effective means to mitigate inter-cell interference and improve reliability through interference avoidance, especially for those services with high QoS requirements. This type of coordination calls for the timely orchestration of network layer action sequences.

When the measurement report from a mobile station indicates strong neighboring cell interference, NRC reacts by instantiating a group of relevant control stations to engage in the coordinated transmission and reception. Consider the example illustrated in Fig. 8.16, where Ship $\mathcal{A}$, served by Station A, is experiencing interference from Stations B and C. Thus, included in the coordination group are Stations A, B, and C. Stations A and B are actively participating in joint transmission to the mobile station on a specific PDTCH resource, both using the scrambling code mask($\mathcal{M}$)of Station A – the anchor serving station of Ship $\mathcal{A}$, so that the operation is transparent to the ship. The key difference from the scenario in Fig. 8.14 is that, now, $\mathbf{c}_{\mathrm{B}\to\mathcal{A}} = \mathbf{c}_{\mathrm{A}\to\mathcal{A}}$ (i.e., both intended for Ship $\mathcal{A}$) and $\dot{\boldsymbol{\chi}}_{\mathrm{B}} = \dot{\boldsymbol{\chi}}_{\mathrm{A}}$(the same scrambling code), as a result of coordination through NRC. Station C is instructed (by NRC) to mute on the corresponding resource, termed *selective muting*, for being the least relevant in the group in terms of the distance and interference strength to Ship $\mathcal{A}$. What Ship $\mathcal{A}$ receives is a superposition of the same burst $\mathcal{A}$, $\dot{\mathbf{c}}_{\mathrm{A}\to\mathcal{A}}$, from both paths, $\boldsymbol{\alpha}_{\mathrm{A}\to\mathcal{A}}$ and $\boldsymbol{\alpha}_{\mathrm{B}\to\mathcal{A}}$, i.e.,

$$\begin{aligned} \boldsymbol{y}_{\mathcal{A}} &= \boldsymbol{\alpha}_{\mathrm{A}\to\mathcal{A}} \cdot \dot{\boldsymbol{c}}_{\mathrm{A}\to\mathcal{A}} + \boldsymbol{\alpha}_{\mathrm{B}\to\mathcal{A}} \cdot \dot{\boldsymbol{c}}_{\mathrm{B}\to\mathcal{A}} \\ &= (\boldsymbol{\alpha}_{\mathrm{A}\to\mathcal{A}} + \boldsymbol{\alpha}_{\mathrm{B}\to\mathcal{A}}) \cdot \dot{\boldsymbol{c}}_{\mathrm{A}\to\mathcal{A}}. \end{aligned} \tag{8.36}$$

The received signal is then descrambled using the scrambling code, $\dot{\chi}_{\mathrm{A}}$, as usual,

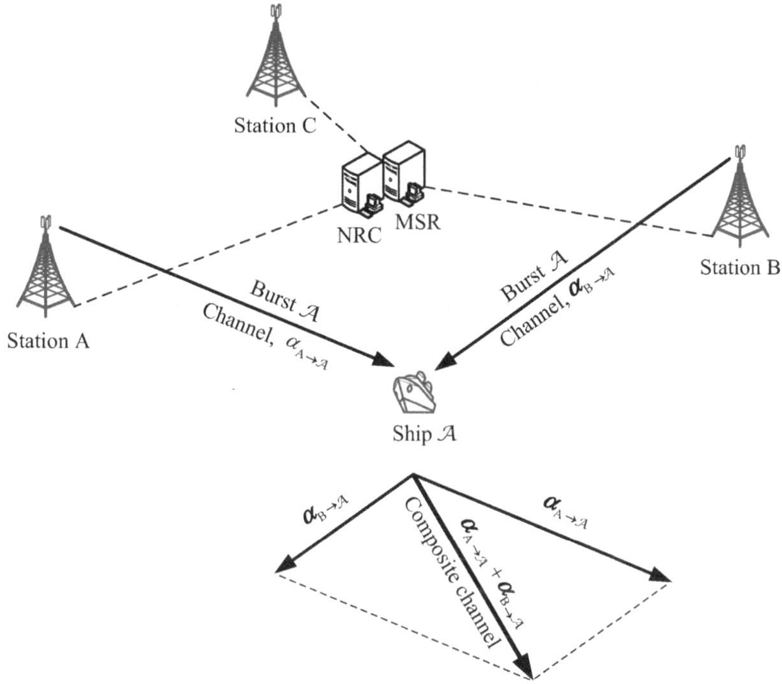

Fig. 8.16 Graphical illustration of a composite channel under the coordinated multipoint transmission orchestrated by NRC

$$\begin{aligned} \boldsymbol{y}_{\mathcal{A}} \cdot \dot{\boldsymbol{\chi}}_{\mathrm{A}}^{*} &= (\boldsymbol{\alpha}_{\mathrm{A}\to\mathcal{A}} + \boldsymbol{\alpha}_{\mathrm{B}\to\mathcal{A}}) \cdot \dot{\boldsymbol{c}}_{\mathrm{A}\to\mathcal{A}} \cdot \dot{\boldsymbol{\chi}}_{\mathrm{A}}^{*} \\ &= (\boldsymbol{\alpha}_{\mathrm{A}\to\mathcal{A}} + \boldsymbol{\alpha}_{\mathrm{B}\to\mathcal{A}}) \cdot \boldsymbol{c}_{\mathrm{A}\to\mathcal{A}} \cdot \dot{\boldsymbol{\chi}}_{\mathrm{A}} \cdot \dot{\boldsymbol{\chi}}_{\mathrm{A}}^{*} \\ &= \boldsymbol{\alpha} \cdot \boldsymbol{c}_{\mathrm{A}\to\mathcal{A}}. \end{aligned} \tag{8.37}$$

Therefore, from the perspective of Ship $\mathcal{A}$, the effective downlink channel is just the composite channel, $\boldsymbol{\alpha} = \boldsymbol{\alpha}_{\mathrm{A}\to\mathcal{A}} + \boldsymbol{\alpha}_{\mathrm{B}\to\mathcal{A}}$, and interference is prevented.

A caveat of this coordinated transmission scheme is that the addition of $\boldsymbol{\alpha}_{\mathrm{A}\to\mathcal{A}}$ and $\boldsymbol{\alpha}_{\mathrm{B}\to\mathcal{A}}$ can be constructive when they are in phase and can also be destructive when $\boldsymbol{\alpha}_{\mathrm{A}\to\mathcal{A}}$ and $\boldsymbol{\alpha}_{\mathrm{B}\to\mathcal{A}}$ are out of phase, e.g., 180° apart, causing fading. To fully exploit the combining gain at the mobile station, Stations A and B must pre-compensate the phase difference between $\boldsymbol{\alpha}_{\mathrm{A}\to\mathcal{A}}$ and $\boldsymbol{\alpha}_{\mathrm{B}\to\mathcal{A}}$, i.e.,

$$\begin{aligned} \boldsymbol{y}_{\mathcal{A}} &= \boldsymbol{\alpha}_{\mathrm{A}\to\mathcal{A}} \cdot \dot{\boldsymbol{c}}_{\mathrm{A}\to\mathcal{A}} + \boldsymbol{\alpha}_{\mathrm{B}\to\mathcal{A}} \cdot \left(\dot{\boldsymbol{c}}_{\mathrm{A}\to\mathcal{A}} \cdot e^{j\varphi}\right) \\ &= \left(\boldsymbol{\alpha}_{\mathrm{A}\to\mathcal{A}} + \boldsymbol{\alpha}_{\mathrm{B}\to\mathcal{A}} \cdot e^{j\varphi}\right) \cdot \dot{\boldsymbol{c}}_{\mathrm{A}\to\mathcal{A}}, \end{aligned} \tag{8.38}$$

where $\varphi = \angle\left(\boldsymbol{\alpha}_{\mathrm{A}\to\mathcal{A}} \cdot \boldsymbol{\alpha}_{\mathrm{B}\to\mathcal{A}}^{*}\right)$, so that $\boldsymbol{\alpha}_{\mathrm{A}\to\mathcal{A}}$ and $\boldsymbol{\alpha}_{\mathrm{B}\to\mathcal{A}} \cdot e^{j\varphi}$ are now in phase. This process requires the knowledge of both downlink channels, involving the coordination from the mobile station (channel feedback). Selective muting is simpler and more practical in this sense.

It is a common misconception that TDD negates the need for channel feedback per the reciprocity principle. However, the reciprocity principle only applies to the radio link, not the radio channel that includes two different transceiver RF paths in uplink and downlink (see Chap. 5 for the definitions of radio link and radio channel). Therefore, RF calibration is needed before uplink and downlink can have the same radio channel.

On the uplink (not shown in Fig. 8.16), Ship $\mathcal{A}$ transmits over the assigned PUTCH resource and is received simultaneously by Stations A, B, and C, each observing distinct uplink channels, $\boldsymbol{\alpha}_{\mathcal{A}\to\mathrm{A}}$, $\boldsymbol{\alpha}_{\mathcal{A}\to\mathrm{B}}$, and $\boldsymbol{\alpha}_{\mathcal{A}\to\mathrm{C}}$. Quite differently from the downlink, the uplink burst$\boldsymbol{x}_{\mathrm{A}\to\mathcal{A}}$ is received *separately* by Stations A, B, and C: $\boldsymbol{\alpha}_{\mathcal{A}\to\mathrm{A}} \cdot \dot{\boldsymbol{c}}_{\mathcal{A}\to\mathrm{A}}$ by Station A, $\boldsymbol{\alpha}_{\mathcal{A}\to\mathrm{B}} \cdot \dot{\boldsymbol{c}}_{\mathcal{A}\to\mathrm{A}}$ by Station B, and $\boldsymbol{\alpha}_{\mathcal{A}\to\mathrm{C}} \cdot \dot{\boldsymbol{c}}_{\mathcal{A}\to\mathrm{A}}$ by Station C. They can then be combined coherently as the following:

$$
\begin{aligned}
\boldsymbol{c}'_{\mathcal{A}\to\mathrm{A}} =& (\boldsymbol{\alpha}_{\mathcal{A}\to\mathrm{A}} \cdot \dot{\boldsymbol{c}}_{\mathcal{A}\to\mathrm{A}}) \cdot \dot{\boldsymbol{\chi}}_{\mathrm{A}}^{*} \cdot \boldsymbol{\alpha}_{\mathcal{A}\to\mathrm{A}}^{*} \\
&+ (\boldsymbol{\alpha}_{\mathcal{A}\to\mathrm{B}} \cdot \dot{\boldsymbol{c}}_{\mathcal{A}\to\mathrm{A}}) \cdot \dot{\boldsymbol{\chi}}_{\mathrm{A}}^{*} \cdot \boldsymbol{\alpha}_{\mathcal{A}\to\mathrm{B}}^{*} \\
&+ (\boldsymbol{\alpha}_{\mathcal{A}\to\mathrm{C}} \cdot \dot{\boldsymbol{c}}_{\mathcal{A}\to\mathrm{A}}) \cdot \dot{\boldsymbol{\chi}}_{\mathrm{A}}^{*} \cdot \boldsymbol{\alpha}_{\mathcal{A}\to\mathrm{C}}^{*} \\
=& \left(|\boldsymbol{\alpha}_{\mathcal{A}\to\mathrm{A}}|^2 + |\boldsymbol{\alpha}_{\mathcal{A}\to\mathrm{B}}|^2 + |\boldsymbol{\alpha}_{\mathcal{A}\to\mathrm{C}}|^2 \right) \cdot \boldsymbol{c}_{\mathcal{A}\to\mathrm{A}},
\end{aligned}
\tag{8.39}
$$

commonly known as *maximum ratio combining* (MRC), thereby enjoying the rich combining gain and, potentially, ample spatial diversity gain as well. MRC in (8.39) is optimal in the sense of maximum symbol SNR, assuming equal noise levels at control station receivers. It is assumed understood that the receiver thermal noise background (including other unaccounted interference) is present but not shown in the above analysis, purposely omitted to simplify the description.

A lighter-weight alternative to (8.39) is "hard combining," in which each coordinating station attempts to decode the received signal from Ship $\mathcal{A}$ independently, and the winner is whoever successfully decodes the burst.

Like on the downlink, Station C, or both Stations B and C, may choose to "mute only," i.e., making no attempt to receive or decode the signal from Ship $\mathcal{A}$ to save the combining complexity. Indeed, the combining process can be rather involved depending on the backhaul network of the infrastructure.

Either way, since the uplink resource, on which Ship $\mathcal{A}$ is transmitting, is not used by any ship in the cells under the control of Stations B and C, interference on the uplink is avoided.

The role of each group member may change as the ship sails along, so may the members of the group. It is the NRC's responsibility to maintain this group when engaging in communication with this mobile station.

Chapter 9
Satellite Air Interface Design

Abstract Mobile stations on the high seas are not reachable by the VDES network on land once vessels have sailed outside the area of VDE-TER coverage from shore stations. Under the maritime MTC architecture, satellite communication is responsible for global coverage to address this issue (cf. Chap. 2). With the newly available satellite downlink spectrum (see Chap. 6), satellite MTC becomes possible on a global scale. It employs the same centralized medium access control strategy as VDE-TER, but via space stations, particularly efficient in addressing the hidden node problem in satellite communications employing distributed medium access control due to the large disparity in the field of view between a space station and mobile stations. Under the unified design criterion, VDE-SAT has the same air interface as VDE-TER (including the layer protocols, the resource frame, and burst structure) and shares the VDE-TER spectrum with VDE-TER to maximize the efficiency of the extremely scarce VHF spectrum. Nevertheless, satellite communication does have considerable differences in radio spectrum regulation, propagation delay, mobility, and, possibly, the link budget. These differences and their potential impact on the design are the highlights of this chapter.

Keywords Satellite air interface · Waveform selection · Bandwidth selection · Doppler shift · Propagation delay · Timing advance · Timing control · TDD timing · VDE-SAT · VDES

9.1 Waveform and Bandwidth

Under the unified air interface design, VDE-SAT and VDE-TER share the same transmission waveform, burst, timing, and resource frame structures, but notable differences are expected and resolved through parameter configuration. This section reiterates the waveform and bandwidth selection and elaborates on the issues pertinent and prominent to the VDE-SAT scenario.

M. M. Wang, J. Zhang, *Machine-Type Communication for Maritime Internet-of-Things*, Wireless Networks, https://doi.org/10.1007/978-3-030-77908-5_9

9.1.1 Waveform

As noted earlier in Chap. 8, a single-carrier waveform is favorable over multi-carrier ones in the current design, mainly due to the power efficiency issue and the issues related to frequency localization and sensitivity to frequency errors inherent to a multi-carrier waveform. We revisit this subject with added implications in the context of satellite communication at sea.

First, a multi-carrier, particularly the OFDM waveform, is an effective solution for combating the multipath effect, a characteristic of a wideband radio link. However, as we recall from Chap. 6, the VDE-SAT spectrum is only 50 kHz in bandwidth (indeed a narrowband spectrum), even with the 100-kHz shared VDE-TER spectrum. For such narrowband satellite communication out on the open sea, the radio link is predominantly characterized by the line-of-sight frequency-flat property. A single-carrier waveform can better exploit this link characteristic than a multi-carrier one in terms of transceiver simplicity, consequently, a simpler and cheaper transceiver than a comparable OFDM transceiver, which is critical for low-cost MTC mobile stations.

Second, the poor frequency localization of an OFDM waveform due to excess sidelobes, as we have seen in Chap. 5, makes cross-link isolation more challenging for full-duplex FDD transceivers aboard LEO satellite stations, along with other interference issues and complications associated with out-of-band emissions as discussed in Chap. 6. Isolation through a spatial separation is not an option for small-dimensioned LEO satellites.

Third, the narrowband VDE-SAT radio spectrum (see Fig. 9.1) would be further divided into multiple even narrower sub-carriers, should a multi-carrier waveform be employed. This further reduced bandwidth, in addition to the inherently large sidelobes of the OFDM spectrum, would make it more sensitive to frequency errors, hence even more susceptible to inter-carrier interference. A fast-moving satellite space station can easily induce a Doppler shift of multiple kHz, depending on the satellite's orbit and traveling directions relative to a mobile station on the earth.

Specifically, consider a VDE-SAT space station orbiting at an altitude of h_{sat} and a velocity of $\mathbf{v}_{\mathrm{sat}}$, as depicted in Fig. 9.2, where $\mathbf{c}_{\mathrm{carrier}}$ is the velocity vector of the carrier or radio wave emitted by the mobile station traveling at a velocity of $\mathbf{v}_{\mathrm{mobile}}$. The frequency shift, or the Doppler shift, that the receiver (the space station in Fig. 9.2) observes is

$$\varsigma_{\mathrm{Doppler}} = \left(\frac{\mathbf{v}_{\mathrm{tx}}}{c_{\mathrm{light}}} \cdot \frac{\mathbf{c}_{\mathrm{carrier}}}{c_{\mathrm{light}}} - \frac{\mathbf{v}_{\mathrm{rx}}}{c_{\mathrm{light}}} \cdot \frac{\mathbf{c}_{\mathrm{carrier}}}{c_{\mathrm{light}}} \right) \cdot f_{\mathrm{carrier}}, \tag{9.1}$$

where $\mathbf{v}_{\mathrm{tx}}$ is the transmitter velocity, $\mathbf{v}_{\mathrm{rx}}$ the receiver velocity, $c_{\mathrm{light}} = |\mathbf{c}_{\mathrm{carrier}}|$ the speed of light, and f_{carrier} the carrier frequency. For a space station orbiting at an altitude of h_{sat}, the orbital speed, $v_{\mathrm{sat}} \triangleq |\mathbf{v}_{\mathrm{sat}}|$, is determined by

$$v_{\mathrm{sat}} = \left(\frac{4\pi}{3} \mathcal{G} M_{\mathrm{earth}} R_{\mathrm{earth}}^2 \right)^{\frac{1}{2}} \cdot \left(1 + \frac{h_{\mathrm{sat}}}{R_{\mathrm{earth}}} \right)^{-\frac{1}{2}}, \tag{9.2}$$

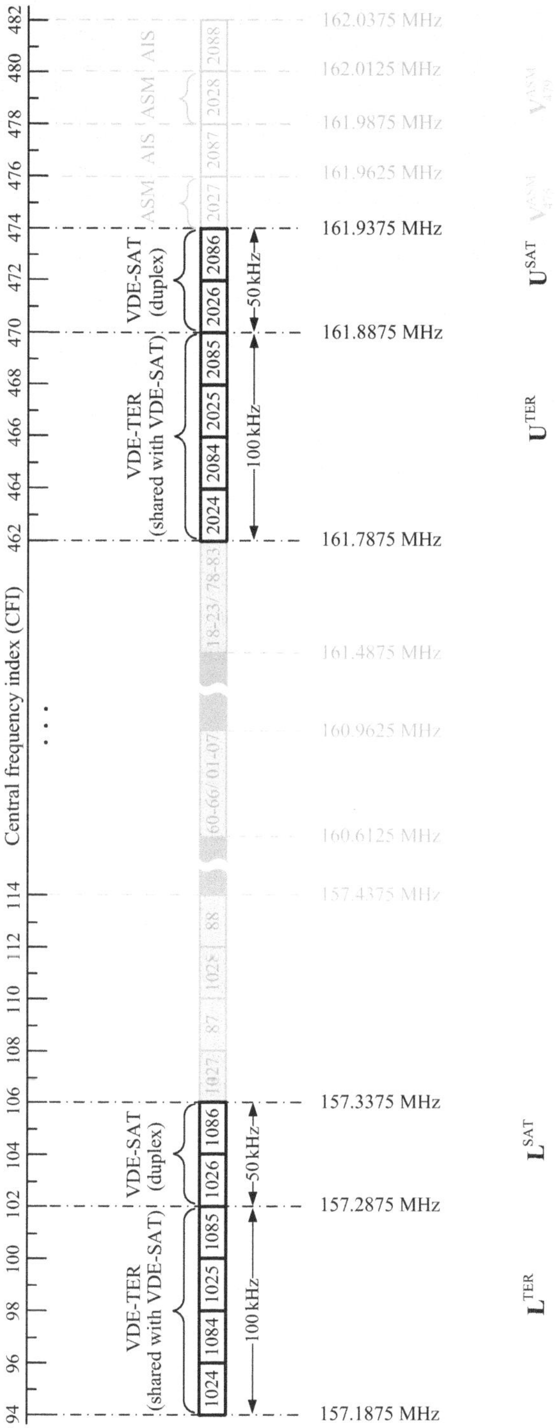

Fig. 9.1 VDE-SAT assigned and shared spectrum with VDE-TER (replotted from Fig. 6.1 in Chap. 6), where $\mathbf{U}^{\mathrm{TER}}$ is shared with VDE-SAT downlink and $\mathbf{L}^{\mathrm{TER}}$ is conditionally sharable under the sharing scheme discussed in Chap. 6

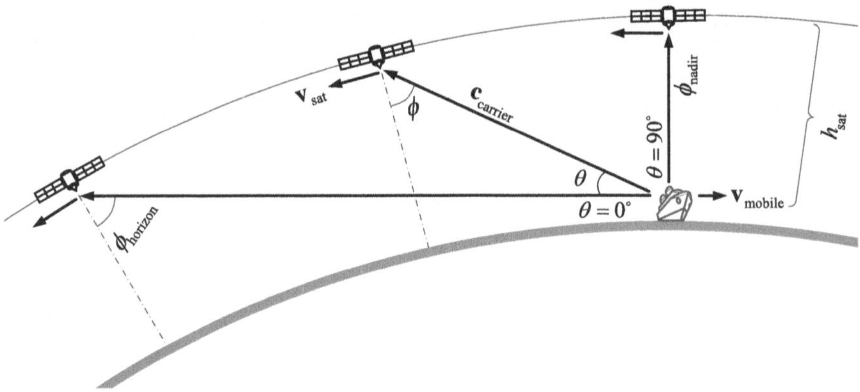

Fig. 9.2 Illustration of the motion between a space station and a mobile station for maximum Doppler shift calculation, where the off-nadir angle, ϕ, is related to the mobile station's eleva-tion angle, θ, through (9.5)

where $\mathcal{G}$ is the gravitational constant, and M_{earth} and R_{earth} are the earth's mass density and radius, respectively.

For a LEO space station orbiting at h_{earth}= 600 km, the corresponding orbital speed is thus v_{sat}= 8 km/s. The speed of a mobile station, $v_{mobile} \triangleq |\mathbf{v}_{mobile}|$, is usually less than 100 knots (~50 m/s), and hence it is reasonable to assert that $v_{mobile} \ll v_{sat}$. Therefore, (9.1) can be comfortably simplified as

$$|\varsigma_{Doppler}| \approx \left| \frac{\mathbf{v}_{sat}}{c_{light}} \cdot \frac{\mathbf{c}_{carrier}}{c_{light}} \right| \cdot f_{carrier}, \tag{9.3}$$

which reaches its maximum when the mobile station appears on the space station's horizon, i.e.,

$$|\varsigma_{Doppler}| \leq \frac{v_{sat}}{c_{light}} \cdot \sin(\phi_{horizon}) \cdot f_{carrier}, \tag{9.4}$$

where $\phi_{horizon}$ is the off-nadir angle directing towards the earth's limb, i.e., the mobile station on the horizon with an elevation angle $\theta = 0°$.

Now, from (6.27) in Chap. 6, we have the following relationship between ϕ and θ:

$$\phi = \arcsin\left(\left(1 + \frac{h_{sat}}{R_{earth}}\right)^{-1} \cdot \cos(\theta)\right), \quad \theta \in [0°, 180°]. \tag{9.5}$$

Thus, the space station's off-nadir angle corresponding to $\theta = 0°$ is

$$\phi_{horizon} = \arcsin\left(1 + \frac{h_{sat}}{R_{earth}}\right)^{-1}. \tag{9.6}$$

For a LEO space station at h_{sat}= 600 km, $\phi_{horizon} \approx 67^{\circ}$. Substituting (9.6) into (9.4) results in the maximum frequency shift that a LEO space station experiences,

$$\max \left| \varsigma_{Doppler} \right| \approx 3.7\,\text{kHz}. \tag{9.7}$$

The result remains the same if the space station is the transmitter and the mobile station is the receiver. Dealing with a frequency offset of this magnitude is quite challenging to frequency-error-sensitive OFDM receivers, especially for low-cost mobile stations operating on the narrowband VDES spectrum (the effect of a frequency error is "amplified" here).

Due to these practical issues, OFDM is not deemed the best fit for VDE-SAT deployed on the VDES spectrum, despite its great efficiency in dealing with multipath and flexibility in spectrum usage, particularly for centralized resource allocation.

9.1.2 Bandwidth

Now let us see how the Doppler shift of this magnitude affects the single-carrier waveform bandwidth selection. As depicted in Fig. 9.3, the satellite-induced frequency shift effectively causes a spectrum shift by $\varsigma_{Doppler}$ at the receiver. It

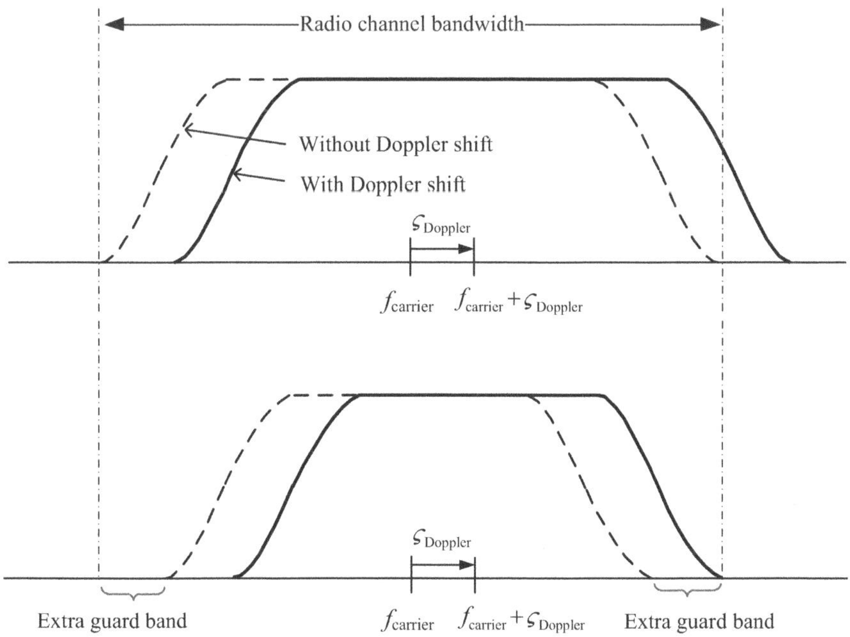

Fig. 9.3 Illustration of the Doppler-shift effect on the radio channel guard bands. The dashed line denotes the received signal spectrum without Doppler shift and the solid line with a Doppler shift, $\varsigma_{Doppler}$

necessitates additional guard bands of at least max $|\varsigma_{\text{Doppler}}|$ on each side to contain the spectrum shift.

Unlike in VDE-TER, where the single-carrier waveform is upper limited by the frequency selectivity of the radio link, it is typically not an issue for line-of-sight predominant satellite radio links at sea. Therefore, under the spectrum-sharing scheme discussed in Sect. 6.5 of Chap. 6, the downlink waveform takes the entire available bandwidth, i.e., $|\mathbf{U}^{\text{TER}} \cup \mathbf{U}^{\text{SAT}}| = 150$ kHz, including the 50 kHz on the dedicated VDE-SAT spectrum, $\mathbf{U}^{\text{SAT}}$, plus the 100 kHz on the shared VDE-TER spectrum, $\mathbf{U}^{\text{TER}}$, as depicted in Fig. 9.1. Resource granularity is less of an issue for downlink (one-to-many) since multiple DTCHs destined to different mobile stations (with similar channel conditions and QoS) can be bundled into a single PDTCH. As such, the default downlink physical channel bandwidth is 150 kHz, except for the system information broadcast channel, PBBCH, which is fixed to $|\mathbf{U}^{\text{SAT}}| = 50$ kHz to avoid interfering with VDE-TER PBBCH. The guard bands help prevent potential interference to the reception of signals (e.g., the ASM signal) from different stations on the adjacent channel.

The uplink remains 50 kHz a piece, regardless of the available bandwidth, for the same reason (i.e., the uplink resource granularity), as noted in Chap. 8. Each of the 50-kHz bands is protected by a 4-kHz guard band of each side of the band against the Doppler shifts that the space station may see on the uplink bursts from different mobile stations, supposedly randomly distributed, $\varsigma_{\text{Doppler}} \in [-3.7, +3.7](\text{kHz})$, depending on the positions (or ϕ) of the mobile stations in the space station's field of view. The guard bands prevent the uplink bursts on the adjacent bands from interfering with each other.

With the guard bands, the square-root raised-cosine pulse roll-off factor gets to be slightly relaxed to 0.25. The corresponding transmit rate is $T_{\text{sym}}^{-1} = 112{,}800$ pulses or channel symbols per second over a 150 kHz downlink spectrum minus the 4-kHz guard band on each side of the band and $T_{\text{sym}}^{-1} = 33{,}600$ symbols per second over a 50-kHz bandwidth, minus the 4-kHz guard band on each side of the band.

9.2 Resource Frame

VDE-SAT utilizes the same resource frame structure as VDE-TER and shares the same unified channelization framework employing the radio resource block concept, as diagrammed in Fig. 8.5, but with several configurational adjustments to reflect the differences between VDE-SAT and VDE-TER, mainly in radio spectrum assignments and regulatory rules, propagation delay and loss, and mobility. We have seen their impact on the bandwidth. We now look at how they affect the resource frame configuration and the link budget.

The VDE-SAT resource frame under FDD is exemplified in Fig. 9.4, much like VDE-TER, except that VDE-SAT shares the VDE-TER spectrum. Like VDE-TER, a physical layer PDU is carried by a physical channel transmitted in the form of a transmission burst on the assigned spectrum in alignment with the slot boundaries synchronized with UTC, as depicted in Fig. 7.2.

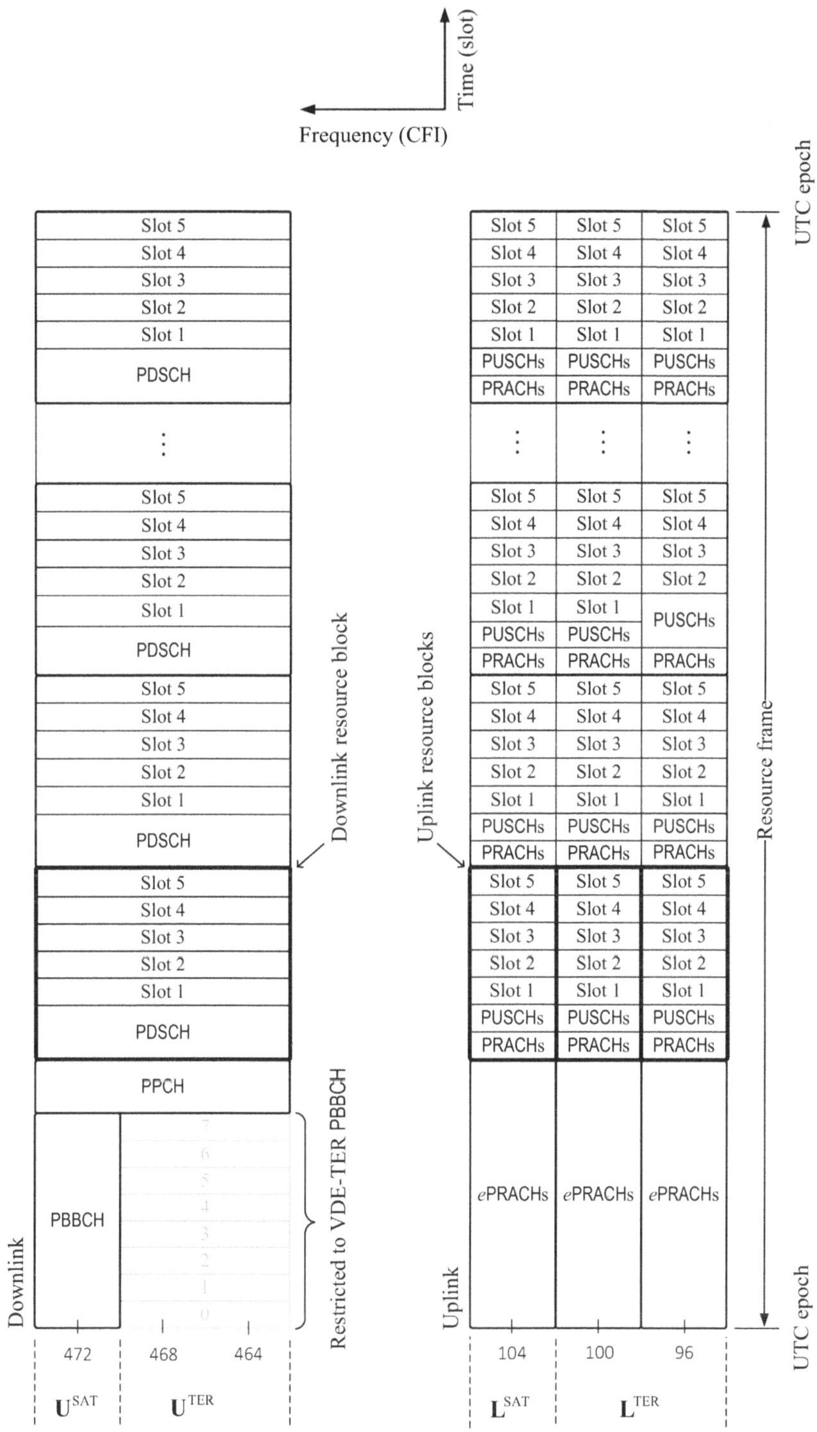

Fig. 9.4 Illustration of VDE-SAT FDD resource frame configuration example that shares the spectrum with VDE-TER. The concept of spectrum sharing is explained in Chap. 6. The dedicated VDE-SAT channels, $\mathbf{L}^{SAT}$ and $\mathbf{U}^{SAT}$, serve as the "anchor carriers." VDE-SAT transmissions on Slots 0 to 7 on a $\mathbf{U}^{TER}$ frame are restricted for protection of VDE-TER PBBCH

9.2.1 Downlink

The first eight slots of a frame on $\mathbf{U}^{\mathrm{TER}}$ are restricted to the sole use of the VDE-TER PBBCH for the protection of the VDE-TER PBBCH from interference, as we recall from Chap. 8 that these eight slots are designated to VDE-TER PBBCHs.

Pertaining to the centralized resource management, the MAC resource allocation function uses Resource Assignment PDU, carried by ACH and transmitted on PDSCH, to convey resource assignment information between transmitting and receiving MAC peers. Each slot in the traffic zone of the resource block pair is individually assigned to a mobile station as the resource for PTCH (PDTCH or PUTCH) indicated by an assignment element included in the MAC Resource Assignment PDU for finest granularity. To allow for the same PDTCH resource granularity as the VDE-TER, there is a need for the VDE-SAT downlink to have more than one DTCH be bundled into a single PDTCH for a slot configured with a bandwidth greater than 50 kHz. In this case, i.e., 150 kHz, a slot may be assigned up to three mobile stations, resulting in three assignment elements per slot. If a downlink slot is assigned to the same mobile station, the mobile station will receive three duplicated assignment elements in a roll.

Unlike VDE-TER, VDE-SAT downlink shares the radio spectrum with the land systems (the primary users). It is critically important to address the conflicting demands of ensuring the resulting interference is within acceptable limits to the victim system while at the same time providing an adequate link budget for the physical signaling channels with sufficient levels of robustness or processing gain.

The link budget evaluation is similar to VDE-TER, except that the space station's transmit power is constrained by the EIRP prescribed in Chap. 6, and the propagation loss is approximated by the free-space model (Friis' equation).

Specifically, the link budget analysis is based on the same burst structure defined in Fig. 8.8 with a specific configuration enumerated in Table 9.1 based on Fig. 9.4. We continue to use the notation convention, denoting the space station's off-nadir angle as ϕ and the mobile station's elevation angle as θ, with a relationship defined in (9.5) from (6.27) of Chap. 6.

Table 9.1 Burst configurations of the VDE-SAT downlink bursts

Physical signaling burst	PBBCH	PPCH	PDSCH	PDTCH
TTI (slot)	8	2	2	Variable
Symbol rate (symbols/sec)	33.6 k	112.8 k	112.8 k	112.8 k
Number of Symbols, $N_{\mathrm{sym}}^{\mathrm{burst}}$	7168	6016	6016	3008 × TTI
`Ramp up/down` (symbols)	None	None	None	None
`Sync word` (symbols)	256	None	None	None
`Guard period` (symbols)	None	None	None	None
`Pilot` pattern	(16, 4)	Configurable		
Payload (symbols)	5184	≤6016	≤6016	≤3008 × TTI

From Chap. 6, the EIRP per unit spectrum of a space station equipped with a typical Yagi antenna is constrained by (6.26). Now, for the VDE-SAT downlink bandwidth of $B_{\text{SAT}}^{\text{DL}}$, the EIRP is $\Lambda_{\text{sat}}(\phi) \cdot B_{\text{SAT}}^{\text{DL}}$, where

$$\begin{aligned} B_{\text{SAT}}^{\text{DL}} &= \left|\mathbf{U}^{\text{SAT}}\right| \\ &= 50 \text{ kHz}, \end{aligned} \tag{9.8}$$

for PBBCH, and

$$\begin{aligned} B_{\text{SAT}}^{\text{DL}} &= \left|\mathbf{U}^{\text{SAT}} \cup \mathbf{U}^{\text{TER}}\right| \\ &= \left|\mathbf{U}^{\text{SAT}}\right| + \left|\mathbf{U}^{\text{TER}}\right| \\ &= 150 \text{ kHz}, \end{aligned} \tag{9.9}$$

for all other downlink channels, PPCH, PDSCH, and PDTCH (see Sect. 9.1.2).

The power collected by a destined mobile station receiver is

$$p_{\text{mobile}}(\theta) = \frac{\Lambda_{\text{sat}}(\phi) \cdot B_{\text{SAT}}^{\text{DL}} \cdot G_{\text{mobile}}(\theta)}{\ell(\theta)}, \tag{9.10}$$

where ℓ is the propagation loss between the space station and the mobile station on the surface of the earth and $G_{\text{mobile}}(\theta)$ the mobile station antenna gain. The assumed average antenna gain of a Category 1 mobile station, the mobile station with the minimum RF capability, is roughly -3 dBi.

Denoting the receiver noise temperature of a mobile station as T_{mobile}, the noise power spectral density is then $N_0^{\text{mobile}} = \kappa T_{\text{mobile}}$ and the downlink receive SNR is determined by

$$\begin{aligned} \eta_{\text{mobile}}(\theta) &= \frac{p_{\text{mobile}}(\theta)}{N_0^{\text{mobile}} \cdot T_{\text{sym}}^{-1}} \\ &= \frac{\Lambda_{\text{sat}}(\phi) \cdot B_{\text{SAT}}^{\text{DL}} \cdot G_{\text{mobile}}(\theta)}{N_0^{\text{mobile}} \cdot T_{\text{sym}}^{-1} \cdot \ell(\theta)}, \quad \theta \in [0^\circ, 90^\circ]. \end{aligned} \tag{9.11}$$

Now, given the receiver mobile station noise temperature T_{mobile} of 30 dBK offshore, and $T_{sym}^{-1} = 33.6$ kHz and 112.8 kHz for the respective downlink bandwidth of $B_{\text{SAT}}^{\text{DL}} = 50$ kHz and 150 kHz, the received VDE-SAT downlink SNR ranges between 4 and 13 dB by a mobile station at an elevation angle between $\theta = 90^\circ$ and 0°, as shown in Fig. 9.6, from a space station under the EIRP constrained by the PFD mask, as replotted in Fig. 9.5 from Chap. 6.

Table 9.2 summarizes the downlink link budgets.

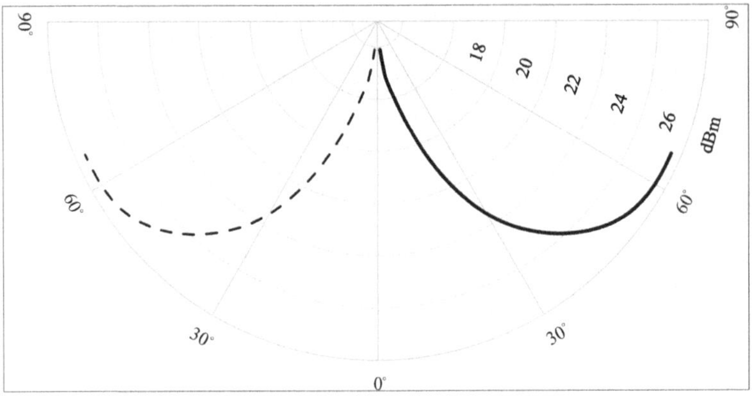

Fig. 9.5 Space station Yagi antenna beam EIRP (per 25 kHz), replotted from Fig. 6.2 in Chap. 6 as a function of the satellite's off-nadir angle ϕ

Table 9.2 VDE-SAT downlink budgets under the burst configuration in Table 9.1

Physical channel	PBBCH	PPCH	PDSCH	PDTCH
SDU (bits)	192	1260	1223	Variable
CRC (bits)	24	24	24	24
Code rate, r	TC: 1/3	TC: 1/3	TC: 1/3	TC: $\geq$1/3
Modulation	QPSK ($m = 2$)	QPSK	QPSK	Adaptive
1%-PER[a] $E_b/N_0, E_b^+/N_0$	1.8 dB	1.2 dB	1.2 dB	Variable
TTI	8 slots	2 slots	2 slots	Variable
Effective code rate, r' ($\mathcal{g}$)	0.021 (17 dB)	0.106 (10 dB)	0.104 (10 dB)	Adaptive
Transmit power, P_{sat}	EIRP per unit spectrum constrained by (6.26)			
Bandwidth, B_{SAT}^{DL}	50 kHz	150 kHz		
Noise temperature, T_{mobile}	30 dBK			
Receive SNR, η_{mobile}	4 dB/13 dB[b]			
Receive SINR, γ_{mobile}	- 1.5 dB	- 1.5 dB	- 1.5 dB	Variable
E_b/N_0 margin, $\Delta E_b/N_0$	11 dB	4 dB	4 dB	Variable

[a]PER: physical layer PDU (or physical channel payload) error rate
[b]Minimum/maximum receive SNR (before inter-cell interference)

9.2.2 Uplink

The uplink frame structure is the same as VDE-TER except that the uplink shares 100 kHz worth of VDE-TER spectrum like the downlink, as diagrammed in Fig. 9.4. Nonetheless, the uplink transmission bandwidth remains to be 50 kHz for a reason explained earlier in Sect. 9.1.2. The burst structure is defined in Fig. 8.9 with a specific configuration enumerated in Table 9.3.

Table 9.3 Burst configurations of the VDE-SAT uplink bursts

Physical signaling burst	PRACH	ePRACH	PUSCH	PUTCH
TTI (slot)	1	10	1	Variable
Symbol rate (symbols/sec)	33.6 k	33.6 k	33.6 k	33.6 k
Number of symbols, $N_{\text{sym}}^{\text{burst}}$	896	8960	896	896 × TTI
`Ramp up/down` (symbol)	14/14	14/14	14/14	14/14
`Sync word` (symbol)	64	640	64	64
`Guard period` (symbol)[a]	20	484	None	None
`Pilot` pattern	Configurable (via BBCH)		Adaptive (via CCCH)	
Payload (symbol)[b]	≤784	≤7808	≤804	≤896 × TTI-92

[a]Detailed in Sect. 9.3
[b]Equality holds when no `Pilot` is present. The actual payload depends on the pilot pattern

Unlike the downlink where the transmit power is constrained by the EIRP per unit bandwidth, the uplink transmit power for a mobile is fixed, e.g., $P_{\text{mobile}} = 100$ mW for a regular Category 1 mobile station. The received signal power is

$$p_{\text{sat}}(\phi) = \frac{\Lambda_{\text{mobile}}(\theta) \cdot G_{\text{sat}}(\phi)}{\ell(\phi)} \tag{9.12}$$

by a space station, where $\Lambda_{\text{mobile}}(\theta) \triangleq P_{\text{mobile}} \cdot G_{\text{mobile}}(\theta)$ is the EIRP of the mobile station and $G_{\text{sat}}(\phi)$ the antenna gain of the space station, replotted in Fig. 9.5 from Chap. 6. The SNR at a receiving space station is

$$\eta_{\text{sat}}(\phi) = \frac{p_{\text{sat}}(\phi)}{N_0^{\text{sat}} \cdot T_{\text{sym}}^{-1}}, \tag{9.13}$$

where $T_{\text{sym}}^{-1} = 33.6$ kHz for $B_{\text{SAT}}^{\text{UL}} = 50$ kHz uplink channel bandwidth, $N_0^{\text{sat}} = \kappa T_{\text{sat}}$, and $T_{\text{sat}} = 26$ dBK is the noise temperature of a space station receiver. Substituting (9.12) into (9.13) yields

$$\eta_{\text{sat}}(\phi) = \frac{\Lambda_{\text{mobile}}(\theta) \cdot G_{\text{sat}}(\phi)}{N_0^{\text{sat}} \cdot T_{\text{sym}}^{-1} \cdot \ell(\phi)}. \tag{9.14}$$

For an uplink signal from a mobile station with 100 mW transmit power on the surface of the earth, the received SNR by a LEO space station (with a Yagi receive antenna orbiting at a 600-km altitude) ranges between 7 and 16 dB at an off-nadir angle ϕ between 0° and 67°, as shown in Fig. 9.6.

Table 9.4 summarizes the uplink link budgets.

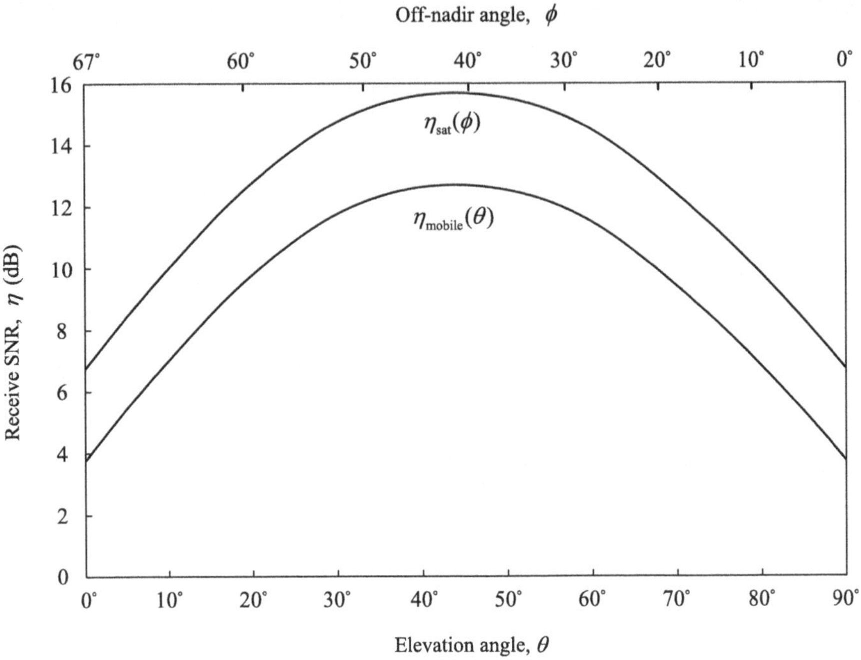

Fig. 9.6 The VDE-SAT downlink and uplink receive SNRs for a LEO satellite space station, orbiting at an altitude of 600 km, with a Yagi antenna and constrained EIRP. The mobile station transmit power is 100 mW

Table 9.4 VDE-SAT uplink budgets under the burst configuration in Table 9.3

Physical channel	PRACH	*e*PRACH	PUSCH	PUTCH
SDU (bits)	48	48	6	Variable
CRC (bits)	12	12	6	24
Code rate, r	TC: 1/3	TC: 1/3	TBCC: 1/3	TC: $\geq$1/3
Modulation	QPSK($m = 2$)	QPSK	QPSK	Adaptive
1%-PERE_b/N_0, E_b^+/N_0	2.6 dB	2.6 dB	1.7 dB	Variable
TTI	1 slot	10 slot	1 slot	Variable
Effective code rate, r' (g)	0.038 (14 dB)	0.004 (24 dB)	0.007 (21 dB)	Adaptive
Transmit power, P_{mobile}	Power controlled			5 mW/1 W[a]
Noise temperature, T_{sat}	26 dBK			
Receive SNR, η_{sat}	7 dB	-6 dB	7 dB	-6 dB/16 dB
Receive SINR, γ_{sat}	-6 dB[b]	-16 dB[c]	-6 dB[b]	Variable
E_b/N_0 margin, $\Delta E_b/N_0$	2.4 dB	2.4 dB	10 dB	Variable

[a]Minimum/maximum transmit power
[b]In the case of a total of 5 power-controlled simultaneous PRACH transmissions
[c]40 power-controlled PRACH transmissions

9.3 Transmission Timing

We have seen that a time buffer on the uplink burst is needed to avoid collisions at the control station between successive uplink transmissions from different mobile stations. So far, we have assumed that the time buffer, particularly the `guard period`, can be saved except for the PRACH burst (Tables 8.8 and 9.3) with the timing advance scheme. In this section, as promised, we first quantitatively determine the relationship between the `guard period` and the propagation time and show that the `guard period` overhead becomes increasingly significant as the propagation time increases. We then describe the timing advance scheme and the MAC timing control protocol that negates the `guard period` and facilitates an efficient synchronization framework for both VDE-SAT and VDE-TER as well as ASM.

9.3.1 Effect of Propagation Delays

To see how the propagation delay affects the time buffer of a VDE-SAT uplink burst, let us take a close look at the propagation time between a space station and a mobile station in the field of view of a space station and the maximum possible difference in the propagation time that a space station sees among all the mobile stations within the view, with the help of Fig. 9.7.

From a space station's perspective, the propagation time or delay can be expressed as a function of the off-nadir angle of a space station, i.e., ϕ,

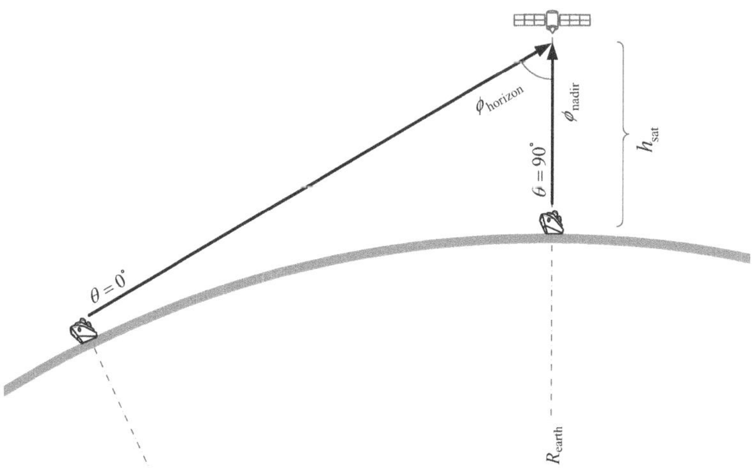

Fig. 9.7 Illustration of the maximum propagation time difference that a space station sees between two mobile stations

$$\delta(\phi) = \frac{\cos(\phi + \theta)}{\sin \phi} \cdot \frac{R_{\text{earth}}}{c_{\text{light}}}. \tag{9.15}$$

From Fig. 9.7, the minimum uplink propagation time occurs when the space station communicates with a mobile station at an angle of

$$\begin{aligned}\phi_{\text{nadir}} &\triangleq \arg\min_{\phi} \delta(\phi) \\ &= 0^\circ,\end{aligned} \tag{9.16}$$

i.e., toward the nadir, which corresponds to the mobile station's elevation angle of $\theta = 90^\circ$, i.e., toward the zenith, whereas the maximum propagation delay is from the mobile station in the direction of

$$\begin{aligned}\phi_{\text{horizon}} &\triangleq \arg\max_{\phi} \delta(\phi) \\ &= \arcsin\left(1 + \frac{h_{\text{sat}}}{R_{\text{earth}}}\right)^{-1},\end{aligned} \tag{9.17}$$

corresponding to $\theta = 0^\circ$ from (9.5), i.e., from the mobile station on the horizon.

Obviously, $\delta(\phi_{\text{nadir}}) = \frac{h_{\text{sat}}}{c_{\text{light}}}$ is the minimum propagation delay attributable to the orbital altitude of the space station, denoted as

$$\delta_{\min} \triangleq \delta(\phi_{\text{nadir}}), \tag{9.18}$$

and $\delta(\phi_{\text{horizon}})$ is the maximum propagation delay attributable to the distance to the horizon, denoted as

$$\delta_{\max} \triangleq \delta(\phi_{\text{horizon}}). \tag{9.19}$$

The largest uplink propagation time difference that a space station sees is, therefore, the difference between (9.19) and (9.18),

$$\begin{aligned}\Delta &\triangleq \delta_{\max} - \delta_{\min} \\ &= \frac{R_{\text{earth}}}{c_{\text{light}}} \cdot \text{ctan}(\phi_{\text{horizon}}) - \frac{h_{\text{sat}}}{c_{\text{light}}}.\end{aligned} \tag{9.20}$$

Substituting (9.17) into (9.20) and some straightforward manipulations, (9.20) becomes

$$\Delta = \left(\left(\left(\frac{h_{\text{sat}}}{R_{\text{earth}}}\right)^2 + 2\frac{h_{\text{sat}}}{R_{\text{earth}}}\right)^{\frac{1}{2}} - \frac{h_{\text{sat}}}{R_{\text{earth}}}\right) \cdot \frac{R_{\text{earth}}}{c_{\text{light}}}. \tag{9.21}$$

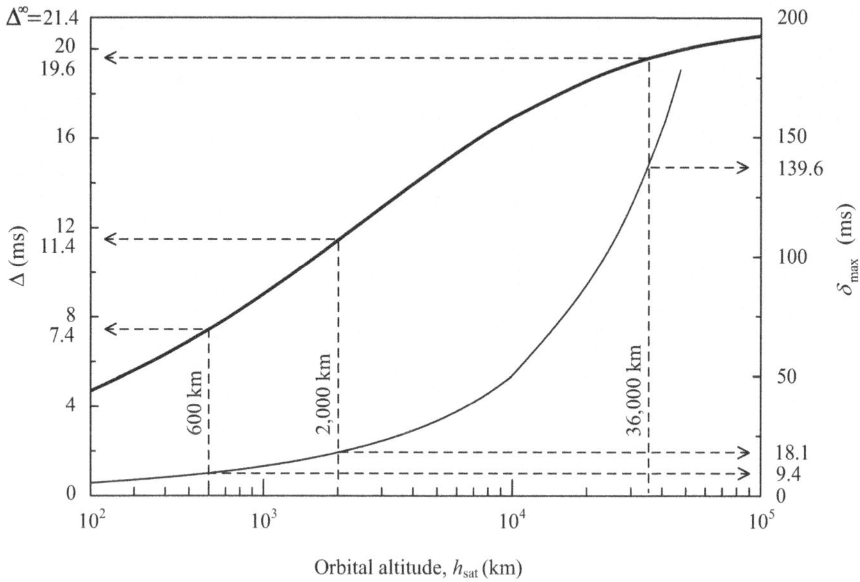

Fig. 9.8 Maximum uplink propagation time, δ_{max}, and maximum difference among mobile stations, $\Delta \triangleq \delta_{\text{max}} - \delta_{\text{min}}$, seen by a space station receiver at various orbital altitudes

This difference increases as h_{sat} increases, and asymptotically approaches a constant, i.e.,

$$\begin{aligned} \Delta^{\infty} &\triangleq \lim_{h_{\text{sat}} \to \infty} \Delta \\ &= \frac{R_{\text{earth}}}{c_{\text{light}}} \\ &\approx 21.4 \text{ ms}. \end{aligned} \tag{9.22}$$

As graphed in Fig. 9.8, the uplink propagation time difference for a LEO satellite space station at $h_{\text{sat}} = 600$ km can be as large as 7.4 ms, meaning that the propagation time difference among different mobile stations seen by a space station receiver can be up to 7.4 ms. The difference extends to ~11 ms for a MEO space station orbiting at 2000 km and ~20 ms for a GEO space station at 36,000 km.

The time buffer of an uplink transmission burst is needed to contain this difference plus transceiver timing errors.

The UTC synchronization error is typically less than 50μs for space stations, twice for mobile stations like ships, and even more for low-cost mobile stations, i.e., $|\varepsilon_{\text{sat}}^{\text{UTC}}| < 50$ μs for satellite and $|\varepsilon_{\text{mobile}}^{\text{UTC}}| < 200$μs for mobile stations. The time buffer of an uplink burst must then satisfy

$$T_{\text{buffer}} > \Delta + 2\left(|\varepsilon_{\text{mobile}}^{\text{UTC}}| + |\varepsilon_{\text{sat}}^{\text{UTC}}|\right). \tag{9.23}$$

Therefore, for a LEO space station orbiting at h_{sat}= 600 km, the propagation delay difference plus timing errors is within 7.9 ms. An 8-ms time buffer should do the job, at the cost of about 30% of a slot and more as the orbit moves further up to a higher altitude, e.g., up to 75% for a GEO space station.

Moreover, this excess overhead is under the assumption that a mobile station possesses a UTC. In reality, UTC is a luxury that most cost-and-energy-limited mobile stations cannot afford. Apart from the overhead, this large difference in propagation delay also creates an excessive timing ambiguity at the receiver, i.e., T_{buffer} as in (9.23), complicating the receiver design and degrading the receiver burst detection and timing performance. Indeed, a good timing scheme to minimize the uplink burst's `guard period` (resulting from multiplexing on the uplink among mobile stations) is crucial to the overall system spectral efficiency and uplink burst detection performance. A transmission timing framework to minimize the effect of the extensive propagation delay is the focus of this section.

There are two possible solutions. A straightforward solution is simply to increase the scheduling granularity of transmission resources, i.e., increases the TTI of an uplink transmission burst. For instance, forcing a minimum of 3-slot TTI reduces the time buffer overhead to 10%, and a 10-slot TTI knocks down the overhead to 3%. The drawbacks are clear: (1) timing ambiguity remains the same, and (2) reduced degrees of freedom in uplink resource assignment, causing waste of resources for short messages. Short-burst data services constitute a primary part of maritime IoT services. Hence uplink resource granularity is a major issue in physical layer design that affects the overall system capacity, as iterated earlier.

Another solution is to employ the *timing advance* approach. A mobile station advances its uplink transmission burst by an amount just enough to compensate for the propagation delay so that, in the eyes of a space station, the burst arrives at the space station without delay. Consequently, all uplink bursts arriving at the space station from different mobile stations align with the UTC slot boundaries, ergo, without overlapping. The time buffer can then be eliminated, and the receiver performance can be improved as well, "killing two birds with one stone."

9.3.2 Timing Advance

Similar to VDE-TER, the uplink spectrum is shared or time-multiplexed among mobile stations. During the uplink transmission period, the uplink bursts received by the space station are from multiple time-multiplexed mobile stations with varying time delays, i.e., $\delta(\phi) \in [\delta_{min}, \delta_{max})$, depending on the off-nadir angle, $\phi \in [\phi_{nadir}, \phi_{horizon}]$ (see Fig. 9.7), thereby potentially causing overlap/collision in time, ergo co-channel interference among uplink bursts at the space station. For the 600-km orbit LEO space station, $\delta_{min} \approx$ 2 ms due to the 600-km orbital altitude, and $\delta_{max} \approx$ 10 ms due to the 2830-km distance to the horizon, according to (9.18) and (9.19). Therefore, a time buffer of $T_{buffer} = 8$ ms needs to be set aside for each uplink burst to avoid collisions between successive uplink bursts, literally throwing away 30% of a slot.

However, if we can somehow align all uplink bursts' arrival time at the *serving* space station, the uplink `guard period`, at least 30% of a slot overhead, can then be avoided. Ideally, the detection window of the uplink bursts, resulting from the arrival-time ambiguity of individual uplink bursts, is nullified at the space station, reducing the detection complexity and improving the uplink burst detection performance.

Both benefits are achieved by advancing the mobile station's transmit time by such an amount just enough to offset its propagation delay, $\delta(\phi)$, as illustrated in Fig. 9.9. This delay is a function of the mobile station position relative to the space station (i.e., a function of ϕ) and time-varying. Hence it needs to be estimated on the fly. We start with the simple scenario in which a mobile station has UTC. In the current design, at the control station, downlink transmission bursts are naturally synchronized with UTC, i.e., the maritime MTC network slot timing, as noted in Sect. 7.3 of Chap. 7.

9.3.2.1 Timing with UTC

A mobile station first detects the arrival time of the downlink signal (e.g., PBBCH and PDSCH), $t_{\text{mobile}}^{\text{DL-slot}}$. The disparity between this detected time and the network slot timing is attributable to the one-way propagation delay, δ_{mobile}, a function of the distance between the mobile station and the control station, according to

$$\begin{aligned} t_{\text{mobile}}^{\text{DL-slot}} &= t^{\text{UTC-slot}} + \delta_{\text{mobile}} \\ &= t_{\text{sat}}^{\text{UTC-slot}} + \delta_{\text{mobile}}, \end{aligned} \tag{9.24}$$

where $t_{\text{sat}}^{\text{UTC-slot}}$ is the network slot time maintained at the control station.

With UTC, the network slot timing, $t^{\text{UTC-slot}}$, is known, so is δ_{mobile} from (9.24). The mobile station then advances its uplink transmission timing by $\Delta^{\text{TA}} = \delta_{\text{mobile}}$, relative to the UTC slot timing so that the one-way propagation delay on the uplink is compensated. This pre-compensation ensures the arrival time of the uplink signal in alignment with the network transmission timeline at the control station.

However, the reality is not nearly as neat. The timing advance error, i.e., the accuracy of the uplink burst alignment at a control station receiver after timing advance, is determined by the estimation accuracy of δ_{mobile}, which is in turn affected by (1) the accuracy of UTC timing, $t_{\text{mobile}}^{\text{UTC-slot}}$ and $t_{\text{sat}}^{\text{UTC-slot}}$, maintained at a mobile station and a space station; (2) detection and tracking accuracy of the downlink slot timing, $t_{\text{mobile}}^{\text{DL-slot}}$, by the mobile station; and (3) mobility of the space and mobile stations.

The network slot timing maintained at the control station can be expressed as

$$t_{\text{sat}}^{\text{UTC-slot}} = t^{\text{UTC-slot}} + \varepsilon_{\text{sat}}^{\text{UTC}}, \tag{9.25}$$

and the network timing slot maintained at the mobile station is

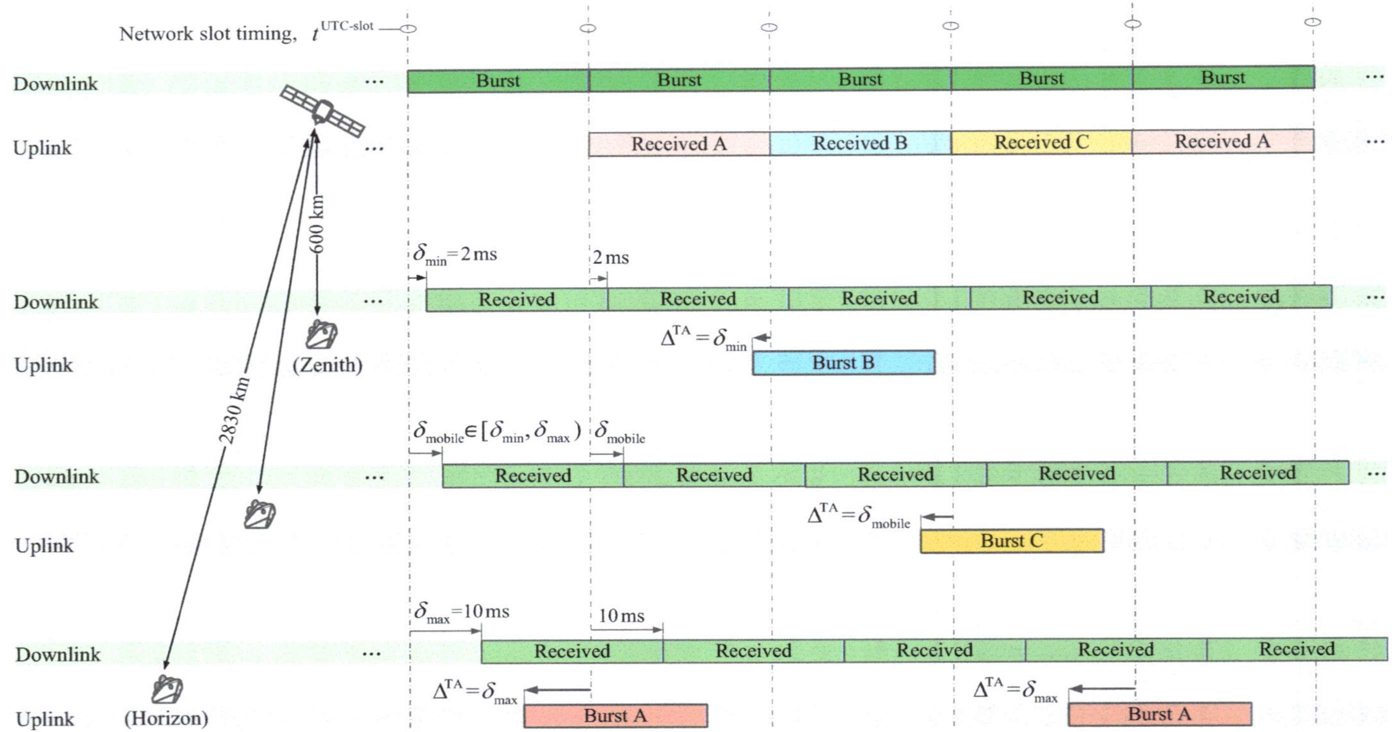

Fig. 9.9 Illustration of the uplink transmission timing advance concept at various mobile positions with various propagation delays, δ_{mobile}. The downlink transmission time at a space station is always aligned with UTC, representing the network slot timing detailed in Chap. 7

$$t_{\text{mobile}}^{\text{UTC-slot}} = t^{\text{UTC-slot}} + \varepsilon_{\text{mobile}}^{\text{UTC}}, \tag{9.26}$$

where $\varepsilon_{\text{sat}}^{\text{UTC}}$ and $\varepsilon_{\text{mobile}}^{\text{UTC}}$ are the UTC timing errors at a space station and a mobile station, respectively.

The downlink slot timing detected by the mobile station is expressed by

$$\begin{aligned} t_{\text{mobile}}^{\text{DL-slot}} &= t_{\text{sat}}^{\text{UTC-slot}} + \delta_{\text{mobile}} + \varepsilon_{\text{mobile}}^{\text{DL}} \\ &= t^{\text{UTC-slot}} + \delta_{\text{mobile}} + \varepsilon_{\text{sat}}^{\text{UTC}} + \varepsilon_{\text{mobile}}^{\text{DL}}, \end{aligned} \tag{9.27}$$

where $\varepsilon_{\text{mobile}}^{\text{DL}}$ is the downlink detection and tracking error at a mobile station and assumed to be well within one pulse interval, i.e., $\left|\varepsilon_{\text{mobile}}^{\text{DL}}\right| < T_{\text{sym}}/2$. Here, $T_{\text{sym}} \approx 30\mu\text{s}$ under a 50-kHz channel bandwidth and ~9μs under 150 kHz.

The one-way propagation delay by the mobile station is estimated by taking the difference between the detected downlink timing and the local UTC timing, i.e.,

$$\begin{aligned} \delta'_{\text{mobile}} &= t_{\text{mobile}}^{\text{DL-slot}} - t_{\text{mobile}}^{\text{UTC-slot}} \\ &= \delta_{\text{mobile}} + \left(\varepsilon_{\text{sat}}^{\text{UTC}} - \varepsilon_{\text{mobile}}^{\text{UTC}} + \varepsilon_{\text{mobile}}^{\text{DL}}\right). \end{aligned} \tag{9.28}$$

When the mobile station is due to transmit at the assigned slot, it advances the uplink burst timing ahead of the local UTC slot timing by the one-way propagation delay estimated from (9.28), $\Delta_{\text{mobile}}^{\text{TA}} = \delta'_{\text{mobile}}$. The actual transmission timing at the mobile is thus

$$\begin{aligned} t_{\text{mobile}}^{\text{UL}} &= t_{\text{mobile}}^{\text{UTC-slot}} - \Delta_{\text{mobile}}^{\text{TA}} \\ &= t^{\text{UTC-slot}} - \delta_{\text{mobile}} + 2\varepsilon_{\text{mobile}}^{\text{UTC}} - \varepsilon_{\text{sat}}^{\text{UTC}} - \varepsilon_{\text{mobile}}^{\text{DL}}, \end{aligned} \tag{9.29}$$

and the time when the burst arrives at the space station is

$$\begin{aligned} t_{\text{sat}}^{\text{UL}} &= t_{\text{mobile}}^{\text{UL}} + \delta_{\text{mobile}} \\ &= t^{\text{UTC-slot}} + 2\varepsilon_{\text{mobile}}^{\text{UTC}} - \varepsilon_{\text{sat}}^{\text{UTC}} - \varepsilon_{\text{mobile}}^{\text{DL}}. \end{aligned} \tag{9.30}$$

The *timing advance error*, i.e., the difference between the received uplink burst timing and the space station slot timing, is

$$\begin{aligned} \varepsilon_{\text{TA}} &= t_{\text{sat}}^{\text{UL}} - t_{\text{sat}}^{\text{UTC-slot}} \\ &= 2\left(\varepsilon_{\text{mobile}}^{\text{UTC}} - \varepsilon_{\text{sat}}^{\text{UTC}}\right) - \varepsilon_{\text{mobile}}^{\text{DL}}. \end{aligned} \tag{9.31}$$

It can be readily shown that this timing advance error is upper-bounded by

$$|\varepsilon_{\text{TA}}| \le 2\left(\left|\varepsilon_{\text{mobile}}^{\text{UTC}}\right| + \left|\varepsilon_{\text{sat}}^{\text{UTC}}\right|\right) + \left|\varepsilon_{\text{mobile}}^{\text{DL}}\right|, \tag{9.32}$$

where, as noted in Sect. 9.3.1, $\left|\varepsilon_{\text{sat}}^{\text{UTC}}\right| < 50\mu\text{s}$ for a space station and $\left|\varepsilon_{\text{mobile}}^{\text{UTC}}\right| < 200\mu\text{s}$ for a mobile station.

The timing advance accuracy/ambiguity given in (9.32) is thus $|\varepsilon^{TA}| < 510\mu s$, which warrants a time buffer of $T_{buffer} = 2|\varepsilon^{TA}| \approx 1$ ms, which leads to a `guard period` of

$$T_{guard} = T_{buffer} - T_{ramp} \tag{9.33}$$

which is about 600μs or 20 uplink symbols (about a 2% overhead) to absorb the timing advance error.

Clearly, in (9.32),

$$2\left(\left|\varepsilon_{mobile}^{UTC}\right| + \left|\varepsilon_{sat}^{UTC}\right|\right) \gg \left|\varepsilon_{mobile}^{DL}\right|, \tag{9.34}$$

i.e., the timing advance error is dominated by the UTC synchronization error. This error becomes even larger when UTC is not available.

9.3.2.2 Timing Without UTC

In the event that UTC is interrupted due to a GPS outage or simply a lack of a GPS receiver (most likely the case for Category 1 mobile stations), a mobile station has to rely on the control stations to provide network timing before transmitting on the uplink traffic channel. As such, a mobile station first acquires the network timing from the downlink signal, which includes a one-way propagation delay, δ_{mobile}. Without UTC as a reference, $t_{sat}^{UTC\text{-}slot}$ is unknown and δ_{mobile} cannot be obtained from (9.24).

Nevertheless, knowing that $\delta_{mobile} \geq \delta_{min}$, the best that a UTC-less mobile station can do at this point is to assume $\delta_{mobile} = \delta_{min}$ and use

$$t_{mobile}^{slot} = \left(t_{mobile}^{DL\text{-}slot} + \varepsilon_{mobile}^{DL}\right) - \delta_{min} \tag{9.35}$$

as the local slot timing reference as an estimate of $t_{sat}^{UTC\text{-}slot}$, which includes the downlink detection error, $\varepsilon_{mobile}^{DL}$.

The mobile station advances its uplink transmission by $\Delta_{mobile}^{TA} = \delta_{min}$ with respect to t_{mobile}^{slot}, i.e.,

$$\begin{aligned} t_{mobile}^{UL} &= t_{mobile}^{slot} - \Delta_{mobile}^{TA} \\ &= t_{sat}^{UTC\text{-}slot} + \delta_{mobile} - 2\delta_{min} + \varepsilon_{mobile}^{DL}, \end{aligned} \tag{9.36}$$

which arrives at the space station at

$$\begin{aligned} t_{sat}^{UL} &= t_{mobile}^{UL} + \delta_{mobile} \\ &= t_{sat}^{UTC\text{-}slot} + 2(\delta_{mobile} - \delta_{min}) + \varepsilon_{mobile}^{DL}. \end{aligned} \tag{9.37}$$

Substituting (9.37) into (9.31), the timing error seen by the space station is thus

$$\begin{aligned}\varepsilon^{\mathrm{TA}} &= t_{\mathrm{sat}}^{\mathrm{UL}} - t_{\mathrm{sat}}^{\mathrm{UTC\text{-}slot}} \\ &= 2(\delta_{\mathrm{mobile}} - \delta_{\min}) + \varepsilon_{\mathrm{mobile}}^{\mathrm{DL}}.\end{aligned} \tag{9.38}$$

Since $\delta_{\min} \leq \delta_{\mathrm{mobile}} \leq \delta_{\max}$ and $\left|\varepsilon_{\mathrm{mobile}}^{\mathrm{DL}}\right| < \frac{1}{2}T_{\mathrm{sym}}$, we maintain that

$$-\frac{1}{2}T_{\mathrm{sym}} < \varepsilon^{\mathrm{TA}} < 2\Delta + \frac{1}{2}T_{\mathrm{sym}}, \tag{9.39}$$

which requires a time buffer of

$$\begin{aligned}T_{\mathrm{buffer}} &= 2\Delta + T_{\mathrm{sym}} \\ &\approx 14.8 \text{ ms}\end{aligned} \tag{9.40}$$

or a `guard period` of

$$\begin{aligned}T_{\mathrm{guard}} &= T_{\mathrm{buffer}} - T_{\mathrm{ramp}} \\ &\approx 14.4 \text{ ms}.\end{aligned} \tag{9.41}$$

where $\Delta = \delta_{\max} - \delta_{\min}$, i.e., the maximum propagation time difference from (9.20). For a LEO space station orbiting at a 600-km altitude, it means a time buffer of 2×7.4 ms, which would take up 60% of a slot, and gets even worse for higher-orbit space stations – a penalty for not having a UTC.

It becomes evident that, in order not to bankrupt the uplink, a mobile station must have better timing before transmitting on the uplink. The MAC timing control protocol is designed just for that purpose by ensuring that mobile stations have the right transmit timing such that the uplink bursts are aligned with the network slot boundary at the space station, regardless of the propagation time differences among transmitting mobile stations, even in the absence of UTC.

The implementation of this scheme under the current MTC framework, i.e., the MAC timing control protocol, is described next.

9.3.3 *The MAC Timing Control Protocol*

Based on the above analysis, a MAC Timing Control Protocol can be devised to keep the overhead under control and timing error in check.

As such, mobile stations, with or without UTC, first acquire and then lock on to the downlink slot timing. For those who have UTC, $t^{\mathrm{UTC\text{-}slot}}$ is known and δ_{mobile} is set as

$$\delta_0 = t_{\text{mobile}}^{\text{DL-slot}} - t^{\text{UTC-slot}}. \tag{9.42}$$

For those who do not, neither $t_{\text{sat}}^{\text{UTC-slot}}$ nor δ_{mobile} is known. A good estimate of δ_{mobile} to start with is

$$\delta_0 = \delta_{\min}. \tag{9.43}$$

In either case, an initial estimate of the network slot timing, $t^{\text{UTC-slot}}$, for use as $t_{\text{mobile}}^{\text{UTC-slot}}$, is

$$t_0^{\text{slot}} = t_{\text{mobile}}^{\text{DL-slot}} - \delta_0. \tag{9.44}$$

The mobile station then transmits a PRACH burst on a PRACH resource,

$$\Delta_0^{\text{TA}} = \delta_0, \tag{9.45}$$

in advance of the estimated slot boundary, t_0^{slot}.

Now let us refer to Fig. 9.10. The space station estimates the timing error, $\varepsilon_0^{\text{TA}}$, from the received PRACH and apprises the mobile station of this error through the MAC Timing Control PDU, as defined in Table 9.5, transported via DTCH and transmitted on a PDTCH. The mobile station updates the one-way propagation delay with

$$\delta_1 = \delta_0 + \frac{1}{2}\varepsilon_0^{\text{TA}}, \tag{9.46}$$

based on the observed $\varepsilon_0^{\text{TA}}$, and the slot timing with

$$t_1^{\text{slot}} = t_{\text{mobile}}^{\text{DL-slot}} - \delta_1. \tag{9.47}$$

With the updated slot timing and propagation delay, the mobile station advances its uplink transmit timing by

$$\Delta_1^{\text{TA}} = \delta_1, \tag{9.48}$$

ahead of t_1^{slot} for both PUSCH and PUTCH transmissions so that they are in alignment with the network slot timing at the space station.

To track the propagation delay, the space station continuously monitors and evaluates the timing error, $\varepsilon_{\text{TA}}^{i}(i \geq 1)$, every time an uplink burst is received and updates the mobile station, as deemed necessary. If so instructed, the mobile station updates the propagation delay

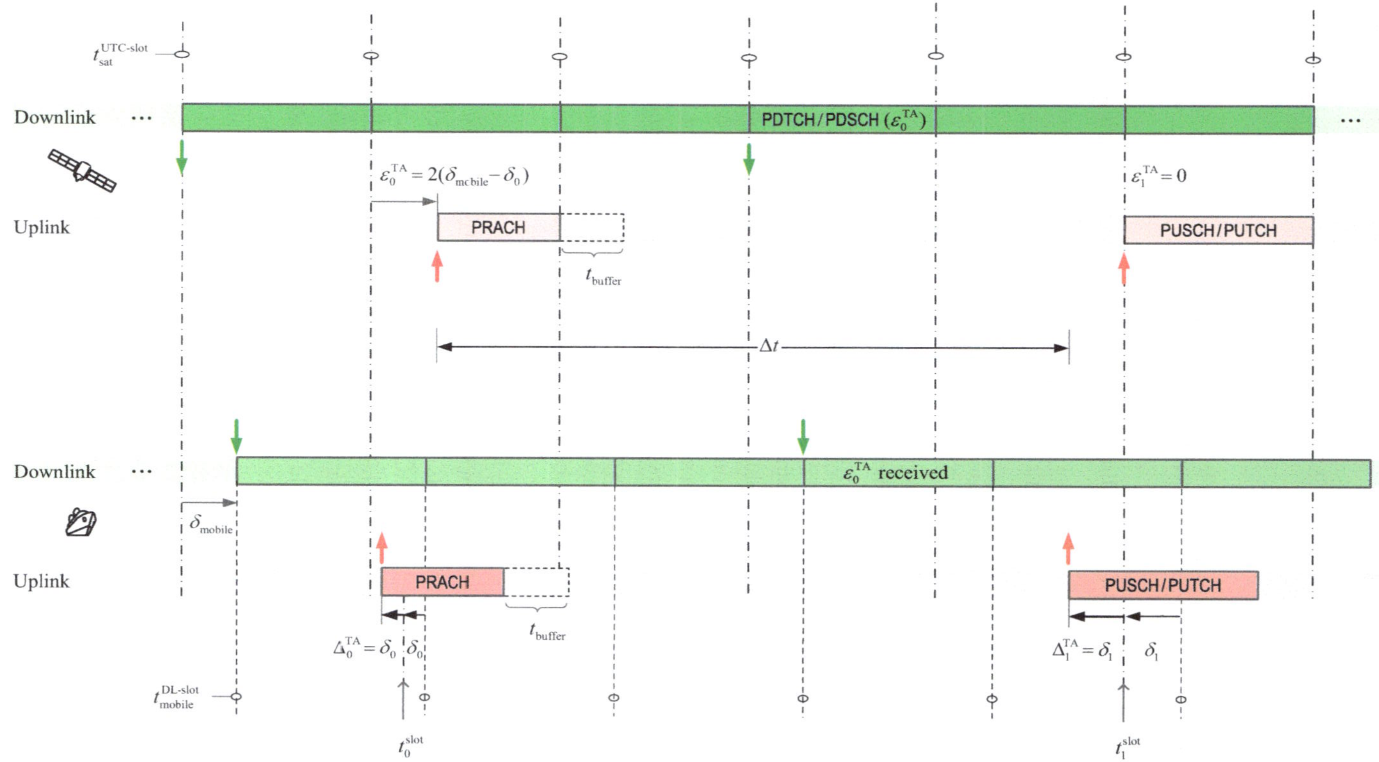

Fig. 9.10 Illustration of network timing acquisition and advancing of a mobile station, in the absent of UTC, through a control station, where $\delta_0 = \delta_{\min}$ as an initial propagation delay estimate for δ_{mobile}. Ideally, $\delta_1 = \delta_{mobile}$. Obviously, the maximum time buffer needed, T_{buffer}, is when $\delta_{mobile} = \delta_{max}$

Table 9.5 MAC Timing Control PDU

Header		SDU	
`Type`	`Control function`	`MMSI`	`Timing offset`
`1`	`0001010`	30 bits	10 bits

$$\delta^i = \delta^{i-1} + \frac{1}{2}\varepsilon_{\text{TA}}^{i-1}, \tag{9.49}$$

and advances the timing of the pending uplink bursts by $\Delta_{\text{TA}}^i = \delta^i$ with respect to

$$t_{\text{slot}}^i = t_{\text{slot}}^{\text{DL}} - \delta^i \tag{9.50}$$

to pre-compensate for the propagation delay.

Typically, $\varepsilon_{\text{TA}}^i \ll \varepsilon_{\text{TA}}^0$, for $i > 0$. Therefore, for the control station MAC timing control protocol to get its message across to its mobile station peer more efficiently, for small timing offsets, the adjustment information is communicated to a mobile station via the `timing adjustment` field of the assignment element of the MAC Resource Assignment PDU, defined in Table 8.2a, transported via ACH over PDSCH. It is designed to offer a small-scale adjustment and hence is more time and resource-efficient than the full-blown one defined in Table 9.5 when dealing with smaller timing errors. The relationship between these two forms of MAC timing adjustment mechanism is complementary rather than overriding. That means that when receiving both forms of updates from the MAC timing control protocol of the control station, a mobile station executes both cumulatively.

Yet, there are two small details with this scheme:

First, from (9.41), a `guard period` of $T_{\text{guard}} \approx 14.4$ ms corresponds to 54% of a one-slot TTI PRACH under the frame configuration in Fig. 9.4; it would be too wasteful and significantly reduce the PRACH resource capacity.

"Luckily," we have an *e*PRACH resource reserved at the beginning of a frame for PRACHs with extended TTI (10 slots) for low-power mobile stations, referring back again to Fig. 9.4. This extended PRACH resource is also ideal for this scenario since a `guard period` of $T_{\text{guard}}^{e\text{PRACH}} = 14.4$ ms is just a 6% overhead for this *e*PRACH, very much acceptable. A `guard period` of $T_{\text{guard}}^{\text{PRACH}} = 0.6$ ms can then be retained for the regular PRACHs for mobile stations with UTC. For space stations with different orbital altitudes, the `guard period` of this *e*PRACH is configured through the `frame configuration` field of the System Bulletin Board.

Second, whether a `guard period` can be eliminated from PUSCH or PUTCH depends on the following factors: (1) uplink timing detection error at the space station, $\varepsilon_{\text{sat}}^{\text{UL}}$; (2) MAC timing adjustment resolution (unit time per adjustment), τ_{TA}; and (3) mobility of the space station between adjustments, $\varepsilon_{\text{motion}}$.

Negating the `guard period` ($T_{\text{guard}} = 0$) means $T_{\text{buffer}} \leq T_{\text{ramp}}$. The 2-bit `timing adjustment` field in the assignment element (Table 8.2a) provides a time resolution of a quarter of the time buffer, i.e., $\tau_{\text{TA}} = T_{\text{ramp}}/4$, as depicted in Fig. 9.11.

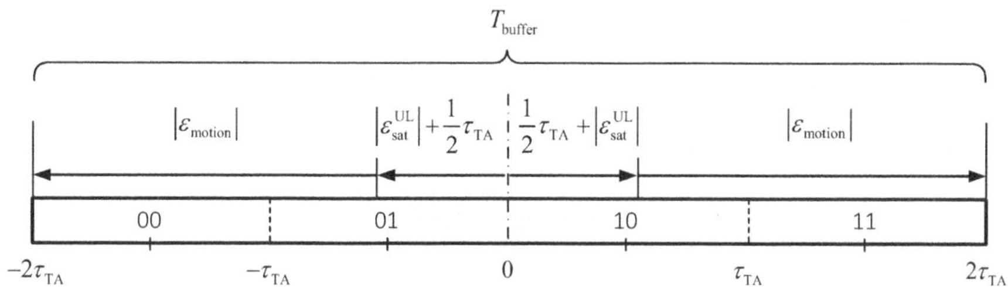

Fig. 9.11 Illustration of timing adjustment, where the time buffer of a burst is divided into four zones, numbered `00`, `01`, `10`, and `11`

For example, a timing adjustment code of `11` indicates the received burst timing is late by τ_{TA} to $2\tau_{\mathrm{TA}}$ (Zone 11) and the mobile station increases its uplink burst timing advance by $3\tau_{\mathrm{TA}}/2$, while `01` signals an early timing in the range from 0 to τ_{TA} (Zone 01) and the mobile station reduces its timing advance by $\tau_{\mathrm{TA}}/2$.

Under this setting, the timing adjustment accuracy/ambiguity obtained from an uplink burst, i.e.,

$$\varepsilon_{\mathrm{TA}} = \pm\left(\left|\varepsilon_{\mathrm{sat}}^{\mathrm{UL}}\right| + \tau_{\mathrm{TA}}/2\right), \tag{9.51}$$

consumes $2\left|\varepsilon_{\mathrm{sat}}^{\mathrm{UL}}\right| + \tau_{\mathrm{TA}}$ of the time buffer, attributable to the timing adjustment resolution and the base station detection error, leaving $T_{\mathrm{buffer}} - \left(2\left|\varepsilon_{\mathrm{sat}}^{\mathrm{UL}}\right| + \tau_{\mathrm{TA}}\right)$ for mobility.

Since there is an elapsed time, Δt, between the measurement of δ_{mobile} and the application of δ_{mobile}, (see Fig. 9.10) the relative movement between the mobile and space station within this duration may introduce a change in ϕ, hence, a change in δ_{mobile} (a function of ϕ), denoted as $\varepsilon_{\mathrm{motion}}$.

As illustrated in Fig. 9.12, the timing error introduced by the space station mobility over a period of Δt is

$$\varepsilon_{\mathrm{motion}}(\phi) = \frac{1}{c_{\mathrm{light}}}\left|\left|\mathbf{c}_{\mathrm{carrier}} \cdot \delta(\phi) + \mathbf{v}_{\mathrm{sat}} \cdot \Delta t\right| - \left|\mathbf{c}_{\mathrm{carrier}} \cdot \delta(\phi)\right|\right|, \tag{9.52}$$

where $\mathbf{c}_{\mathrm{carrier}}$ is the velocity of the RF carrier and $\mathbf{v}_{\mathrm{sat}}$ the velocity of the space station. Since $|\mathbf{v}_{\mathrm{sat}} \cdot \Delta t| \ll |\mathbf{c}_{\mathrm{carrier}} \cdot \delta(\phi)|$, (9.52) can then be comfortably simplified as

$$\begin{aligned}\varepsilon_{\mathrm{motion}}(\phi) &\approx \frac{\left|\left(\mathbf{v}_{\mathrm{sat}} \cdot \mathbf{c}_{\mathrm{carrier}}\right)\right|}{c_{\mathrm{light}}^2} \cdot \Delta t \\ &= \frac{v_{\mathrm{sat}}}{c_{\mathrm{light}}} \cdot \sin\phi \cdot \Delta t.\end{aligned} \tag{9.53}$$

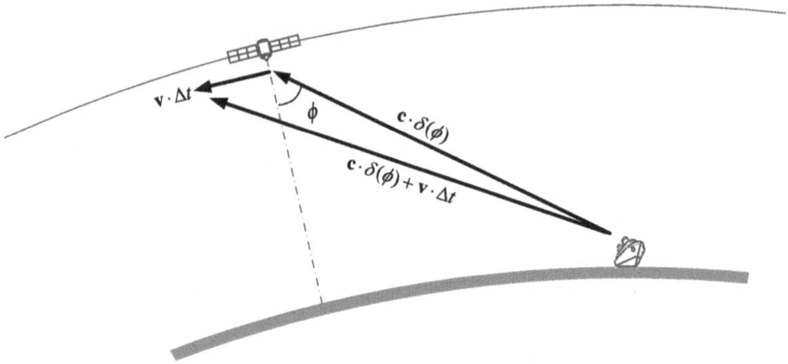

Fig. 9.12 Calculation of the timing advance error introduced by the mobility of a space station

Noting that $0 \le \phi \le \phi_{\text{horizon}} < 90°$, practically, we can assume

$$\dot{\varepsilon}_{\text{motion}} \triangleq \frac{\varepsilon_{\text{motion}}}{\Delta t} \le \frac{v_{\text{sat}}}{c_{\text{light}}} \cdot \sin \phi_{\text{horizon}}. \tag{9.54}$$

Given the LEO space station orbital speed of $v_{\text{sat}} = 8$ km/s, and $\phi_{\text{horizon}} = 67°$ from (9.6), $\dot{\varepsilon}_{\text{motion}} \le 0.65\ \mu\text{s/slot}$.

Given a time buffer of $T_{\text{buffer}} = T_{\text{ramp}} = 416\mu\text{s}$ and an uplink burst detection accuracy of $\left|\varepsilon_{\text{sat}}^{\text{UL}}\right| < 15\ \mu\text{s}$, we obtain $\varepsilon_{\text{motion}} = \pm 141\mu\text{s}$ (see Fig. 9.12), meaning the timing adjustment remains valid within $\varepsilon_{\text{motion}}/\dot{\varepsilon}_{\text{motion}} = 215$ slots.

Under the same setting, the MAC Timing Control PDU in Table 9.5 is designed to bring the uplink burst timing to the center of the time buffer (Fig. 9.11) for any timing gap larger than $2\tau_{\text{TA}}$. The gap can be as large as the *e*PRACH `guard period`, $T_{\text{guard}}^{e\text{PRACH}}$; hence, the `timing offset` field of the PDU should provide a dynamic range up to $T_{\text{guard}}^{e\text{PRACH}}$. A 10-bit of bit width covers this range with a resolution of $\Gamma_{\text{TA}} = 14\mu\text{s}$, and an accuracy/ambiguity of

$$\begin{aligned} \varepsilon_{\text{TA}} &= \pm \left(\left|\varepsilon_{\text{sat}}^{\text{UL}}\right| + \Gamma_{\text{TA}}/2 \right) \\ &\approx \pm\, 22\ \mu\text{s}, \end{aligned} \tag{9.55}$$

allowing for a maximum $\varepsilon_{\text{motion}} = \pm\ 186\mu\text{s}$. Hence, this timing remains valid within 290 slots.

When operating under indirect UTC timing, the timing advance derived from the most recent timing adjustment update may become stale as time elapses, due to, e.g., a lack of uplink activities for over, e.g., 200 slots, remembering that a timing adjustment update only ensues from the uplink burst successfully received by the control station. In that case, a timing update can be obtained by sending a MAC Random Access PDU (Sect. 8.3.2 in Chap. 8) through PRACH.

A flowchart that summarizes the timing process, with and without UTC, is plotted in Fig. 9.13.

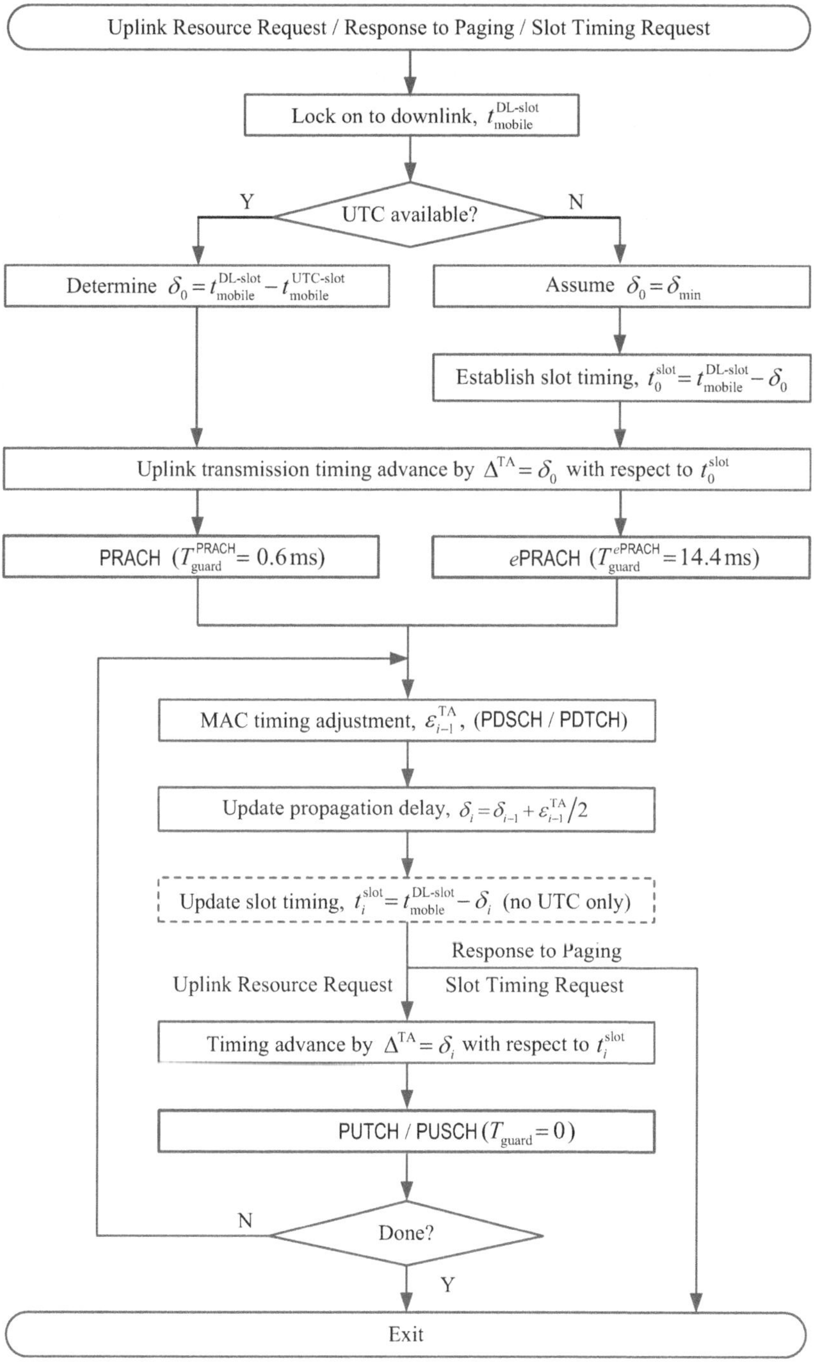

Fig. 9.13 Flowchart of the timing acquisition and advance scheme for mobile stations' Response to Paging, Slot Timing Request, and Uplink Resource Request for uplink transmissions, in a MAC Random Access PDU over PRACH or *e*PRACH, where $i \geq 1$

9.3.4 Timing for VDE-TER and ASM

The same unified timing mechanism derived for VDE-SAT applies to VDE-TER transmissions as well, only that the propagation delay issue is not nearly as prominent as in VDE-SAT and hence can be well contained within the derived timing framework.

With UTC under VDE-TER, the initial timing advance error is upper-bounded by (9.32),

$$|\varepsilon_{\mathrm{TA}}| \leq 2\left(\left|\varepsilon_{\mathrm{mobile}}^{\mathrm{UTC}}\right| + \left|\varepsilon_{\mathrm{shore}}^{\mathrm{UTC}}\right|\right) + \left|\varepsilon_{\mathrm{mobile}}^{\mathrm{DL}}\right|, \tag{9.56}$$

where $\left|\varepsilon_{\mathrm{shore}}^{\mathrm{UTC}}\right|$, i.e., the shore station UTC timing error, is within 50μs for a shore station and $\left|\varepsilon_{\mathrm{mobile}}^{\mathrm{DL}}\right| \leq T_{\mathrm{sym}}/2 = 13\mu\mathrm{s}$. Thus, $|\varepsilon_{\mathrm{TA}}| \leq 513\mu\mathrm{s}$, meaning a time buffer of

$$\begin{aligned} T_{\mathrm{buffer}} &= 2\varepsilon_{\mathrm{TA}} \\ &\approx 1\,\mathrm{ms} \end{aligned} \tag{9.57}$$

is required to contain the uncertainty.

Without UTC, the initial timing advance error is bounded by (9.39), with $\Delta = \delta_{\max}$ for VDE-TER. Thus,

$$-\frac{1}{2}T_{\mathrm{sym}} < \varepsilon^{\mathrm{TA}} < 2\delta_{\max} + \frac{1}{2}T_{\mathrm{sym}}. \tag{9.58}$$

For a one-way delay over a 50-NM (or 93-km) communication range, $\delta_{\max} = 310\mu\mathrm{s}$. With $T_{\mathrm{sym}} = 26\mu\mathrm{s}$ for VDE-TER, (9.58) necessitates a time buffer of

$$\begin{aligned} T_{\mathrm{buffer}} &= 2\delta_{\max} + T_{\mathrm{sym}} \\ &\approx 646\ \mu\mathrm{s}. \end{aligned} \tag{9.59}$$

We see that, under VDE-TER, UTC-based timing advance potentially has larger timing error, implying that, unlike in VDE-SAT, mobile stations without UTC are free to use PRACH, no longer required to use *e*PRACH for initial access. Nevertheless, the same time buffer of 1 ms is reserved for both *e*PRACH and PRACH for the sake of simplicity. Deducting the `ramp-down` period leads to a `guard period` of

$$T_{\mathrm{guard}}^{e\mathrm{PRACH}} = T_{\mathrm{guard}}^{\mathrm{PRACH}} = 610\ \mu\mathrm{s}. \tag{9.60}$$

The uncertainty introduced by mobility is also expected to be insignificant compared to VDE-SAT, considering that the shore station is static (fixed). Assuming the speed of a mobile station, v_{mobile}, is less than 100 NM/h, we have

$$\begin{aligned}\dot{\varepsilon}_{\text{motion}} &< v_{\text{mobile}}/c_{\text{light}} \\ &\approx 4.4\text{ns per slot.}\end{aligned} \quad (9.61)$$

As noted earlier in Chap. 7, ASM transmissions also share the same UTC-based network slot timing. Therefore, the same MAC timing control protocol diagrammed in Fig. 9.13 applies.

In the absence of UTC, the mobile station may acquire the network slot timing by explicitly sending a Slot Timing Request in a MAC Random Access PDU to its control station peer via RACH on *e*PRACH (over VDE-SAT or VDE-TER) or PRACH (VDE-TER only). The control station will respond with a MAC Timing Control PDU via DTCH on PDTCH, from which the mobile station derives the slot timing, inferred by the `timing offset` field of the PDU with respect to the *e*PRACH/PRACH burst timing. The accuracy, ε_{TA}, provided by the MAC Timing Control PDU in (9.55) is absorbed by the ε_{TA} term in (7.21) of Chap. 7.

In addition, the ASM time buffer given in (7.22) leaves 600μs for absorbing the mobility-induced error, meaning that the timing offset is valid as long as it is received within ~400 slots (after the timing request is sent) when acquiring timing from a space station.

9.4 TDD and Half-Duplex FDD

Normally, the primary reason to choose TDD over FDD is that it does not need a paired spectrum, as noted in Chap. 2. It is obviously not the case for VDE-SAT since the VDE-SAT spectrum is a traditional paired FDD spectrum (see Fig. 9.1). The reason to allow TDD on the VDE-SAT spectrum is mainly because of the potential cross-link interference at a space station transceiver, as mentioned in Chap. 6.

So far, we have seen how to use the "building blocks" (the resource block and frame) to establish an FDD system; we now demonstrate how the same building blocks are used to build a time-division-duplexing or TDD system for VDE-SAT as an example.

9.4.1 TDD

In spite of the difference in duplexing between the uplink and downlink, the logical association between the physical channels of the resource block pairs remains the same, except that the physical transmissions of the downlink and uplink resource blocks of a resource block pair are no longer frequency-duplexed but rather time-duplexed. In FDD, the time and frequency resources for uplink and downlink of a frame are separated by frequency, while in TDD, they are separated by time. Just as noted in Chap. 6, time-duplexing the downlink and uplink provides temporal (rather than frequency) separation to avoid cross-link interference between downlink and uplink.

Timing in TDD is similar to that in FDD, but with added complication resulting from the time-division-duplexed transmissions between the uplink and downlink on the *same* radio frequency spectrum, implying that not only the mobile stations but also the control station must share the same spectrum and multiplex on the same radio link. As we all know too well, sharing means conflicting. The design, therefore, must ensure no temporal overlapping (to preserve orthogonality) between uplink and downlink, not only at the control station (as in FDD) but at the mobile stations as well, which further complicates the timing design. Specifically, under FDD, a control station has a dedicated radio spectrum, and hence there is no conflict there, whereas, under TDD, the very same radio spectrum is shared by both a control station and mobile stations, meaning all stations, regardless control or mobile, share the same radio link in a time-division-multiplexing or TDM fashion.

Figure 9.14 shows a direct application of FDD timing discussed earlier to TDD, only that the spectrum is shared among downlink and uplink in a time-duplexing manner. We have two observations:

First, with the same FDD timing advance mechanism in place, the overlap issue at the transitions between uplink to downlink at a space station is taken care of automatically since the uplink bursts are aligned with the network slot timing at the control station, as shown in Fig. 9.14 (top) – another benefit of timing advance under TDD.

Second, let us take a close look at what happens at the downlink to uplink transition from a mobile station's perspective. Referring to Fig. 9.14 (bottom), we see an inevitable temporal overlap or collision between the downlink burst and the uplink burst at a mobile station, as a combined result of (1) the uplink transmission timing advance and (2) downlink propagation delay. When time-duplexing on the same spectrum with the uplink, the downlink burst protrudes into the period earmarked for uplink transmissions at a mobile station receiver due to the propagation delay from the space station to the mobile station. At the same time, the uplink burst intrudes into the period designated for downlink due to the uplink timing advance by the mobile station. The intrusion and protrusion give rise to co-frequency interference at the mobile station, blocking it from receiving downlink transmissions while transmitting on the uplink. Recall that TDD relies on "temporal separation" between the downlink and uplink to provide isolation against the cross-link interference. Any temporal overlap breaches the isolation.

A solution to this issue is to earmark another time buffer at the transition from the downlink to the uplink, large enough to absorb the downlink propagation delay *plus* the time for uplink timing advance (which equals the uplink propagation delay). Denoting this time buffer as the `time gap`, we have

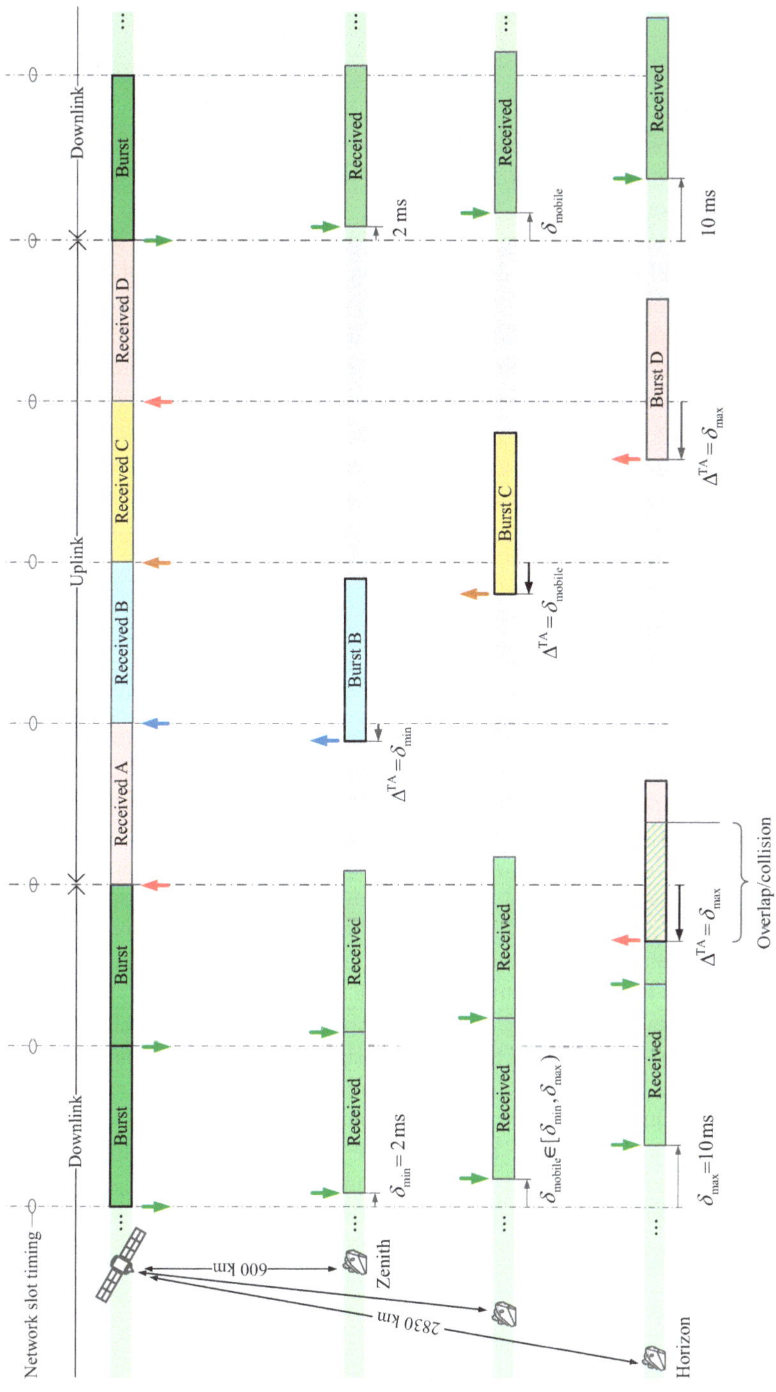

Fig. 9.14 Illustration of temporal collision between downlink and uplink bursts at the downlink to uplink transition under TDD

$$\begin{aligned} T_{\text{gap}} &= \underbrace{\delta_{\max}}_{\text{max uplink timing advance}} + \underbrace{\delta_{\max}}_{\text{max downlink propagation delay}} \\ &= 2\delta_{\max}, \end{aligned} \tag{9.62}$$

where $T_{\text{gap}} \approx 20$ ms for a LEO space station orbiting at $h_{\text{sat}} = 600$ km, from (9.19) or Fig. 9.8. We thus devote a slot as the `time gap` at every downlink to uplink transition, as diagrammed in Fig. 9.15.

There is still one subtle issue with the TDD timing framework in Fig. 9.14, however. The problem is that it takes a transition time of $\tau_{\text{R-T}}$ for a TDD transceiver to switch from receive to transmit and $\tau_{\text{T-R}}$ from receive to transmit (not the case for FDD – no switching between downlink and uplink). They are typically around 100μs, thereby necessitating a time gap at a transition for the control station to switch from receive to transmit and vice versa.

However, these two `time-gaps` (one at the downlink-to-uplink transition and another at the uplink-to-downlink transition) can be consolidated into a single time gap by increasing the uplink timing advance by an additional amount of $\tau_{\text{R-T}}$ to leave a time gap for switching at the uplink to downlink transition, as illustrated in Fig. 9.16. Therefore, a `time gap` with a lump sum of

$$\begin{aligned} T_{\text{gap}} &\geq \underbrace{2\delta_{\max} + \tau_{\text{R-T}}}_{\text{at mobile station}} + \underbrace{\tau_{\text{R-T}}}_{\text{at space station}} \\ &= 2(\delta_{\max} + \tau_{\text{R-T}}) \end{aligned} \tag{9.63}$$

guarantees no overlap at a space station or a mobile station and ample switching time at both transitions. The first part of (9.63) is attributable to the downlink propagation delay ($\delta_{\min} \leq \delta_{\text{mobile}} \leq \delta_{\max}$) and the uplink timing advance (δ_{mobile}) and switching time from receive to transmit ($\tau_{\text{R-T}}$) at the mobile station, whereas the second part is attributable to the switching time from receive to transmit ($\tau_{\text{R-T}}$) at a space station. Note that explicit reservation for a switching time from transmit to receive is not needed at either station under the current setting because the propagation delay leaves ample time between transmit and receive.

A one-slot worth of `time gap` provides a maximum propagation delay ($\delta_{\max}$) up to 13 ms per (9.63), sufficient for a LEO space station at a 600-km orbit. One slot is thus reserved as a `time gap` at every downlink to uplink transition. A mobile station advances its uplink burst by δ_{mobile} plus an additional $\tau_{\text{R-T}} = 100\mu\text{s}$, i.e., $\Delta_{\text{TA}} = \delta_{\text{mobile}} + \tau_{\text{R-T}}$. This way, 100μs is made available for the control station to switch at the uplink-to-downlink transition and at least $T_{\text{gap}} - (2\delta_{\max} + \tau_{\text{R-T}}) \approx$ 6.6 ms for a mobile station at the downlink-to-uplink transition, as shown in Fig. 9.16.

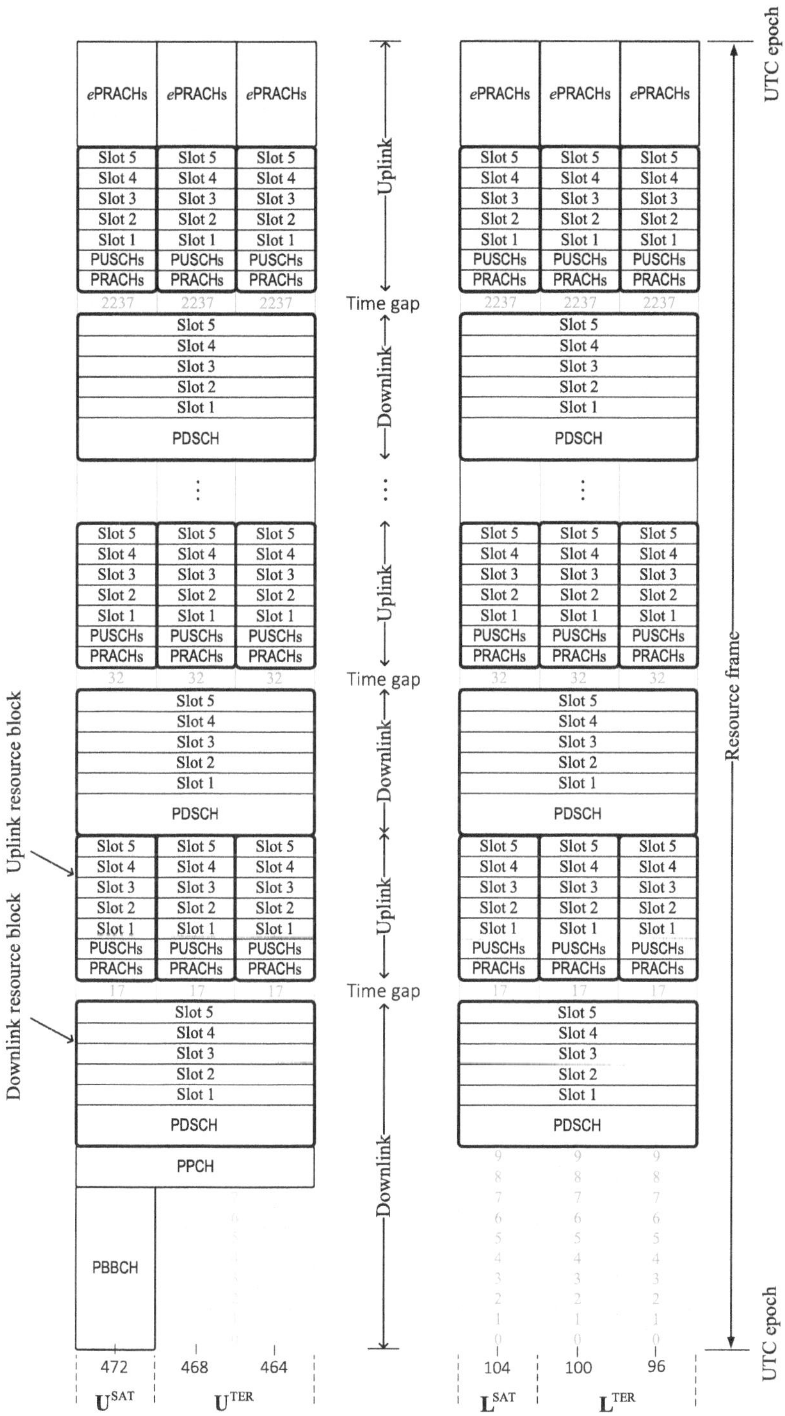

Fig. 9.15 Illustration of resource frame configuration example under TDD. Here, uplink and downlink switch every resource block for ease of illustration

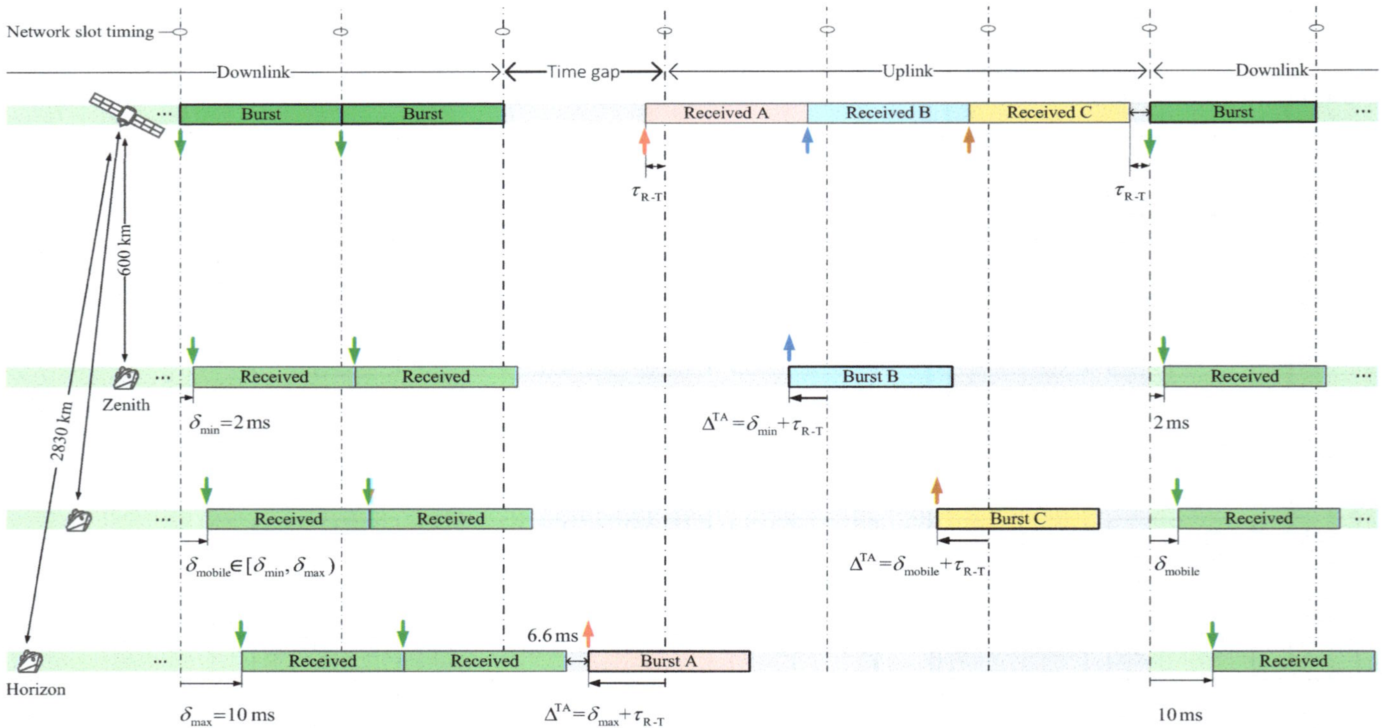

Fig. 9.16 Illustration of TDD transmission timing framework, where downlink transmissions by satellite are time-aligned with the UTC-based network slot timing, and a one-slot (80/3 ms) time gap is reserved at the downlink-to-uplink transition for absorbing *both* downlink and uplink propagation delays as well as the *switch time*, i.e., $2(\delta_{max} + \tau_{R-T})$

For orbital altitudes higher than 1100 km, more than one slot is needed. Indeed, the need for such a time gap for buffering two-way delays is purely a cost incurred by sharing a radio channel between the downlink and uplink. In fact, the need for such a time gap is one of the most significant factors that put TDD at a disadvantage with its full-duplex FDD counterpart. It ultimately limits the switching frequency between downlink and uplink, which has a direct implication on system overhead and latency.

9.4.2 Half-Duplex FDD

As noted in Chap. 6, the simplest (but spectral-efficiency-wise costly) way to prevent cross-link interference in FDD is to deploy the half-duplex FDD, under which the space station and mobile stations take turns to transmit on the downlink and uplink frequencies, i.e., $\mathbf{U}^{\mathrm{SAT}}$ and $\mathbf{L}^{\mathrm{SAT}}$, respectively, thereby negating the need for a transceiver to transmit and receive simultaneously to simplify the transceiver design.

However, this seemingly simple solution is not as simple as it seems. The temporal overlapping problem between the uplink and downlink at a transceiver, not a problem for full-duplex FDD, now becomes an issue. Although not a head-on inband collision as in TDD, where uplink and downlink share the same spectrum, cross-link interference occurs due to out-of-band emission leakage when an uplink burst overlaps *temporally* with a downlink burst, much like in TDD, which beats the purpose of going FDD in the first place.

The good news is that the same timing scheme, whatever works for the TDD to avoid temporal overlap at the transitions between uplink and downlink, works for the half-duplex FDD, despite the fact that the uplink and downlink transmissions are now on separate frequency channels. Like TDD, half-duplex FDD uses temporal separation between downlink and uplink to avoid temporal overlap between uplink and downlink transmissions for cross-link interference avoidance. Therefore, they share the same resource frame structure and timing framework as illustrated in Figs. 9.17 and 9.18.

So far, we have seen that the timing advance scheme consolidates the temporal overlap or collision issue resulting from multiplexing with great flexibility and efficiency, especially under TDD when radio resources are not only shared among mobile stations but also between mobile stations and a control station. Nonetheless, it is evident that, under TDD, time duplexing between downlink and uplink introduces a downlink and uplink transmission latency that could be significant depending on the switching frequency. Ideally, we want the switching frequency as high as possible to minimize the latency. However, switching incurs the need for `time gaps`, and the overhead limits the switching frequency. The same goes for

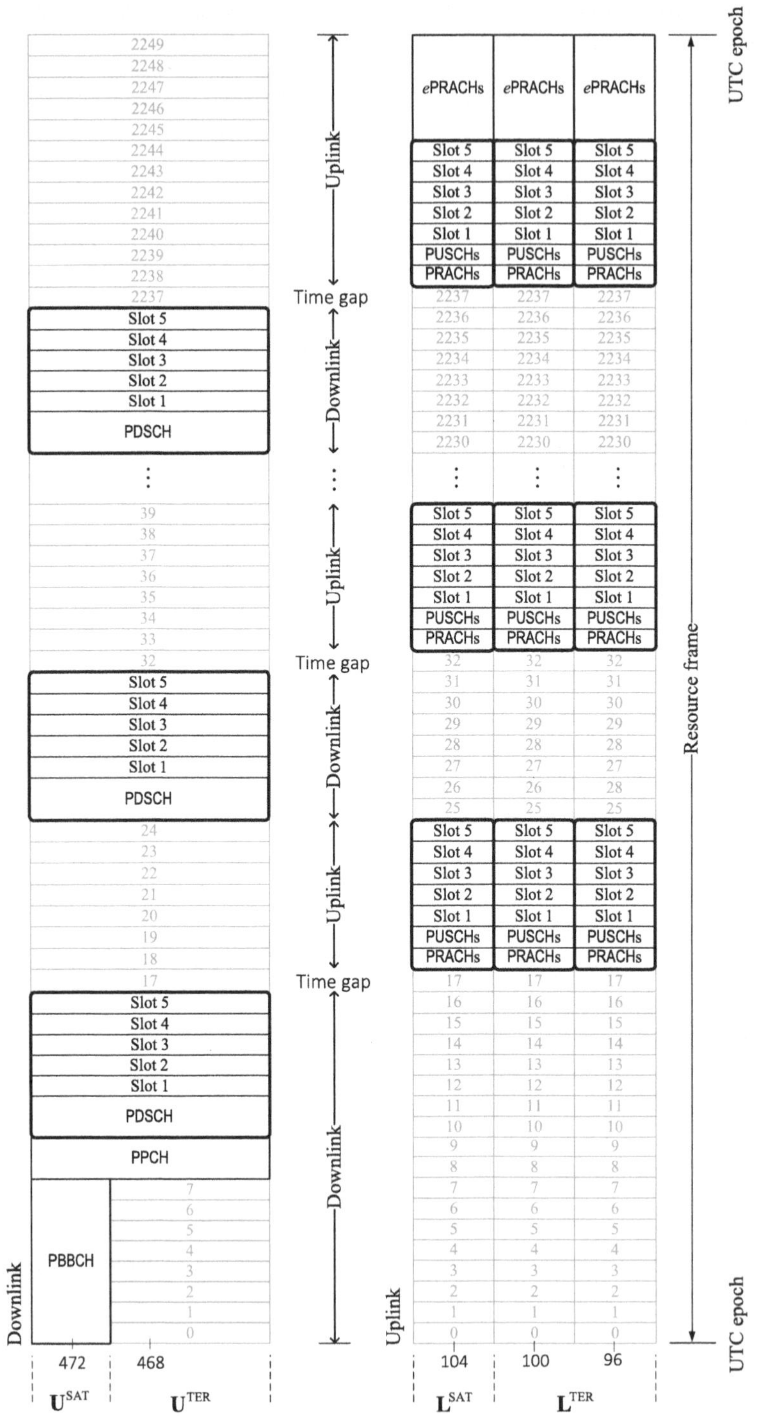

Fig. 9.17 Illustration of VDE-SAT half-duplex FDD transmissions that share the same timeline and resource configuration as TDD

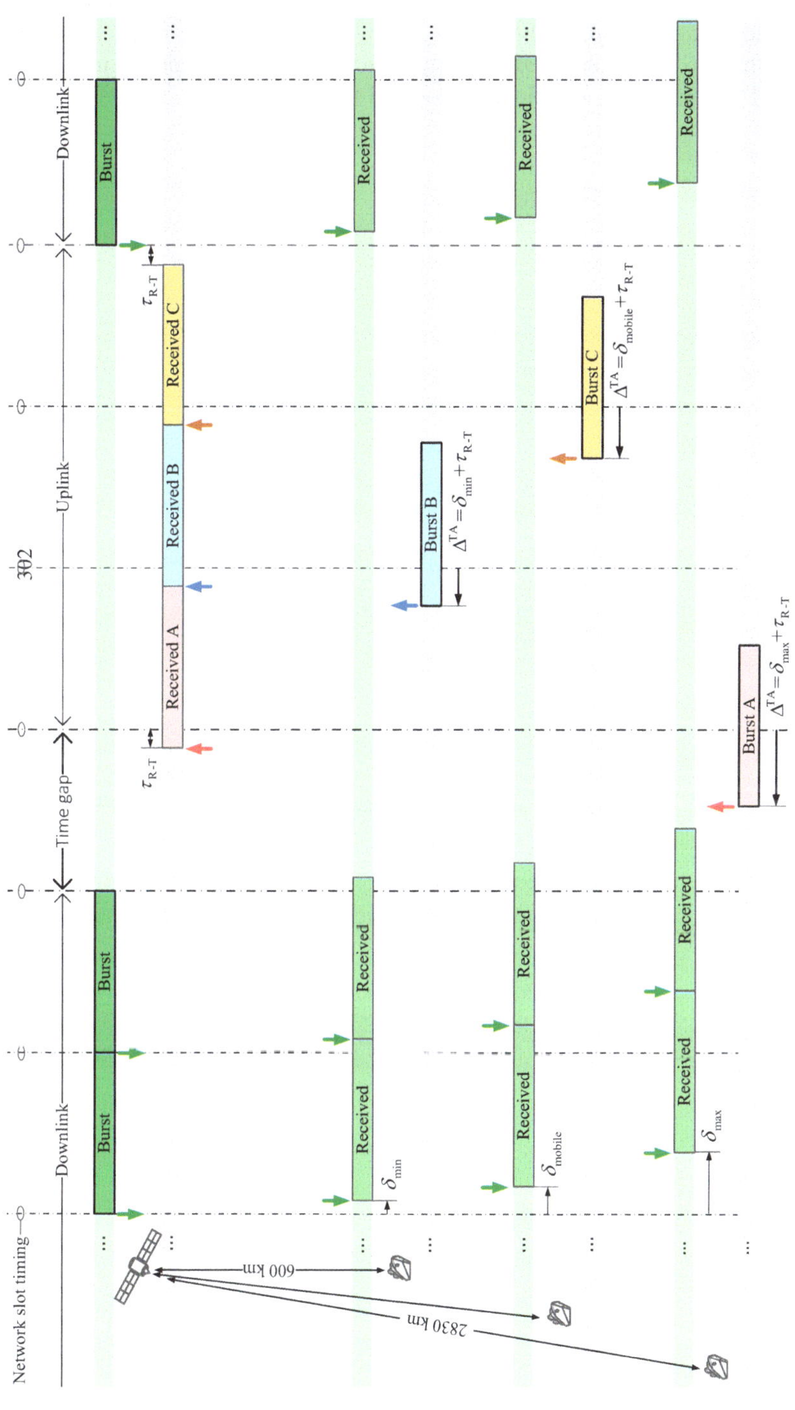

Fig. 9.18 Illustration of the half-duplex FDD transmission timing framework derived from Fig. 9.16

half-duplex FDD – an appreciable disadvantage compared to FDD, in addition to the other disadvantages noted in Chap. 6.

Nevertheless, half-duplex FDD is likely to be the default transmission method for VDE-SAT because of its simplicity unmatched by FDD or TDD.

Chapter 10
Concluding Remarks

We now come to the conclusion of this book. In meeting the broad requirements of maintaining a robust and vibrant maritime MTC system, the challenges are centered on one key theme: global ubiquitous connectivity and service continuity, uniting communications, applications, and physical elements that enable the global maritime IoT to support a full range of maritime services, covering the full range of environmental diversity over the full range of geographic areas.

Indeed, the maritime IoT spectrum assignment in the international maritime mobile communication band has laid the foundation for the globalization of a practical, cost-effective maritime MTC technology, providing low-bandwidth connectivity for short burst data services. It should be made clear that this technology is not poised to serve broadband human-type communication needs at sea.

The international standardization of the technology on this spectrum, i.e., VDES, is essential to compatibility, interoperability, repeatability, and sustainability. It is a process of developing and implementing technical standards based on the consensus of different parties that include stakeholders, manufacturers, interest groups, standards organizations, and government agents. The development and standardization thus require the culmination and synergy of various technological efforts, essential for maritime MTC that involves a comprehensive space and earth integrated solution. This new paradigm provides a unique opportunity for the terrestrial and maritime communities as well as the satellite community to join forces to establish a harmonized and fully fledged space-earth communication network. In the past, these technologies evolved almost independently from each other.

However, as noted at the beginning of the book, maritime communications have been relegated to a lower priority tier for many decades, mainly because of the limited market and thin profit margins compared to its land counterpart. This situation has inevitably resulted in the depletion of genuinely experienced professionals, retardation in communication technologies, and failure to attract and retain a sufficient number of contributors, which places the development and standardization of maritime MTC technology at a considerable disadvantage. Consequently, it would not be a complete surprise to see a handful of companies or organizations

M. M. Wang, J. Zhang, *Machine-Type Communication for Maritime Internet-of-Things*, Wireless Networks, https://doi.org/10.1007/978-3-030-77908-5_10

in control of the standardization, driven by the interests of their individual agendas rather than the broader holistic goals and the integrated objectives outlined in Chap. 1. A maritime MTC technology that can truly address the maritime IoT requirements and stand the test of time requires no less.

What could happen next is that some immature VDES technology is rushed into an international standard that lacks (1) clearly defined goals and requirements; (2) comprehensive system architecture, including network structure and well-defined protocols that support the structure; (3) ability to adapt to diverse applications and services and interwork with different systems and networks; (4) unification and cohesion across the compositional components of the system; (5) spectral efficiency—crucial to the system capacity on such scarce and precious VHF spectrum; (6) verification and optimization; and (7) forward-thinking and long-term planning, thereby ultimately failing to meet the maritime MTC requirements.

This book is intended to provide guidelines and designs to avoid these potential pitfalls. The unique requirements and challenges for maritime MTC are identified and individually addressed through a concrete system design, based on the recently authorized spectrum in the international VHF maritime mobile band and foundationed on a comprehensive maritime MTC architecture. From the top down, the design promotes a service-centric network platform via centralized maritime network management that makes it easier and securer to represent maritime service and application requirements so that software-defined networking solutions can be seamlessly deployed across the terrestrial and satellite segments of the maritime MTC infrastructure; ubiquitous connectivity and service continuity can be better supported over a universal services platform capable of supporting diverse services under a single, unified framework that carries all three networks, i.e., proximity networks, satellite networks, and terrestrial networks. From the bottom up, the radio access network encompasses three unique air interfaces with dedicated frequency bands on the VHF spectrum to achieve distinct goals and requirements. By leveraging the unified protocol and resource framework, the discussed design concept not just can balance its coverage, transmission latency, system overhead/spectral efficiency but can provide (1) proximity, terrestrial, and satellite networking as well as seamless spectrum sharing, (2) support for various categories of mobile stations and applications, and (3) efficient interworking with IP networks, all under a unified operationalized framework.

The question then arises to what extent the ubiquitous connectivity and service continuity "manifesto" is realistic. The uncomfortable truth is that this promise, in the real world, might only hold at the national or regional level; service continuity remains in question as soon as a mobile station moves out of its national network's footprint. Indeed, control ownership and data privacy are among the primary concerns challenging the observability and centralized controllability of the network architecture, compromising the network's ability to effectively mobilize the global maritime infrastructure assets, invalidating the very concept of the maritime MTC, and ultimately plaguing the globalization of maritime IoT.

Today's maritime communication systems comprise many isolated national and regional level systems that are typically proprietary infrastructures and protocols, where the network resources within the infrastructure are dedicated to a single organization and accessed and controlled locally. They do not communicate to the outside maritime world, strange as it may seem, not even the neighboring countries. That means the network resource, configuration, and state information stay inside and are not shared. Therefore, the maritime IoT network will likely continue as a collection of such non-communicating private networks until such time as there is an assurance of security from an internationally federated maritime service platform, under which information can be collected from and safely shared among these networks.

A network structure that resonates more with this reality is the one that leverages the cloud service delivery approach under a multi-cloud/nation maritime MTC network structure. Figure 10.1 exemplifies such a two-tier multi-nation maritime MTC network architecture.

Under this approach, the proprietary maritime MTC infrastructure (satellite or terrestrial) remains as a private cloud. Each owns its virtually isolated maritime MTC solutions customized to its physical environment and its native service providers and clients. It shares the underlying proprietary infrastructure as a service with the global maritime MTC network through a gateway that functions as the maritime application server and network controller of the national network and communicates with the central IMO-certified maritime service platform that serves as an internationally federated maritime application server and network controller via virtual network functions that run on top of the private network infrastructure. Multiple such central service platforms could be deployed in each region on an all-IP distributed architecture, serving various regions of the world to handle the required capacity and reduces overall latency. More importantly, they help load-balancing and avoid a single-point-of-failure.

This structure provides the architectural foundations that enable network information and service data to be accessed and shared appropriately and securely across organizational, regional, and national boundaries. It also enables multiple nations/organizations to share the capital and operating expenses of deploying maritime MTC infrastructure worldwide.

When different operators share the same spectrum, co-frequency interference between systems is bound to occur. For instance, VDE-SAT networks deployed by different operators may create significantly overlapped coverage, ergo co-frequency interference that may degrade or even disrupt services. Hence, another importance of the embodiment of this structure that cannot be overlooked is the enabling of the interference coordination among private infrastructures that share the same air interface spectrum. Again, this is where the integrated multi-cloud structure is not optional but crucial.

In a nutshell, this structure promotes the concept of platform-based services on the cloud that hides the underlying infrastructure from the outside world and yet allows a service provider to efficiently extend its service beyond the boundaries of physical infrastructure and continuously deliver carrier-grade trans-region connectivity and services to the transiting and visiting vessels.

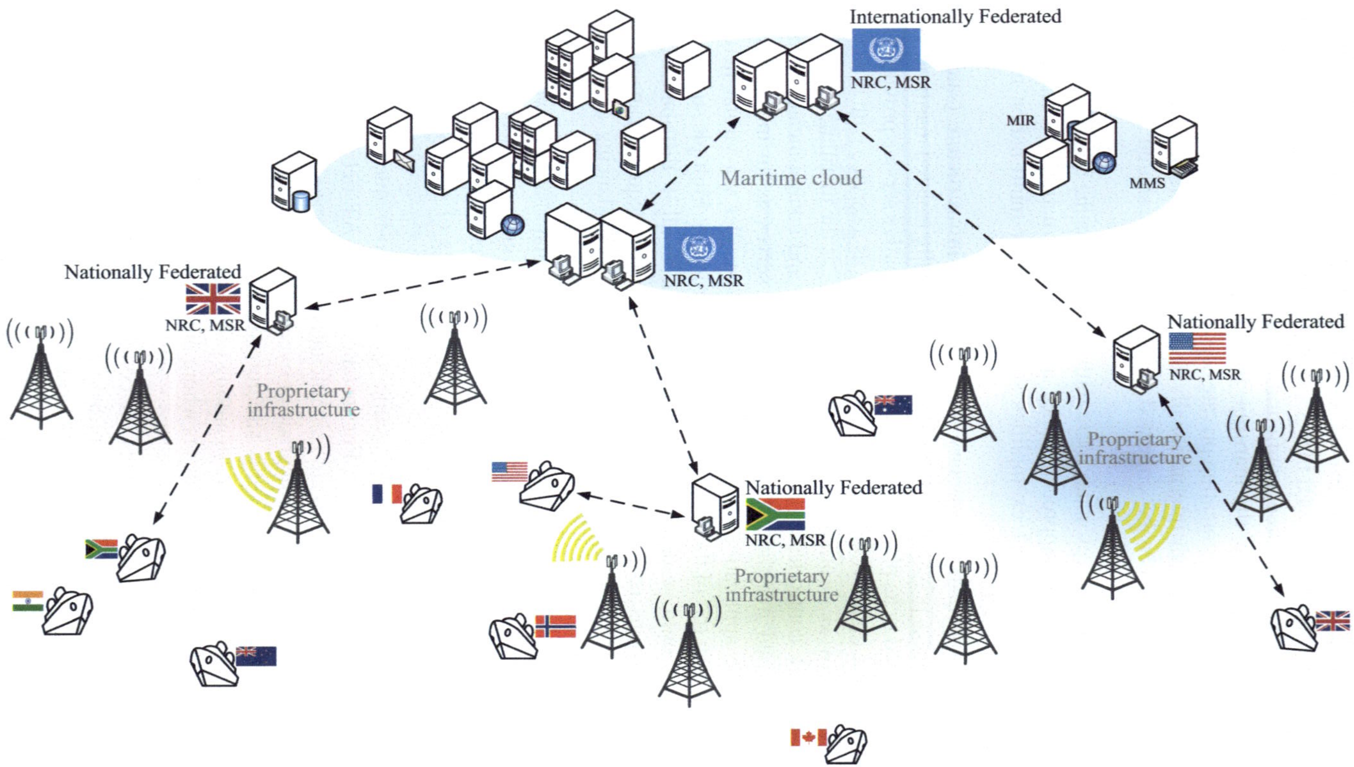

Fig. 10.1 Graphical illustration of a multi-nation maritime MTC network, where national-level networks are typically proprietary networks (serving as private clouds) and are connected to the internationally federated maritime service platform, i.e., MSR and NRC, to enable global MTC for maritime IoT

Index

M. M. Wang, J. Zhang, *Machine-Type Communication for Maritime Internet-of-Things*, Wireless Networks, https://doi.org/10.1007/978-3-030-77908-5

The manufacturer's authorised representative in the EU is Springer Nature Customer Service Centre GmbH, Europaplatz 3, 69115 Heidelberg, Germany. If you have any concerns regarding our products, please contact ProductSafety@springernature.com

Printed and bound by CPI Group (UK) Ltd, Croydon, CR0 4YY

30/09/2026

02225504-0008